JN255739

基礎から学ぶ
クルーズビジネス

池田　良穂　著

KAIBUNDO

はじめに

かつて人を運ぶ客船は海運の花形でした。大西洋や太平洋を横断する航路には，各国が威信をかけて建造した大型客船が就航していました。こうした客船は，単に人の輸送というだけでなく，人々の夢も運ぶ憧れの存在でした。

しかし，日本からヨーロッパまで客船で行くと1ヶ月以上もかかりました。この船の最大の欠点である「遅いスピード」を一気に解消したのが飛行機です。船で行く1日が，飛行機では1時間にまで短縮されました。飛行機が世界中を結ぶようになって，長距離の航路に就航する大型客船は次々と姿を消しました。1960年代から1970年代のことです。

こうして長距離の旅客輸送という舞台から消えた客船が，クルーズという客船による周遊旅行の分野で復活を遂げて，今では巨大な海事産業に成長しています。1970年代初めには，わずか50万人だったクルーズ客船の乗客は，2016年には2,500万人あまりに達し，産業規模は約14兆円になっています。この経済規模は，貨物輸送の花形であるコンテナ船の約20兆円に迫る大きさです。貨物船のビジネスは世界の経済成長にほぼ比例して拡大していますが，クルーズはそれよりもはるかに高い成長率で，世界中で拡大しています。海事産業の中で，唯一，需要が急拡大している成長分野です。使用される船は急速に大型化されて，もっとも大きなクルーズ客船は，23万総トンで，6,000人もの乗客を乗せ，2,000人もの乗組員によって運航されています。この船は，かつて大洋を横断していた最大の客船の3倍もの大きさです。東京駅より長く，3倍以上の高さがある巨大な建造物が，海の上を走っているのです。

さて，クルーズビジネスは，人や貨物を運ぶ海運業というよりは，人を楽しませる観光業といった方がよいでしょう。人々はクルーズ客船で観光地を巡るだけでなく，船上でも大いに楽しむようになりました。海上を移動中に，美味しい食事を堪能し，ショーを観たり，音楽を聴いたり，ショッピングを楽しみ，プールで泳ぎ，そしてカジノもできます。まさに，海上を移動しながら船上で楽しむ「動く統合型リゾート」なのです。

この本は，クルーズのガイドブックではなく，クルーズというビジネスを学ぶための教科書です。2010年に神戸夙川学院大学の観光文化学部にクルーズコースが設置され，非常勤講師として「クルーズビジネス論」という授業を担当した時に書き下ろした教科書をベースとして，2018年から始まった大阪経済法科大学の国際学部の授業のために大幅に改訂したものです。

本書では，クルーズの歴史，成長する現代クルーズの特徴，需要予測，経済性，使用される船のハードとソフト，港湾との関係について学べるようになっています。また，クルーズ客船に乗ったことのない読者でも，クルーズの楽しみ方を理解できるように，実際のクルーズで乗客に配られた資料をそのまま事例研究として掲載しています。これらの資料を使って，クルーズの疑似体験をすることができるようになっています。海外でのクルーズの資料は，英語の原文のまま掲載していますので，辞書を片手にクルーズの内容が理解できるまで読み込んでください。そして，なぜ，クルーズが世界中でブームになっているのか，その理由を見つけだしてください。

2018 年 3 月 8 日

池田　良穂

目　次

第1章 クルーズとは

1.1　クルーズの定義

クルーズ（Cruise）とは，英語で「船で巡遊，巡航する」という意味で，広辞苑には「周遊船旅行」とある。「Cruise」に「r」をつけたクルーザー（Cruiser）という単語は，軍艦の一種の「巡洋艦」および宿泊設備をもつレジャー用ヨットを指す。

多数の乗客を乗せて「クルーズ」を行う客船を「クルーズ客船」（Cruise ship）と呼び，100総トン程度の小型船から，22万総トンの大型船まで，さまざまな大きさと種類がある。こうしたクルーズ客船による船旅が，本書が扱う「クルーズ」である。

クルーズアドバイザー認定委員会によるクルーズの定義は次のとおりである[1]。

①船に乗ること自体が旅行の主目的の1つであること。すなわち，「船」そのものが主要な目的地（デスティネーション）であること。

②航空機や鉄道などの代替・振替の輸送機関としての船旅でないこと。

③単なる輸送機関としてではなく，船内でのレジャーや滞在，洋上ライフを楽しむことが乗船の主目的となっていること。

④原則的に船内での宿泊が伴うこと。

図1-1　大型の現代的クルーズ客船（16万総トン）

図1-2　クルーザー（宿泊機能付の帆船）

1.2　定期客船との違い，他の交通機関との違い

定期客船（Passenger liner）とは，ニューヨーク（アメリカ）とサウサンプトン（イギリス），ロサンゼルス（アメリカ）と横浜などのように，大洋を挟む両側の港を一定のスケジュールで運航される船で，民間航空機網が整備されるまでは，大西洋や太平洋などの大洋横断航路に就航していた。定期客船は，基本的には大洋を渡る人を移動させるのが目的の「輸送用客船」であるのに対し，クルーズ客船は観光地を巡ることを目的とした「観光用宿泊滞在型客船」である。

クルーズ客船は，飛行機，鉄道，バスなどの移動目的の輸送機関とは根本的に違った特性をもっている。その特性は，船内の広い空間が生み出している。すなわち快適な寝室のベッドでの安眠，豪華なレストランでの食事，多彩な催し物，人々との触れ合い，さらにスポーツまで楽しみながら，いながらにして移動することのできる乗り物はクルーズ客船以外にはない。船が寄る港が観光の目的地であるが，前述のように，船自体も目的地の1つなのである。

図 1-3　大西洋横断定期客船「タイタニック」

1.3　リゾートホテルとの相違点

クルーズ客船は「動くリゾートホテル」とも呼ばれる場合があるが，リゾートホテルとの間には，多くの相違点がある。

まず「動く」こと。リゾートホテルでは，朝起きても同じ景色だが，クルーズ客船では景色が変わる。日本のクルーズ業界では「目覚めれば新しい街」というのがクルーズの1つのキャッチフレーズになっている。また，ホテルごと移動するのだから，クルーズ期間中をとおして客室を自分の部屋，プライベート空間として継続的に利用できることも大きなメリットとなる。スーツケースのような大きな荷物を移動する必要もなく，きわめて「楽」に，離れた観光地を手ぶらで周遊できる。

このように旅客側には多くのメリットがある反面，運航する側には種々のハンデがある。すなわちクルーズ客船は，いったん出港すると，陸とは完全に隔離される。したがって，陸上からの物理的な支援なしにすべて自己完結しなくてはならない。食事，電気などのエネルギー，各種のエンターテイメントまで，すべてを船内にある資源だけを使って運営しなくてはならない。

また，安全に関する違いもある。安全を担保するために，リゾートホテルは各国が定める「建築基準法」に従って設計，建築されるが，クルーズ客船は国際海事機関（IMO）が定める国際規則（SOLAS規則など）に則って設計，建造され，運航される。船は，一度洋上に出ると外部からの援助を受けることが難しいので，陸上以上に厳しい規則となっている。

1.4　観光旅行の中での位置づけ

客船によるクルーズ観光は，19 世紀末から始まっているので決して新しい観光旅行ではないが，船以外の交通機関を使って海外に行くことのできなかった昔と違って，現代では長距離の移動は飛行機が中心となっているので，「船による観光」は観光全体のマーケットの中ではニッチ・マーケット（Niche market，すきま産業）といえる。

特に，「飛行機の 1 時間が船の 1 日」といわれるほど，その移動速度に違いがあり，多くの人々の観光旅行は，飛行機で移動して，現地で宿泊して観光をするというスタイルが一般的になった。日本からヨーロッパまで，飛行機だと片道 12 ～ 14 時間だが，船なら往復には最低でも 1 ヶ月はかかる。すなわち，一般の仕事をもつ人には，とても船でのヨーロッパ旅行はできず，比較的時間に余裕のある退職した高齢者に限られることとなる。かつて，クルーズ客船の多くは，こうした長距離の周遊ルートを航海していた。日本にたまに寄港した世界一周のクルーズ客船が，「動く豪華養老院」とまでいわれた所以である。

しかし，1960 年代にアメリカで誕生した「現代クルーズ」（定義は後述）は，この船のデメリットを完全に克服して，短期間で「優雅な船旅」を誰もが楽しめるというコンセプトを構築した。居住地からクルーズ起点港までは飛行機で移動し，そこから船旅に適した水域だけを船で巡る「フライ & クルーズ」という観光ビジネスモデルである。そして，現代クルーズは，海上を移動するホテルとしての宿泊機能だけでなく，ふんだんな食事，多彩なエンターテイメント，ショッピング，そして寄港地での観光を複合的にまとめた，あらゆる人々のためのバケーション商品，すなわちユニバーサル・バケーションとして，多くの人々を魅了するようになった。

1.5　クルーズ料金の特徴

前述のようにクルーズは複合観光商品であり，一般的にクルーズ料金には，船の移動費用，船上でのすべての食事，各種施設の使用，各種エンターテイメントへの参加などの料金を含んでいる。場合によっては乗下船する港までの往復の飛行機代まで含まれている（日本の場合には含まれていない場合が多い）。こうした料金体系から，「オール・インクルーシブ（All Inclusive）」，すなわち「すべてを含んだ料金」をうたい文句にしている。乗客にとって，他に費用が発生するのは，船上でのアルコール飲料，特殊なサービス，サービス要員へのチップ，ショッピング，そして寄港地でのオプショナルツアーや買い物などである。

最近は，スタークルーズのように食事を別料金にしているクルーズも現れているが，これは欧州のクルーズフェリーの流れを取り入れたもので，クルーズの主流は今でも「オール・インクルーシブ」，すなわち「すべてこみこみ」がその料金コンセプトとなっている。

かつての定期船のように船内に等級差（1 等，2 等など）をつけて旅客を区別する船は，一部の特殊な船を除いてない。このような等級差のない船を「モノクラス船」と呼ぶ。モノクラス船でもクルーズ料金の違いはあるが，これは使用するキャビンの大きさとグレードの違いによるものと考えてよい。

1.6　クルーズの内容

クルーズという観光の内容を，利用者の立場からみてみよう。

1.6.1　乗船

クルーズ客船へ乗船するのは，港の客船ターミナル。国内クルーズであればチェックイン，国際クルーズであればチェックインの後，パスポートチェックなどの出国検査がある。

スーツケースなどの大きな荷物は，ターミナルでピックアップしてくれて，船室（キャビン）まで届けてくれるのが普通だ。

船内に入ると，船室に案内される。まずは，船室内に用意された「船内新聞」に目を通すことが必要だ。特に「ボートドリル」と呼ばれる避難訓練は全員参加が義務付けられているので，その時間は要チェックだ。また，この新聞には，どのレストラン，バー，ラウンジが，いつオープンしているかも記載されている。

もう 1 つ大事なのが，デッキプランと呼ばれる船内の配置図だ。船内のどこになにがあるかが一目でわかるように色分けしてある。

出港時間になると，デッキでの出港パーティが開催される。シャンペンなどの飲み物が配られたり，日本船では紙テープが配られたりして，岸壁の見送りの人々との間で別れのテープが結ばれる。汽笛が鳴り響き，いよいよ出港。船は港を後にする。

出港前にボートドリルが行われ，これは乗客が万一の時に自分の乗るボートまでの避難順路を確認するための大事な訓練だ。船室に用意されているライフジャケットを身に着けて，船室のドアの内側に表示されている指定のボートの場所（マスター・ステーション）まで駆けつける。最近になって，ライフジャケットはボート付近に用意されて，ボートドリルの時にはライフジャケットを着用する必要がない船も現れている。

夕食のアナウンスがあれば，指定のダイニングルームへと向かう。乗船当日の服装は，乗船時のままの普段着（カジュアル）で OK だ。

夕食の後，シアターやラウンジでの各種催し物に参加するのは自由だ。いつどんな催し物があるかは，船室に配られている船内新聞に記載されている。たっぷり遊んで小腹が空いたなら，ダイニングルームのミッドナイトビュッフェに出かけよう。寝る部屋も船内なので，帰宅の心配もなしに，ゆっくりと船上の夜を楽しむことができる。

①　クルーズ客船ターミナルの受付カウンター

②　まずはキャビンに

③　ボートドリルには必ず参加

CRUISE COMPASS

DAY 1 *Freedom of the Seas*SM
DEPART MIAMI, FLORIDA
Sunday, June 11, 2006

TODAY'S FORECAST Thundershowers, 85°F

SUNRISE 6:29 am
SUNSET 8:10 pm

TOMORROW At Sea

WELCOME TO THE FIRST DAY OF THE MOST AMAZING VACATION OF YOUR LIFE.
This week will be filled with boatloads of activities, wonderful adventures and unforgettable experiences. In other words, there is a lot to do and see. The list below highlights the things you won't want to miss. Each night you will receive a Cruise Compass and Daily Planner in your stateroom so you can get a jump-start planning your adventures for the next day. Remember that our crew is here to ensure you receive the best service and have the best vacation, so feel free to ask any questions you may have. Get out there!

TODAY'S TIPS
- Please note that luggage delivery is a lengthy process. Thank you for your patience and understanding.
- Familiarize yourself with the safety routine onboard and your Assembly Station, your assigned meeting place in case of emergency. View our safety film on your stateroom television, channel 41. There will be a compulsory mustering at 4:45 pm.
- A nibble of this, a bite of that, try it all at the Windjammer buffet on Deck 11 aft from 11:30 am to 3:30 pm.
- Once we set sail, stop by and browse our duty-free shops and boutiques.
- Tonight's the perfect night to dine at Chops Grille or Portofino.

TONIGHT'S ENTERTAINMENT

Your Cruise Director, **Ken Rush**, proudly presents the

WELCOME ABOARD SHOW
Starring
The Comedy of
Steve Bruner

7:00 pm (Pre-Dinner Show) for Second Seating Guests
9:00 pm for Main Seating Guests
Arcadia Theatre, Decks 3 & 4

Please be reminded that the saving of seats and video taping of shows is strictly prohibited. Also, children in the first three rows of the theatre must be accompanied by a parent or guardian.

DON'T MISS THE EXCITING FREEDOM EXPO!
Join us on the Royal Promenade from 2:30 pm. Let us show you what we have in store for you this cruise vacation. Over $800 in prizes to be won in the free raffle. The raffle will be drawn at 4:00 pm. (You must be present to win).

FAMILY WELCOME ABOARD SHOW
We invite all families to meet our Youth Staff for a special welcome aboard show and introduction to our award-winning Adventure Ocean Youth Program.
Doors open at 7:45 pm, Showtime at 8:15 pm, Studio B, Deck 3

PARTY AROUND THE WORLD PARADE
Don't miss the first of the two beautiful parades! Bring your cameras!
11:15 pm, Royal Promenade, Deck 5

FREEDOM-ICE.COM
Don't miss this incredible ice show. Due to seating limitations, complimentary tickets are needed for this spectacular event. Ticket distribution is tomorrow from 9:30 am – 11:00 am in the On Air Club. Shows will take place on Monday, Wednesday and Thursday.

DINNER IS SERVED Casual Attire
Main Seating – 6:15 pm in Leonardo's, Isaac's & Galileo's Dining Rooms, Decks 3, 4 & 5
Second Seating – 8:45 pm in Leonardo's, Isaac's & Galileo's Dining Rooms, Decks 3, 4 & 5

THE TOP 10 THINGS TO DO ON A ROYAL CARIBBEAN CRUISE VACATION
1. Bring your cruise vacation to new heights – literally. Soaring 200-ft above the sea, the Rock Climbing Wall is something to see but, more importantly, conquer.
2. Indulge yourself, you deserve it. Book an appointment today at our Freedom Day Spa.
3. Let our entertainers amaze you in the Arcadia Theatre.
4. With so many shore excursions like SCUBA diving and parasailing, we have just the thing for everyone seeking new adventures. Book today.
5. Whether you like salsa, rock or ballroom dance, one of our many bars and lounges will surely suit your dancing needs.
6. Napa Valley in the middle of the Caribbean? You bet! Be sure to visit Vintages, our wine bar at sea.
7. The Royal Promenade is lined with boutiques and specialty shops with everything from designer duds to exquisite jewelry. So get out there and shop 'til you drop.
8. Practice your figure 8's on one of the only ice rinks at sea.
9. An intimate dinner awaits you. Enjoy fine Italian cuisine at Portofino or the perfect steak at Chops Grille. Call and make your dinner reservations now.
10. Top off your evening in Olive or Twist or one of our other many exciting bars and lounges.

Get ready for fun! Check out the Daily Planner for each day's activities.

④　船内新聞を見て，まずは当日の予定を決定

⑤　出港，日本船では紙テープが投げられる

⑥　時間になったら，ダイニングルームで指定のテーブルに

⑦　誕生日には，みんなで祝福

図1-4　クルーズ客船への乗船〜夕食

1.6.2　寄港日

翌朝，水平線から昇る朝日を浴びながらデッキをウォーキングするのも悪くない。プロムナードデッキと呼ばれる船の周りを一周できる通路が設けられた甲板があるのが普通だ。船尾などのデッキでは，インストラクターによる朝の体操教室なども開催される。昨日の乗船当日から，ついつい食べ過ぎた人には，適度の運動が必要だ。モーニングコーヒーも用意されているので，汗を流した後，海を眺めてゆったりとした時間を過ごすのも楽しい。

朝食は，ビュッフェ・レストランで好きなものを適当にとって食べてもよいし，メイン・ダイニングルームで食事をしてもよい。日本船であれば，和食か洋食を選択できるのが普通だ。朝食も終わる頃，最初の寄港地に船は近づく。海から入る時の町の風景は，いつもと違って新鮮だ。

下船ができるというアナウンスがあればいよいよ上陸だ。小さな港で，船が岸壁に着岸できない場合には，テンダーボートと呼ばれるボートでの上陸となり，これもなかなか体験しがたいので楽しい。

岸壁には，オプショナルツアーのバスがたくさん並んで乗客を待っている。オプショナルツアーは，事前に船上で申し込んでおくのが普通だ。オプショナルツアーに参加せず，自分で近くの町や観光地を見て回ることも自由だ。岸壁から町まで遠い場合には，無料の送迎バス（シャトルバス）が運行されていることもある。もちろん，船から降りずに船上で楽しむ人も少なくない。停泊中でも船上では，飲食のサービスは提供されている。

定刻に，船は港を出港する。一般的には，朝に入港して，夕食前には出港というパターンが多い。くれぐれも最終帰船時間には遅れないことが肝心だ。海外で寄港し，観光に下船する時には，ボーディングカードと呼ばれるカードがパスポート代わりで，パスポートは船が預かってイミグレーション（出入国管理）などを行うのが普通だ。それでは，パスポートも持たない状態で，出港に遅れて，船に置いてきぼりにされたらどうするか。その時には港にあるクルーズ会社の代理店に助けを求めるしかなさそうだ。

船上では，船長主催のカクテルパーティが開催される。できれば正装（タキシードやイブニングドレス）で楽しむことをおすすめしたいが，現代のクルーズ客船ではあまりうるさいことはいわなくなった。ただし，夕刻以降のダイニングルームでは，短パン，水着はご法度だ。パーティでは，会場の前に船長が立ち，一人ひとりの乗客と握手をして歓迎の意を表するという，定期客船時代の慣習を残している船もある。1 週間以上のクルーズでは，乗船翌日にウェルカムパーティ，下船の前々日にフェアウェルパーティとして行われる場合が多い。船長以下，船を運航する士官が紹介される場合も多いので，なかなか貴重な機会となる。

パーティの後は，ダイニングルームでの食事となる。テーブルはあらかじめ指定されているのが普通で，一緒にテーブルを囲むテーブルメイトに挨拶をして着席しよう。ウェイターがメニューを配ってくれるので，好きな料理を選んで注文する。

食事の後は，シアターやラウンジでプロによるショー，歌などが楽しめる。カジノもオープンするが，18 歳未満は入室できないことが多い。ただし，日本の船は，現金を掛けるカジノは法律上できないので，あっても雰囲気を楽しむだけのゲームだ。

① 着岸できない港では，沖に停泊してテンダーボートで上陸

② テンダーボートが着く桟橋では，クルーズスタッフがお出迎え

③ 船長主催のカクテルパーティでは，船長以下の主要スタッフが紹介される

④ 深夜のミッドナイトビュッフェでは，シェフがお出迎え

図1-5　寄港地～カクテルパーティ～夜食

1.6.3　終日航海日

1日，どこにも寄港せずに航海する日があるのは乗客にとって嬉しい。寄港日ばかりだと，陸上の観光に疲れきってしまうので，たまの航海日にはゆっくりと船上生活を楽しむことができるからだ。船内では，いろいろな催し物が，船内新聞に記載のスケジュールどおりに開催されている。どれに参加するかは乗客の自由だ。1日，プールサイドのチェアで横になって，読書をしたり，うつらうつらしていても誰も咎めはしない。忙しく船上生活を満喫するのも，ゆったりとした時間を楽しむのも，退屈するのも乗客の自由だ。「クルーズはお気に召すまま」「Enjoy as you like」といった言い方もされるが，まさにクルーズという観光のあり方をよく表している。

最近の標準的な1週間クルーズだと，2日間くらいの航海日があるのが普通だ。

① 航海日のプールサイド

② プールでのイベントに集まった乗客

③ 船上での乗客同士のコミュニケーションも楽しい

④ フェアウェルショーでは，プロのエンタテイナーとクルーズスタッフが勢ぞろいして，再会を約束

図 1-6 船上での楽しみ

1.6.4 下船

下船の前日には，荷物のパッキングをして，大きな荷物は指定された時間にドアの前に出しておく。カラータグが配布されていれば，それを付けておくことを忘れてはならない。下船すると，ターミナルのバゲージ置き場に，このカラーごとに区分けされて，スーツケースがずらっと並べられている。最近の大型船では，数千個のスーツケースとなるから，色を忘れてしまうと自分のスーツケースを探すのに苦労することとなる。ただし，パスポートは絶対に，この荷物の中に入れてはいけない。荷物を取る前に，パスポート検査があるからだ。

下船前夜の大事なことが，チップを渡すことだ。日本船や，海外の高級船の一部では，このチップ制度がないが，一般的には専属でサービスをしてくれた船員にチップを渡すのがしきたりとなっている。パンフレットや船内新聞にチップの相場が記載されているので，いくら渡したらよいかと悩む必要はない。サービス要員は，給料の多くの部分がこのチップとなっているので，特別の不満がなければ快くチップを渡そう。船室にチップを入れる封筒が用意されているのが普通で，それに現金を入れて，ダイニングルームのテーブルの担当者（ウェイター，アシスタントウェイター，チーフウェイターの3人），キャビン担当者に直接渡す。

港に着いても下船するまでに時間がかかるので，下船のアナウンスがあるまで船内でゆっくりと過ごそう。帰りの飛行機の時間にあまり余裕がない場合には，優先して先に下船できるサービスもある。その場合には下船前夜までにフロントで申し出ておくことが必要で，スーツケースなどの荷物を自分で運ぶことが求められることもある。

以上は，クルーズ客船の旅の一例を示したにすぎない。クルーズ客船にもいろいろなタイプがあり，それぞれのクルーズを実施している[2]。

1.7　クルーズ旅行の欠点

クルーズという観光にも欠点はある。その中でも最も大きな欠点は「船酔い」であろう。海が荒れると，船は大きく揺れて，乗客は「船酔い」を起こす。この船酔いは動揺病の一種であり，車酔い，エレベーター酔い，飛行機酔いなどと同種の「病気」だが，船の場合には症状が重くなりがちで，嘔吐にまで至り，寝たままの状態を余儀なくされる場合も多い。起きると気持ちが悪くなり，食事も喉を通らなくなる。

日本における観光に関するアンケート調査のなかには，約50％の人が「船酔いがいやなので客船には乗りたくない」と答えているものもあるほど。すなわち，「船酔い」がクルーズのマーケット規模を半減させているといえる。したがって，船酔い対策はクルーズ運航会社にとってはきわめて重要となる。この船酔い発症のメカニズムおよび対策については詳しく後述する。

クルーズ旅行のもう1つの欠点は，一度船が洋上に出ると，場合によっては数日間も完全に船内に閉じ込められることとなり，陸上のホテルに滞在している時に比べて自由度がきわめて低くなることだ。特に小型船ほど自由度が少なくなる。そのため最近の傾向としては，船を大型化して船内での乗客の選択の自由度をできるだけ上げるようになってきている。大型化すると，複数の違ったレストラン，ラウンジ，ショップ，シアター，スポーツ施設などを船内に配置することができ，いくつもの選択肢の中から乗客が自由に選ぶことができるようになる。その結果，プライベート空間から一歩外に出ると多彩なエンターテイメントが楽しめるという，クルーズならではの快適空間が創造され，これが現代クルーズの人気を高めた要因となっている。

コラム①　陸上のクルーズ

クルーズという言葉は，船以外にも使われている。海外では，タクシーが町の中を客を探してゆっくりと流したり，パトカーがゆっくりと巡回したりすることもクルーズと呼ぶ。さらに盛り場を飲み歩くのもクルーズというそうだ。千鳥足でいったりきたりする様は，まさに海上のクルーズと似ていなくもない。

【参考文献】

[1] クルーズアドバイザー認定委員会：クルーズ教本（各年度版），（社）日本外航客船協会

[2] 池田良穂：静かな海と楽しい航海，朔北社，1994 年 4 月

演習

巻末資料「事例研究：「にっぽん丸」のクルーズ」に基づき，神戸在住のあなたが，東京で乗船，福井で下船するとして，クルーズ旅行の計画（資金計画，部屋の選択，持参する荷物など），クルーズ期間中の行動計画（毎日のスケジュール，寄港地での観光計画など）を立てなさい。

≪問題≫

1-1　クルーズとはどのような観光か。

1-2　大洋を渡る定期客船がなくなった理由はなにか。

1-3　定期客船の船旅とクルーズ客船の船旅の違いはなにか。

1-4　クルーズ客船とリゾートホテルの違いはなにか。

1-5　現代クルーズのビジネスモデルとはどのようなものか。

1-6　クルーズ料金の特徴はなにか。

第2章 クルーズ発展史

2.1 商業クルーズの発祥

世界で最初に本格的商業クルーズ客船が現れたのは 19 世紀の終わり頃のことで，その最初の船は元 P&O の「セイロン」という 2,210 総トンの客船だったと言われる。同船は，イギリスの事業家に購入されて，富裕層のためのクルーズ客船として運航され，1881 年（1883 年という説もある）には世界一周クルーズも実施し，これが最初の世界一周クルーズとされている。

1926 年には，スウェーデンでヨットタイプのクルーズ客船「ステラ・ポラリス」が建造された。まだ本格的な商業用飛行機は出現しておらず，船が唯一の海上を渡る交通手段であった頃であり，欧州での新造クルーズ客船による本格的な商業クルーズ事業の始まりとみなすこともできる。大洋を渡る定期客船が，渡航客が減少するシーズンオフにクルーズ客を乗せて世界各地を周遊することは決して珍しくなかったが，クルーズ専用船はたいへん珍しいものであった。顧客は欧州の富裕層であり，マーケットとしてはたいへん小さいものであった。

この「ステラ・ポラリス」は，第 2 次大戦後も欧州で稼動し，その後，日本に売却され，伊豆でフローティングホテル「スカンジナビア」として利用されたことから，日本の人々にも馴染みの深い船となった。しかし，2007 年に同ホテルは閉鎖，海外に売却されて日本を離れたが，老朽化が進んでいたため曳航中に沈没した。この海外売船については，国内の有志による保存のための活動も行われたが，残念ながら実現せず，最終的には海の藻屑と消えた。

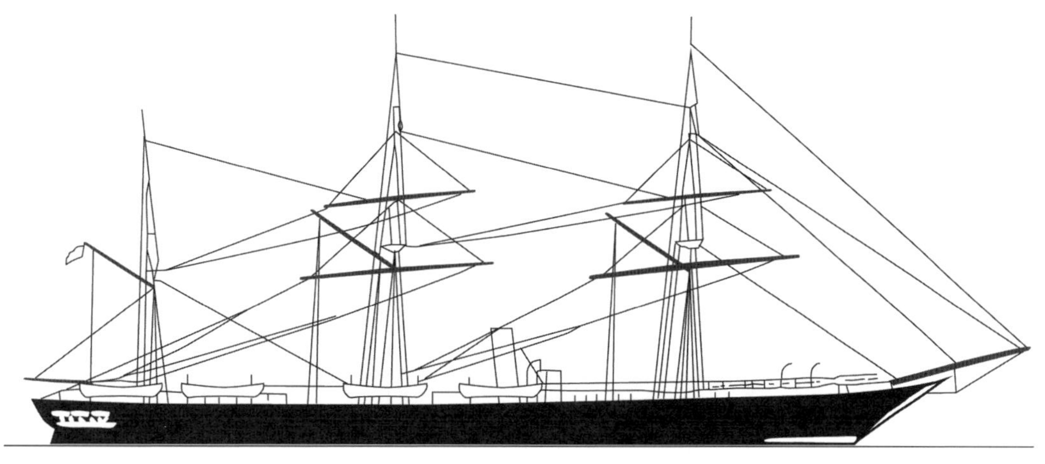

図 2-1　世界最初の商業クルーズ客船「セイロン」

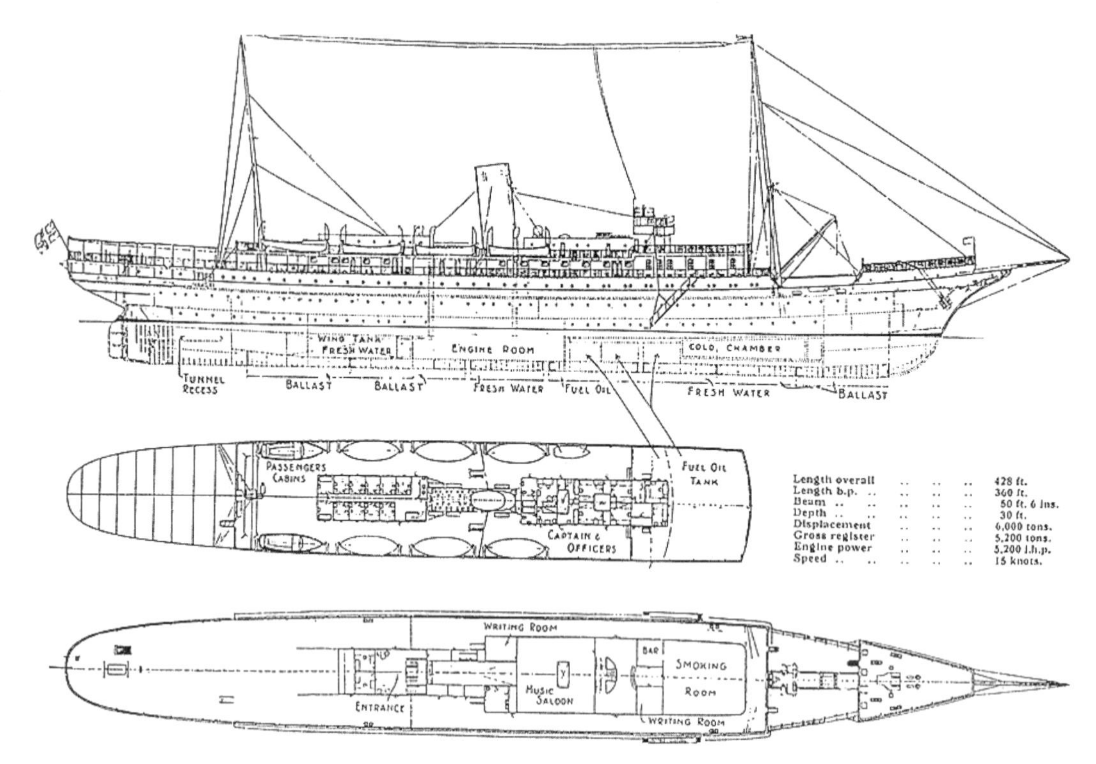

図 2-2　ベルゲン・スティーム・シップ社が 1926 年に建造したクルーズ客船「ステラ・ポラリス」

図 2-3　晩年，伊豆でフローティングホテル「スカンジナビア」として使用された「ステラ・ポラリス」

コラム② 飛行機の登場

ライト兄弟の飛行機の初飛行は 1903 年のこと。「ステラ・ポラリス」が完成した 1926 年に，リンドバーグが単独大西洋横断飛行に成功している。1930 年代になって，ようやく乗客数が 10 ～ 30 人程度の旅客機が登場している。

2.2　定期客船の終焉とクルーズへの進出

1960年代から1970年代のはじめにかけて，世界的な民間航空機網の発達によって，大西洋や太平洋などの大洋を渡る定期客船は次々と撤退した。それらの多くはクルーズに転用されたが，そのほとんどの事業は失敗に帰した。

この理由として以下のことが考えられる。まず定期客船はいくつかの等級に分かれており，基本的には等級差のない「モノクラス」の観光クルーズには非常に使いにくいハードであり，また，ソフト面でも，船旅自体を楽しむことを目的とするレジャー客に対応できるものではなかったことがある。また，大都会と大都会を結ぶ航路に就航していた定期客船が，そのマーケット基盤であった大都会を起点としたクルーズにこだわったこと，数週間から数ヶ月におよぶ比較的長い期間のクルーズを行ったこと，たくさんの定期客船がクルーズ・マーケットに投入されて供給過多となったことなども，この失敗の要因であった。

フレンチ・ライン，イタリア・ライン，ユナイテッド・ステーツ・ライン，日本郵船などは，定期客船の運航廃止とともに早々に客船事業から撤退したのに対し，P&O，ホランド・アメリカ・ライン，チャンドリス・ライン，ホーム・ライン，シトマー・クルーズ，コスタ・ライン，スウェディッシュ・アメリカ・ライン，ノルウェージャン・アメリカ・ラインなどはクルーズを継続したもののうまくは事業展開できず，最終的に生き残った会社は，北米の現代クルーズで成功した会社に引き取られるか，グループ化する形で存続したものであった。

このことは，現在，世界的な規模で発展しているクルーズは，定期客船から派生したクルーズとは一線を画した新しいビジネスであることを物語っており，定期客船ビジネスの延長線上に現代のクルーズを捉えることはできないことを示している。

図2-4　欧州～豪州航路の定期客船として活躍後，クルーズ客船として使われたP&Oの「オリアナ」

図2-5　スウェディッシュ・アメリカ・ラインの「グリップス・ホルム」。大西洋横断定期航路引退後，クルーズ客船となり，世界一周クルーズで日本にも寄港した。

2.3 現代クルーズの発祥―カリブ海に登場した新しいクルーズ

2.3.1 現代クルーズの発祥

前述したように1960年代は，客船にとって暗黒の時代といえた。大洋を渡る大型定期客船は，高速の航空機に旅客を奪われて，次々と撤退し，その多くがクルーズに転用されたものの，ほとんどは事業が成り立たずに姿を消した。

こうした状況の中で，カリブ海で新しいクルーズが芽吹き始めていた。このビジネスは，フロリダ半島の一大リゾート地であるマイアミを基点とし，15,000～20,000総トン，旅客定員700～1,000人程度の客船を使った定期的なカリブ海クルーズから始まった。最初は地元のマーケットを中心としていたが，やがて飛行機とのタイアップを行い，北米全土から飛行機でマイアミに旅客を運び，そこからカリブ海のクルーズを楽しんでもらうというスタイルを確立していった。いわゆる「フライ＆クルーズ」の発祥である。飛行機との競争に敗れて廃業の憂き目にあった定期客船会社では，なかなか発想できない新しいビジネスモデルであった。

こうして，従来のクルーズとは性格のまったく異なる，新しいクルーズ産業が産声をあげたのである。この新しいクルーズを，筆者は「現代クルーズ」と呼び，それまでの伝統的なクルーズとは完全に違ったビジネスモデルとして分類している。

2.3.2 現代クルーズのパイオニア

この現代クルーズのパイオニアは，ノルウェージャン・カリビアン・ライン（NCL，後にノルウェージャン・クルーズ・ラインに改名。略称のNCLは変わらず）である。元々，マイアミ基点の旅客船業を営んでいたアリソン氏が，ノルウェーの海運企業人のクロスター氏と手を結んで設立した新しいクルーズ会社で，欧州の旅客カーフェリーをクルーズ仕様に改装して「サンワード」（SUNWARD）と命名し，マイアミ基点のバハマクルーズを始めて成功した。この成功をうけて，15,000総トン級のクルーズ客船4隻の建造を発注して，そのうち3隻を1969年から次々にカリブ海クルーズに投入した。ただし，第4船については，さすがに事業の急拡大に躊躇したようで，当時アラスカクルーズに進出していたP&O社に売却した。第1船の「スターワード」（STARWARD）は

図2-6　NCLの初期のクルーズ客船の1隻「スターワード」

200台の乗用車甲板を有していた。これはマイアミ周辺の乗客が港に車を放置してクルーズをすることを嫌がったため，ガレージ代わりに船内に設置したものだが，港のパーキング場が整備されて，やがてこの乗用車甲板は姿を消した。この「3隻の同型船の運航」が，実は「フライ＆クルーズ」とともに，現代クルーズ成功のキーポイントであった。それは，大量の乗客を扱うことによって旅客1人あたりの運航コストを削減し，クルーズ料金を低廉化して，だれでも気軽に乗船できるレジャーへと変身させたのである。それまでのクルーズは，かつての大洋を渡る大型定期客船の1等の豪華な雰囲気を継承したもので，暇な富裕層の道楽旅行とみられ，特に長期のクルーズは「動く高級養老院」とまで，若干のやっかみをもって呼ばれていた。この「高級な船旅」を，一般庶民にも十分に手の届くレジャーにできた要因の1つが，この複数客船の効率的な運航であった。さらに航空機によるマイアミへのアクセスが，クルーズの「旅行期間の短縮化」となり，仕事をもつ現役の人でもクルーズを楽しむことが可能となった。

図2-7　マイアミ港に並ぶクルーズ客船群。1980年代の初めには，まだ2万総トン程度の小型クルーズ客船が主流であった。

2.3.3　よきライバルによる相乗効果

この現代クルーズのパイオニアであるNCLに続いて，同じく北欧船主が中心になって設立したロイヤル・カリビアン・クルーズ・ライン（RCCL，現RCI）が，やはり3隻のクルーズ客船を建造して，ほぼ同じスタイルのクルーズをマイアミを基点にして始めた。第1船の「ソング・オブ・ノルウェー」は，フィンランドのバルチラ造船所で建造されたが，鋭く突出した船首，煙突の中腹の展望ラウンジ，最上デッキに広がるサンデッキなど，それまでの定期客船にはない斬新なデザインで人々を驚かせた。

さらに，1970年代になって，NCLの創業者の1人であったアリソン氏が，クロスター氏と意見が対立してスピンアウトし，カーニバル・クルーズ・ラインを設立した。同社は，中古客船3隻を購入してクルーズ客船に改装し，同じくマイアミ基点のクルーズに投入した。コンセプトはNCLやRCCLと同じで，3隻のクルーズ客船を投入したことも同じであった。また，カーニバル・クルーズ・ラインの3隻は，改造船ではあったがNCLやRCCLの船よりも大型で，旅客定員は40～50%あまりも多かった。これも旅客1人あたりのコストを削減し，より安い料金でのクルーズ商品を実現する

図 2-8　RCCL の初代 3 姉妹船の 1 隻「サン・バイキング」。煙突中腹の展望ラウンジがユニーク。

図 2-9　カーニバル・クルーズ・ラインの改造クルーズ客船「カーニバル」。3 隻の中古客船を改造してクルーズを開始した。

こととなり，クルーズの大衆化を一気に推し進めることとなった。

このように 1970 年代には，新しいクルーズのビジネスモデルを構築した 3 社がよきライバルとしてマイアミ起点のカリブ海クルーズの需要を急速に開拓していった。その基本コンセプトは，①年間を通した定点定期クルーズ，②飛行機とクルーズを組み合わせたフライ & クルーズ，③ 1 週間程度の短期クルーズ，④ハイレベルなサービスをリーズナブルプライスで提供，という 4 点に絞られる。これこそが，現代クルーズが急成長したキーポイントといえる。

コラム③　現代クルーズを育てた 3 社

1960 年代後半～ 1970 年代前半にカリブ海で新しいクルーズビジネスを開始した 3 社（NCL，RCCL，カーニバル）の共通点は，いずれもマイアミ港を起点とする定点定期クルーズを運航し，飛行機とタイアップしたフライ & クルーズを中心として販売を行い，複数隻の客船の運航によってコストを削減して，高級なレジャーであるクルーズを，一般庶民をメインターゲットとしてリーズナブルな価格設定で提供したことにあった。しかも，3 つのライバルが，ほぼ同時期に同じスタイルのクルーズを実施したことが，よい相乗効果を挙げた。毎週末の金・土・日曜日には，マイアミから 3 社のクルーズ客船が乗客を満載してカリブ海へと出港していった。

2.3.4　旅行代理店がマーケットを開拓

年間を通じて，週末の金・土・日曜日の 3 日間には，毎日，クルーズ客船が同じ時間に出港するようになって，旅行代理店にとっては販売が非常に便利になった。また，マイアミへの飛行機についてもクルーズ会社が手配をして，クルーズ料金に含まれるようにしたので，旅行代理店は，クルーズ会社に 1 本電話連絡をするだけですべての旅行の手配が終了できた。また，その後のファックスの普及も，小さな旅行代理店にとっては予約業務の簡易化に大きく寄与したという。

この手軽さが，全米にクルーズ専門の旅行代理店（Cruise Only Travel Agency）を林立させ，これが北米の各地域でのクルーズ・マーケットを強力に開拓した。クルーズ会社の首脳に，「最終顧客はもちろん乗客だが，営業上の最重要顧客は旅行代理店」と言わせるほど，クルーズという旅の内容をよく知って，かつクルーズに愛着をもつ人が経営する旅行代理店を最重要顧客として大事にしたマーケティングが展開された。マイアミでは，停泊中に旅行代理店を対象とした船内見学を行い，体験招待クルーズにも力を入れた。旅行代理店の教育こそ，クルーズ・マーケット拡大の最大の武器であったのである。やがて，主要都市の旅行代理店にはクルーズのパンフレットが常時前面に並ぶようになった。

カリブ海でのクルーズが活況を呈するようになり，古い船会社でもカリブ海クルーズに進出するようになったが，そのほとんどは，この3つの会社の新しいビジネスモデルの本質には気づかないまま，旧態然とした伝統的クルーズ事業を展開して，失敗をしている。

コラム④　クルーズだけを扱う旅行会社

クルーズ専門旅行会社には，家族だけで経営する小さなものも少なくなく，地元の固定客を集客し，代理店主の夫婦も一緒にクルーズを楽しむといったこともたくさんあったという。クルーズ会社のマージンは15%程度と大きく，さらに15人の集客に1人の添乗があるなど，クルーズファンの経営者には魅力的なものであった。こういうインセンティブが十分に機能して，クルーズが全米のレジャーマーケットに浸透していった。

2.4　伝統的クルーズの動向

伝統的クルーズにも，1960年代末から新しい動きが始まっていた。その1つの事例が，1968年に完成した大型客船「クイーン・エリザベス2」で，イギリスの老舗定期客船会社であるキュナード・ラインの船であった。同船は，イギリスのグラスゴーの造船所で建造されていたが，建造の途中で大型定期客船の役割が終焉をむかえつつあることがはっきりとしてきたため，クルーズ客船仕様に変更されて完成した。「これこそが正統派のクルーズ」との評価もあったが，等級分けされたレストランや，旅客定員のわりには狭い公室など，その後で登場する現代クルーズ用の客船とは違った面も多々あった。

クルーズ事業への進出に積極的なノルウェー船主の中にも，NCLやRCCLのような現代クルーズに進出して成功した船主以外に，伝統的クルーズに新造船を投入して進出した会社もある。その1つがロイヤル・バイキング・ライン（RVL）であり，1972～1973年に，2.2万総トンのクルーズ客船3隻をフィンランドのバルチラ造船所で新造して，ワールドワイドな不定期クルーズを開始した。「ロイヤル・バイキング・スター」「ロイヤル・バイキング・スカイ」「ロイヤル・バイキング・シー」の3隻である。高級感溢れるそのハードとソフトは，クルーズのトップマーケットの顧客を魅了して，「クイーン・エリザベス2」（6.6万総トン）を擁するキュナード・ラインと並ぶ高いランキングに常に君臨した。これらが現在のクルーズのトップマーケットで，「ラグジュアリー・マーケット」と呼ばれるクラスである。

しかし，退職後の老齢者をメインターゲットとした，このビジネスモデルのマーケットは，主にリピーター（何度も乗船する顧客）に頼るもので，マーケット自体が現代クルーズのように大きく拡大することはなかった。ロイヤル・バイキング・ラインは1980年代の初めに，3船に27.7mの船体延長工事をして大型化して旅客定員を増やし，さらに1988年には3.8万総トンの「ロイヤル・バイキング・サン」を建造してマーケットの拡大を図ったが，その後登場したクリスタル・クルーズなどの新興勢力との激しいシェア争いの中で衰退していった。

このクルーズ・マーケットのセグメントには，その後，1万総トン以下の小型高級クルーズ客船が多く投入されるようになった。その先駆けが「シーゴッデスⅠ」と「シーゴッデスⅡ」の姉妹船である。ニッチ・マーケットに合ったキャパシティをもち，公室数は限られているが，マンツーマンに近いサービス要員によるきめ細かいサービスを受けながら，ゆっくりと静かにクルーズを楽しみたいという層には大いにうけた。これが後にブティック・クルーズと呼ばれるようになった。

日本郵船は，ロサンゼルスにクルーズ運航子会社クリスタル・クルーズを設立し，1990年には，5万総トン，定員900人の大型豪華クルーズ客船「クリスタル・ハーモニー」を建造して，このトップマーケットに進出した。同船は，当時トップマーケットに君臨していた「クイーン・エリザベス2」（QE2）やロイヤル・バイキング・ラインの客船のシェアを獲得して，キュナード・ラインおよびロイヤル・バンキング・ラインを窮地に追い詰めていく。これは両社が客船の代替時期を逸して，老朽化したクルーズ客船を使い続けた結果，多くの顧客に飽きられたものとみられている。

その後，ロイヤル・バイキング・ラインはNCLに統合されて消滅し，「ロイヤル・バイキング・サン」はキュナード・ラインに売却された。ラグジュアリー・マーケットでの地位を確立したクリスタル・クルーズは，その後2隻の大型クルーズ客船を新造し，一時は3隻体制で運航していたが，最終的には供給過剰となり，初代船「クリスタル・ハーモニー」を郵船クルーズに移籍させ，同船は「飛鳥Ⅱ」となって日本のマーケットで活躍している。

一方，経営危機に陥ったキュナード・ラインは，カーニバルグループの傘下に入り，同グループのラグジュアリー・マーケット部門の中心として再生する。世界最大（建造当時）のクルーズ客船「ク

図2-10　キュナード・ラインが1969年にイギリスで建造した定期ライナー兼クルーズ客船「クイーン・エリザベス2」（70,327総トン）。経営難時には，日本に長期チャーターされ，横浜と大阪でホテルシップとなったこともある。2009年に引退。

図2-11　クリスタル・クルーズの最新鋭船「クリスタル・セレニティ」（68,870総トン）。旅客定員1,100人に対し船員数が650名と多く，ハイグレードなサービスを売りにしている。

イーン・メリー2」(15万総トン) を建造して，名門キュナード・ラインのブランド力を発揮して復活し，老朽船「QE2」を引退させるとともに，9万総トンの「クイーン・ビクトリア」「クイーン・エリザベス」を新造して，ラグジュアリー・マーケットでのシェア拡大戦略を推し進めている。

図2-12　ラグジュアリー・マーケットで一世を風靡したロイヤル・バイキング・ラインの3姉妹船の1隻「ロイヤル・バイキング・スカイ」

図2-13　ラグジュアリー・クラスの中の小型船をブティック・クルーズと呼び，そのパイオニアが「シーゴッデスⅠ」「シーゴッデスⅡ」の姉妹船。旅客定員108人に対し，船員が89人。1984年に建造され，その後キュナード・ラインを経て，現在はシードリーム・ヨットクラブが運航する。

2.5　クルーズ客船の大型化とマーケットの爆発

1970年代，マイアミを起点とするカリブ海の現代クルーズによるマーケットの開拓が急速に進んだ。当初は，北米大陸の厳冬期間の避寒のために南のマイアミに来る観光客が多かったため，冬がオンシーズン，夏がオフシーズンとなっていたが，しだいに夏の需要も拡大して，カリブ海は年間を通じて大きなクルーズ需要があるようになった。

定員700～900人の3隻のクルーズ客船を投入していた先発のNCLとRCCLは，定員1,300人の3隻のクルーズ客船を投入した後発のカーニバルに比べると，コスト面ではるかに劣っていた。船は，一般に大型化するほど，旅客1人あたりの初期投資，運航コストともに減少する。これは「スケールメリット」と呼ばれる。

NCLは当初自動車を搭載していた甲板を廃止して旅客定員を増やし，RCCLは1979年から2隻のクルーズ客船の中央部に26mの新しい船体を挿入して船体を大きくするジャンボ化工事を行って，旅客定員の40%あまりの拡大を図った。

しかし，クルーズの需要拡大は続き，各社ともに大型船の投入に走る。まず1979年，NCLは係船されていた7万総トンの世界最大の定期客船「フランス」(1961年建造) を購入して，クルーズ客船に大改造し，「ノルウェー」と改名して就航させた。旅客定員は2,356人と，NCLのそれまでの船の

2.6 倍に増している。この大型化には数々の難題が控えていた。カリブ海クルーズで寄港する島々の港に入れないというのが最大の難題であったが，沖止めして，船首甲板上に搭載した大型テンダーボート（交通艇）で乗客をピストン輸送するという手法を編み出して，これを解決している。

これに続いて，RCCL が 1982 年に，38,000 総トン，旅客定員 1,575 人の「ソング・オブ・アメリカ」を新造した。この船は，RCCL の初代の 3 姉妹船の拡大版ではあるが，2 層の広大なサンデッキに複数のプールを配置した，陽光のカリブ海クルーズに適したコンセプトを導入していた。

カーニバルは，1981 年に 2.3 万総トンの「トロピカル」を新造し，さらに 1985 ～ 1987 年に 5 万総トン弱の「ホリデー」「ジュビリー」「セレブレーション」の新造の 3 姉妹船の建造を行った。後者の 3 隻は旅客定員が約 1,800 人で，船体は内部容積を最大限にとるため外観は箱型となり，「ティッシュボックス型」などと呼ばれた。

この各社のクルーズ客船の大型化は，急拡大する需要に対応するためのキャパシティの増加，乗客 1 人あたりのコスト削減によるクルーズ料金の低廉化，さらに大きな船体内でのファンクションの多

図 2-14　大西洋横断定期客船「フランス」を改造した「ノルウェー」（70,202 総トン）。NCL は，この改造でカリブ海のクルーズ客船大型化の口火を切った。

図 2-15　RCCL が新造した 38,000 総トンの「ソング・オブ・アメリカ」。1982 年にフィンランドで建造された。

図 2-16　マイアミ港に並んで停泊するカーニバル・クルーズ・ラインの 4.6 万総トンの「ホリデー」（右）と 4.7 万総トン「セレブレーション」。船内スペースをできるだけ広くとるため箱型にした船型からティッシュボックス型と呼ばれた。旅客定員は約 1,800 人。

様化をもたらし，利用者の満足度を大きく向上させた。

各地域でのマーケット拡大には，前述したとおり旅行代理店が大きな役割を果たし，クルーズ各社は旅行代理店の教育に努力した。研修会，マイアミ港での停泊時のクルーズ客船見学，クルーズへの招待などを積極的に行った。こうした口コミに近いシステムが全米規模でのクルーズの浸透に大きく貢献した。当時はまだテレビを使っての大々的・広域的な宣伝活動を行うまでは至っていないのが現状であった。

2.6　さらなる大型化の進展（第 2 次大型化）

アメリカのクルーズ・マーケットは，クルーズ客船の大型化に伴って急速に拡大した。また，各社ともにクルーズ客船の運航隻数を増やし，かつそのほとんどをマイアミ起点の定点定期クルーズに集中することで効率化を図り，コストを削減することができた。北米における航空機の規制緩和の時代であったこともあり，フライ & クルーズも格安の費用で実施することができた。各クルーズ会社ともに，飛行機代金をクルーズ料金に含めて販売するようにしたことも，乗客の手間そして各地の旅行代理店の手間を省くこととなった。さらに当時，航空機料金が下がり，航空券の販売手数料ではやっていけなくなった小規模旅行代理店が，一斉にクルーズの販売に力を入れたことも，北米各地でのクルーズ・マーケットの爆発に大きな貢献をしたといわれる。

大型化のメリットを認識して，7 万総トン級という大型化に最初に着手したのは前述のように NCL であった。これに続いて，RCCL は 1989 年に 7 万総トン型のクルーズ客船「ソブリン・オブ・ザ・シーズ」を新造して，マイアミ起点の 1 週間カリブ海クルーズに投入した。この斬新な内装をもつ大型船は，クルーズに対する乗客の満足度を一気に高めた。北米におけるクルーズ・マーケットの爆発がこの時から始まった。

カーニバルも，7 万総トン型の「ファンタジー」を 1990 年に建造し，マイアミ起点の短期（3 ～ 4

図 2-17　1989 年に建造された 7 万総トン型新造クルーズ客船「ソブリン・オブ・ザ・シーズ」。同船の成功により，7 万総トン以上の超大型クルーズ客船の建造ブームが到来する。

図 2-18　カーニバルは 1990 年に建造した「ファンタジー」を第 1 船に，同型船を 9 隻建造し，3 ～ 4 日間の短期クルーズ，1 週間クルーズに投入した。写真は「エレーション」で，ポッド推進器を採用している。

日間）のバハマクルーズに投入した。それまで，1 週間以内の短期クルーズには，比較的小型の中古クルーズ客船が多く就航しており，この 7 万総トンの新造クルーズ客船は当初疑問視されていたが，クルーズの入門用として注目を集めて，クルーズ・マーケットの底辺の爆発に大きな寄与をすることとなった。

RCCL およびカーニバルは，7 万総トン型クルーズ客船の絶大なマーケット発掘能力を確認して，いずれも同型船の連続建造に着手した。これらのクルーズ客船は「メガクルーズ客船」と呼ばれ，2,000 ～ 2,500 人の乗客を積んだ。

2.7 アラスカでのクルーズ

前節までに紹介したように，1960 年代から 1970 年代にかけてカリブ海で現代クルーズが急速に成長していた頃，同じ北米のアラスカ水域でも新しいスタイルのクルーズが芽生えていた。

氷河によって形づくられた風光明媚なアラスカは古くから人気のある観光地で，フィヨルド，氷河観光のための小型の観光船が就航していた。その中で 5,000 総トン型の宿泊設備をもつ客船「プリンセス・パトリシア」を用船してアラスカクルーズを 1965 年から運航していたのが，プリンセス・クルーズ（アメリカのシアトル）であった。同社は 1972 年，ニューヨーク～バミューダのクルーズ客船として新造されたものの遊休化していたノルウェージャン・クルーズシップ社の「アイランド・ベンチャー」をチャーターして，「アイランド・プリンセス」としてアラスカクルーズに投入した。

一方，イギリスの老舗客船会社 P&O は，オーストラリア航路に就航後，遊休化していた客船「アルカディア」を使ったサンフランシスコ起点のアラスカクルーズを 1970 年に開始していた。同社は 1972 年に，NCL が建造中であったクルーズ客船のうち 1 隻を購入し「スピリット・オブ・ロンドン」としてアラスカクルーズに投入した。ただし，アラスカのクルーズは季節性が強く，夏季に限られるため，冬季には北米西岸起点のメキシコクルーズや，パナマ運河クルーズを行ってからカリブ海域に移動してクルーズを行っていた。

激しい競争を行っていたプリンセス・クルーズと P&O の 2 社は，1974 年に P&O がプリンセス・クルーズを買収し，プリンセス・クルーズの名前を引き継ぐことで決着した。その後，「スピリット・オブ・ロンドン」は「サン・プリンセス」と改名し，さらに「アイランド・プリンセス」の同型姉妹船の「シー・ベンチャー」を購入して「パシフィック・プリンセス」として，その船隊の充実を図った。

1977 年に，クルーズ・マーケットの爆発を誘起する出来事が起こった。それは，プリンセス・クルーズのクルーズ客船を舞台にしたテレビドラマ「ラブボート」の大ヒットであった。これにより，北米におけるクルーズの認知度は大幅に上がり，一般庶民でも参加できるレジャーであることが浸透した。

アラスカクルーズには，中古定期客船を改造した「フェアスター」「フェアシー」「フェアウィンド」を使ったクルーズを実施していたイタリア系のシトマー・クルーズ，さらにオランダのホランド・アメリカ・ラインも小型クルーズ客船「プリセンダム」で参入した。アラスカクルーズは夏季のみなので，ロサンゼルスなどを起点とするメキシコクルーズの開拓，パナマ運河クルーズ，カリブ海クルーズへのシフト，さらにはアジアクルーズへのシフトなどで冬季間をなんとか凌いでいた。

このようにアラスカ水域もクルーズ・マーケットとしては成長をしていたが，カリブ海クルーズのように画期的な新造クルーズ客船が次々と登場して，それが爆発的なマーケット拡大を促すようなことはなかった。この水域に新造船が登場するのは，1980 年代半ばになってからのことである。

その後，アラスカクルーズには，カリブ海のクルーズ会社が相次いで大型クルーズ客船を投入して参入し，バンクーバーおよびシアトルを起点としたクルーズにたくさんの客船が就航するようになった。

2.8　現代クルーズの世界展開

1980 年代から，北米で現代クルーズのビジネスモデルを確立したクルーズ運航会社の世界展開が始まった。まず NCL が，7 万総トンの「ノルウェー」による夏季の欧州クルーズを開始する。これは，客船は毎年 1 週間程度の修理，検査を受ける必要があるが，アメリカでは同船を受け入れるドックがなく，欧州の造船所でのドック入りに合わせて，欧州での現代クルーズのフィージビリティーを探る意味合いもあった。

当時の欧州でのクルーズは，ドイツのクルーズ・マーケット，ギリシャを起点とするエーゲ海マーケット，そして細々とイタリアマーケット，さらにイギリスマーケットがある程度であった。そのうちドイツのマーケットは，ドイツ語による単言語クルーズを実施しており，ハパグロイド・クルーズの「オイローパ」他数隻による高級マーケットと，ドイツの大手旅行業者がソ連の客船をチャーターして運航する大衆クルーズとに 2 極化していたが，30 万人近いクルーズ人口を誇っていた。

またエーゲ海クルーズは，ピレウス起点のクルーズはカボタージュ（国内輸送はその国籍の船だけができるという規則）に守られてギリシャ籍船の独占市場となっており，非常に古い中古船を使ったレベルの低いクルーズが多かった。このようにクルーズとしては低レベルでかつ高価格でも，どの島に行っても素晴らしい遺跡や美しい町並みが見られるという観光資源に恵まれて，十分な乗客を獲得し，欧州のクルーズ・マーケットの中では大きな割合を占めていた。

イギリスは，キュナード・ラインと P&O が定期客船の運航を中止した後にもクルーズを実施していたが，欧州域でのクルーズは全般に振るわず，年間の多くの期間をアメリカ市場へ出稼ぎにいくという状況であった。

こうした状況の中，前述したように「ノルウェー」の欧州域でのクルーズが始まったが，急速なマーケットの成長は見られなかった。

この欧州域で，最初に現代クルーズのビジネスコンセプトを導入して定点定期クルーズの運航を始めたのは，イギリスの旅行会社エアーツアー社であった。地中海のマジョルカ島を起点とした定点定期のクルーズを実施し，旅客をイギリスから格安の飛行機で運んだ。しかも，料金は地中海へのバカンスパック旅行と変わらないものであった。使用船は，現代クルーズのパイオニアとしてカリブ海で活躍した RCCL の中古船が中心であった。こうして，まずイギリスのクルーズ・マーケットの成長が始まった。

アメリカでプリンセス・クルーズを運航して現代クルーズのノウハウをもったP&Oクルーズも，従来の古典的クルーズとは違ったクルーズを展開し始めた。また，カリブ海の現代クルーズの成功者であるNCL，RCLグループ（RCIとセレブリティ），カーニバルグループ（HAL，コスタ，プリンセス）などが，続々と大型クルーズ客船を欧州マーケットに投入した。この結果，長く20万人台だったイギリスのクルーズ人口は1980年代から急成長した（図4-15参照）。

欧州のもうひとつの大きなクルーズ・マーケットであったドイツマーケットでは，カジュアルブランドとしてアイーダ・クルーズが現代クルーズを展開し，これがドイツのクルーズ・マーケットの爆発を誘引した。

北米系クルーズ会社が欧州で行ったクルーズの乗客は，当初，フライ&クルーズでのアメリカ人客の比率が圧倒的に高かったが，その後，現代クルーズが浸透して欧州の乗客の比率の方が高くなった。

2.9 アジアのクルーズ

アジアのうち日本のクルーズについては第3章で詳しく紹介するので，ここでは日本以外のアジアのクルーズについてみていこう。

アジアでも，1970年代に，本格的クルーズの試みがなされている。香港のC.Y.トン氏の率いるオリエント・オーバーシーズ・ライン（OCL）は数隻の中古客船を購入してクルーズ事業に乗り出したし，チャイナ・ナビゲーションはブラジルから中古客船を購入してクルーズ客船に改造して「コーラル・プリンセス」と改名し，日本マーケットをターゲットとした近海クルーズを実施した。しかし，これらの試みは1980年代半ばまでには消滅する。いずれの会社も，日本のクルーズと同様に，一般的なレジャークルーズだけでは十分な集客ができず，団体チャーターや洋上大学といった企画も実行したものの十分な成果を上げることはできなかった。

シンガポールや香港の港は，クルーズ客船誘致には積極的であったが，いずれも長期クルーズの途中寄港誘致にとどまり，マイアミのようなクルーズ起点港になるための定点定期クルーズの誘致に関心を示すことはなかった。

このアジアに現代クルーズのコンセプトを導入したのは，マレーシア資本のスタークルーズであった。同社は，バルト海のクルーズフェリーを購入してクルーズ客船に改造し，香港とシンガポールでの定点定期クルーズを開始した。それまでのアジアのクルーズは，欧米のクルーズ・マーケットをターゲットとしたクルーズ客船によって行われていたが，スタークルーズのターゲットは地元の特に中国系の人々に設定されていたところが大きく異なっていた。また，バルト海のクルーズフェリーを購入してクルーズ客船に改造し，北欧のクルーズフェリーと現代クルーズをミックスしたビジネスモデルを構築した。その最大の特徴は食事をクルーズ料金に含めずに別立てにした点で，従来のクルーズとは大きく異なっていた。これがクルーズ料金を安く設定できることに貢献し，その後のクルーズ界に広がったフリーダイニングなどの動きにも大きな影響を与えた。

スタークルーズは，1998年に7万総トン型の「スーパースター・ヴァーゴ」と「スーパースター・

レオ」の姉妹船を新造して，シンガポールと香港起点のクルーズに投入した。さらに，中古客船の購入，大型クルーズ客船の新造を続けた。

さらに同社は，現代クルーズのパイオニアである NCL を傘下に収め，一躍世界のクルーズのひのき舞台に躍り出た。

2.10　現代クルーズの更なる大型化（第 3 次大型化）

7 万総トン型クルーズ客船が急増した 1990 年代も終わり近くなって，ついにパナマックス（パナマ運河を通過できるぎりぎりの船幅 32.2 m の船）を超える船幅の船が出現した。これが 10 万総トンを超える巨大船で，その第 1 船はプリンセス・クルーズの「グランド・プリンセス」である。この船に続いて，カーニバルは「カーニバル・ディスティニー」を建造した。

一方，RCCL は，「ソブリン・オブ・ザ・シーズ」を第 1 船とする 7 万総トン型を 5 隻建造した後，9 万総トン型，8 万総トン型の連続建造をして，10 万総トン以上では他のライバルを一気に引き離して，14 万総トン型のクルーズ客船の連続建造を始めた。高速定期客船時代の約 2 倍の大きさのクルーズ客船が登場したこととなる。

2009 年 12 月には，22 万総トン，定員 6,000 人の「オアシス・オブ・ザ・シーズ」がフォート・ローダーデールを起点とする 1 週間カリブ海クルーズに登場した。

【参考文献】

［1］山田廸生，池田良穂：世界の客船 81，船と港編集室，1981 年
［2］山田廸生，池田良穂：世界の客船 85，船と港編集室，1985 年
［3］山田廸生，池田良穂：世界の客船 90，船と港編集室，1990 年
［4］山田廸生，池田良穂：世界の客船 93，船と港編集室，1993 年

資料

主要なクルーズ各社の歴史については，山田廸生氏が月刊誌「世界の艦船」（海人社発行）に「世界のクルーズオペレーター」として詳しい紹介をしているので，ぜひ参照されたい。

カーニバル・クルーズ・ライン：世界の艦船，2003 年 6 月号
コスタクルーズ：世界の艦船，2003 年 7 月号
ノルウェージャン・クルーズ・ライン：世界の艦船，2003 年 8 月号
P&O クルーズ：世界の艦船，2003 年 9 月号
クリスタル・クルーズ：世界の艦船，2003 年 10 月号
ラディソン・セブンシーズ・クルーズ：世界の艦船，2003 年 11 月号
スター・クリッパーズ：世界の艦船，2003 年 12 月号
シーボーン・クルーズ・ライン：世界の艦船，2004 年 1 月号
ロイヤル・カリビアン・インターナショナル：世界の艦船，2004 年 2 月号

キュナード・ライン：世界の艦船，2004年3月号
プリンセス・クルーズ：世界の艦船，2004年4月号
ゴールデン・スター・クルーズ：世界の艦船，2004年5月号
シルバーシー・クルーズ：世界の艦船，2004年6月号
ハパグロイド・クルーズ：世界の艦船，2004年7月号
ウインドスター・クルーズ：世界の艦船，2004年8月号
MSCクルーズ：世界の艦船，2004年9月号
フッティルーテン：世界の艦船，2004年10月号
セレブリティ・クルーズ：世界の艦船，2004年11月号
ディズニー・クルーズ・ライン：世界の艦船，2004年12月号
ホランド・アメリカ・ライン：世界の艦船，2005年1月号
スタークルーズ：世界の艦船，2005年2月号

演習

(1) 1960年代に始まった現代クルーズのビジネスモデルを分析し，その成功の要因を挙げなさい。

(2) 7万総トン型の大型クルーズ客船が大きな成功をする中で，キュナード・ラインの「クイーン・エリザベス2」がほぼ同規模の大型船にもかかわらず成功できなかった理由はなにか。どうすれば，成功できる可能性があったのかを論じなさい。

(3) 7万総トン型の大型クルーズ客船をカリブ海に最初に導入したパイオニアであるNCLは，その後，経営危機に陥りスタークルーズに身売りをして再建の道を歩んでいる。同社が経営危機に陥った原因を分析し，その対策を考えなさい。

≪問題≫

2-1　世界で最初の商業クルーズ客船が現れた年代と，船名を挙げよ。
2-2　定期客船がクルーズ客船にかわっても事業的に成功できなかったのはなぜか。
2-3　現代クルーズが現れた場所と年代を挙げよ。
2-4　現代クルーズのパイオニアの3社を挙げよ。
2-5　現代クルーズが北米で急成長した過程における旅行代理店の役割はなにか。
2-6　伝統的クルーズと現代クルーズの違いを挙げよ。
2-7　クルーズ客船が大型化した理由を挙げよ。
2-8　新造の現代クルーズ客船が，7万総トン，10万総トン，20万総トンを超えたのはそれぞれいつごろか。
2-9　アラスカクルーズとカリブ海クルーズの違いはどこにあるか。
2-10　現代クルーズの世界展開はどこから始まったか。

第3章 日本のクルーズの歴史

3.1 クルーズの発祥

日本には昔から「舟遊び」のように，小船を使ったレジャーはあったものの，現代のクルーズとはかけ離れたものであった。

歴史的にみて，日本で最初に本格的なクルーズを行ったのは，朝日新聞社が尾城汽船から「ろせった丸」をチャーターして1906年に実施した，1ヶ月にわたる「日露戦跡を巡る満韓巡遊船」だとされている。乗客は375名であったという。この「ろせった丸」は，1880年にP&Oがイギリスで建造した客船「ROSETTA」で，3,502総トン，14ノットであった。イギリスとオーストラリアを結ぶ航路に就航した後，1900年に日本郵船に購入され，イギリス時代の名前に「丸」がついて「ろせった丸」となった。翌年，東洋汽船に売却され，さらに1905年には尾城汽船に転売されて，チャーター船として利用された。同船は，引退後にはフローティングホテルとして利用された。

欧州で「セイロン」が商業クルーズ客船として活躍を始めてからわずか20年あまり後には，日本でも本格的なクルーズが始まったのである。このことは，広く世界を見てみたいという「観光」へのニーズは，洋の東西を問わず，人間の欲望の一つとしてあることを示している。

3.2 コーラル・プリンセス

戦後の日本で本格的なクルーズをビジネスとして始めたパイオニアといえば，香港船籍のクルーズ客船「コーラル・プリンセス」（Coral Princess）であろう。9,639総トンの小ぶりの船だが，480人の

図3-1 「コーラル・プリンセス」

乗客を乗せて，1971 年から主に日本を起点とするクルーズ事業を開始した。運航していたのは香港のジョン・スワイヤー & サンズ。まだ一般観光客を対象としたレジャークルーズのマーケットがなかったことから，団体チャーターによる研修クルーズが多かった。同船は外国籍船なので，日本国内だけのクルーズはできず，寄港地には必ず海外の港を入れる必要があった。1989 年，折しも日本のクルーズ元年と呼ばれる年に同船は撤退している。

この他にも，「オリエンタル・クイーン」「トロピカル・レインボー」なども，まだ日本にクルーズの時代が来る以前に果敢にクルーズ事業に挑戦した客船として知られている。

図 3-2 「オリエンタル・クイーン」

図 3-3 「トロピカル・レインボー」

3.3 移民船廃止とクルーズ

戦後になっても，商船三井は南米航路の移民船の運航を続けていた。しかし，移民の輸送も飛行機になり，1972 年（昭和 47 年）には「あるぜんちな丸」は移民船としての最後の航海を終え，三菱重工業神戸造船所でクルーズ客船「にっぽん丸」として生まれ変わった。僚船「ぶらじる丸」は翌年に最終航海を終え，その後，団体クルーズなどを数航海行った後，1973 年に引退した。

生まれ変わった「にっぽん丸」は，1972 年 4 月に，東京発の 2 週間香港・マニラクルーズを第 1 弾として，クルーズ客船としてデビューした。

1975 年には，商船三井がクルーズ客船運航のために設立した子会社商船三井客船が，ブラジル客船を購入してクルーズ客船に改装し，「セブンシーズ」として就航させた。同船はパナマ籍で，乗組員は外国人という外国籍船として，日本を起点とするクルーズに稼動した。人件費の安い外国人船員を使えるというメリットはあったが，外国籍のため，日本国内だけを回る国内クルーズには就航できないというハンデがあった。

1976 年に「にっぽん丸」は台湾で解体され，「セブンシーズ」が日本籍を取得し，2 代目の「にっぽん丸」を襲名した（「にっぽん丸」の船史については参考文献 [1] に詳しい）。

商船三井客船は，1980 年に産業巡航見本市船「新さくら丸」を取得し，定員 550 名のクルーズ客船に大改造してクルーズ市場に投入した。

図3-4　南米移民船「あるぜんちな丸」

図3-5　クルーズ客船「にっぽん丸」（旧あるぜんちな丸）

図3-6　南米移民船「ぶらじる丸」

図3-7　クルーズ客船「にっぽん丸」（2代目）（旧セブンシーズ）

3.4　研修クルーズビジネス

1代目，2代目「にっぽん丸」と「新さくら丸」が開拓したのは，国や地方公共団体，各種団体，企業などがチャーターして行う「研修クルーズ」というビジネスモデルであった。なかでも総理府や各地方公共団体による「青年の船」「生産性洋上セミナー」などが有名であった。船上に缶詰状態で各種の研修を行うことができ，かつ各寄港地での交流ができるということで人気があった。

この研修クルーズビジネスの成功により，同様のクルーズを開始する会社も現れた。

沖縄航路の内航客船を運航していた大島運輸は，巡航見本市船「さくら丸」（1962年建造）を購入してクルーズ客船に改装し，1972年に「さくら」と改名して，主にチャータークルーズに使用した。さらに1983年にはクルーズ客船「サンシャインふじ」を新造して，宗教団体によるチャーターを中心とした研修クルーズ・マーケットに投入した。

瀬戸内海を中心として内航客船を運航する老舗会社関西汽船は，沖縄航路の貨客船「若潮丸」をクルーズ客船に改造し，「さんふらわあ7」として1979年に就航させた。

また，長距離フェリーのパイオニアである阪九フェリーグループは，西日本商船（後の日本クルーズ客船）を設立し，遊休化した大型カーフェリー「フェリーライラック」（元すずらん丸）をクルーズ客船に改造して，1980 年に「ゆうとぴあ」として就航させた。さらに 1982 年には，沖縄航路の高速貨客船「だいやもんどおきなわ」を購入して改造し，クルーズ客船「ニューゆうとぴあ」として就航させた。

図 3-8　巡航見本市船を改造した「新さくら丸」（商船三井客船）

図 3-9　巡航見本市船を改造した「さくら」（大島運輸）

図 3-10　大島運輸の「サンシャインふじ」

図 3-11　沖縄航路の貨客船を改造した「さんふらわあ 7」（関西汽船）

図 3-12　北海道航路のフェリーを改造した「ゆうとぴあ」（西日本商船）

図 3-13　沖縄航路の高速フェリーを改造した「ニューゆうとぴあ」（西日本商船）

3.5　クルーズ元年

1980年代末に，日本の海運会社では，北米でのクルーズ市場が急成長していることを受け，クルーズ市場への参入を計画するところが続出した。また，日本のクルーズのパイオニアである商船三井客船は，中古客船を改造したクルーズ客船を運航していたが，新造クルーズ客船の建造に乗り出した。

商船三井客船の「ふじ丸」と昭和海運の「おせあにっくぐれいす」が竣工した1989年を，「日本のクルーズ元年」と呼んでいる。

図3-14　「ふじ丸」（商船三井客船）

図3-15　「おせあにっくぐれいす」（昭和海運）

3.6　海運大手のクルーズ進出

1980年代後半は海運不況で，各海運会社は新規事業を求めていた。当時，前述のように北米のクルーズ・マーケットが急成長しており，カリブ海には5～7万総トン級の大型クルーズ客船が就航し，日本のクルーズ元年である1989年には7万総トンの新造メガクルーズ客船「ソブリン・オブ・ザ・シーズ」が華々しく登場していた。

こうした海外でのクルーズ産業の盛況ぶりから，日本の海運各社は次々とクルーズ事業に乗り出し，下記のとおりクルーズ客船の運航を始めた。

- 1989年　昭和海運　「おせあにっくぐれいす」
  商船三井客船「ふじ丸」
  少年の船協会「少年の船協会21世紀号」（改造船）
- 1990年　川崎汽船「ソング・オブ・フラワー」（改造船）
  クリスタル・クルーズ（日本郵船）「クリスタル・ハーモニー」
  商船三井客船「にっぽん丸」
  日本クルーズ客船「おりえんとびいなす」
  フロンティア・クルーズ（日本郵船，三菱重工業）「フロンティア・スピリット」
  日本旅客船「ジャパニーズ・ドリーム」

- 1991 年　郵船クルーズ「飛鳥」
- 1998 年　日本クルーズ客船「ぱしふぃっくびいなす」
- 2006 年　郵船クルーズ「飛鳥Ⅱ」（改造・就航）

このように日本のクルーズ元年から，わずか 3 年の間に，日本の海運会社は 10 隻ものクルーズ客船隊を運航するに至った。

このうち，商船三井客船の「ふじ丸」と「にっぽん丸」，日本クルーズ客船の「おりえんとびいなす」は，それまでの研修クルーズをベースとして，さらに一般個人客をもターゲットとしたクルーズ・マーケット用に建造された。いずれも 2 万総トン級で，旅客定員は約 600 名。研修用にも使うため 4 人部屋も多く，全室を 2 人部屋として使うと 400 名あまりの乗客しか乗船できない。

一方，クリスタル・クルーズは日本郵船がアメリカのロサンゼルスに設立した子会社で，北米を中心とするクルーズのトップマーケットへの進出であった。当時，このマーケットでは，キュナード・ラインとロイヤル・バイキング・ラインの 2 社が君臨していたが，すでにクルーズ・マーケット全体の数 % というニッチ・マーケットになっていた。ここに進出したクリスタル・クルーズは，比較的古いクルーズ客船を使用していた両社のクルーズと比較すると，5 万総トンという大きな新鋭船に約 1,000 人という限定した数の乗客を乗せて，質の高いサービスを提供したことで，このマーケットで確固たる地位を築くことができた。

昭和海運の「おせあにっくぐれいす」，川崎汽船の「ソング・オブ・フラワー」は，日本市場においてトップマーケットを創生することを目論んで事業展開を行い，小型クルーズ客船でマンツーマンに近い船員構成により高級なサービスを提供するというコンセプトで，1 日あたりのクルーズ料金は 5 ～ 6 万円以上となっていた。世界的には，こうした小型船での超高級クルーズはブティック・マーケットと呼ばれており，1984 年に建造された「シーゴッデスⅠ」「シーゴッデスⅡ」がそのパイオニアとされている。しかし，元々クルーズのマーケット自体が存在しない日本において需要を掘り起こすことはきわめて難しく，いずれも数年の運航の後，早々に撤退した。

また，フロンティア・クルーズは日本郵船と三菱重工業との合弁企業として設立され，世界の秘境や極地などを回る探検クルーズ客船として「フロンティア・スピリット」を建造した。世界的にみても「リンドブラッド・エクスプローラ」など数隻しか稼動していない，特殊クルーズ客船によるニッチ・マーケットへの参入であった。したがって，日本を中心とする未成熟なマーケットにおいて，この特殊でかつ新しいニッチ・マーケットを開拓することはできずに，就航後 3 年で撤退し，ドイツのハパグロイド・クルーズに売船された。

少年の船協会の「少年の船協会 21 世紀号」は，主に学生などの若者をターゲットとした研修クルーズ専用船として，青函連絡船を改造して登場したが，短い間しか運航されずに売却されている。

また，大阪のレストランチェーン「はや」の経営者は，クルーズ運航会社日本旅客船を設立し，青函連絡船を大改造したクルーズ客船「ジャパニーズ・ドリーム」を使った神戸・横浜間の定期クルーズを行った。会員制を取り入れたユニークな経営だったが，小型船のために欠航が多かったことなどもあり，集客は伸びず，これも早々に撤退した。

2001 年，商船三井客船と日本クルーズ客船は，チャーター船専業の日本チャータークルーズを設立し，「ふじ丸」と「おりえんとびいなす」を移籍させたが，2005 年には「おりえんとびいなす」を売却，さらに 2013 年には廃業に至った。

その後，日本では，商船三井客船，日本クルーズ客船，郵船クルーズの 3 社が，それぞれ 1 隻の外航クルーズ客船を運航する状態が現在まで続いている。なお，2017 年，久々に日本籍クルーズ客船が登場した。

図 3-16 「ソング・オブ・フラワー」（川崎汽船）

図 3-17 「クリスタル・ハーモニー」（クリスタル・クルーズ）

図 3-18 「おりえんとびいなす」（日本クルーズ客船）

図 3-19 「飛鳥」（郵船クルーズ）（撮影：関正哲）

図 3-20 「にっぽん丸」（3 代目）（商船三井客船）

図 3-21 「ぱしふぃっくびいなす」（日本クルーズ客船）

図 3-22　焼肉レストランチェーンの「はや」が神戸・横浜間に投入した青函連絡船改造の「ジャパニーズ・ドリーム」

図 3-23　日本郵船系のフロンティア・クルーズが探検クルーズ市場に投入した「フロンティア・スピリット」

図 3-24　郵船クルーズが「飛鳥」の代替船として投入した「飛鳥 II」。同じ日本郵船系のクリスタル・クルーズ（米）が運航していた「クリスタル・ハーモニー」がその前身。

図 3-25　2017 年に常石造船グループのクルーズ会社せとうちクルーズが，瀬戸内海短期クルーズに投入した 3,200 総トンの超高級船「ガンツウ」

コラム⑤　日本の海運界の勘違い？

1980 年代終わりに，日本の海運会社は相次いでクルーズ産業に進出した。ちょうど，カリブ海では 7 万総トン級のメガクルーズ客船が登場し，クルーズ・マーケットが爆発的に成長していた頃であった。海運不況に苦しんでいた日本の海運界にあっては，新しい海事産業としてのクルーズが魅力的に映ったようだ。

しかし，「現代クルーズ」という新しいビジネスモデルが急成長していることを，本当に理解してクルーズ事業に進出した会社はほとんどなかったようだ。再三来日している「クイーン・エリザベス 2」のイメージが，そのままクルーズのイメージとして定着し，すでにニッチ・マーケットとなりつつあった伝統的クルーズ（トラディショナル・クルーズ）が，アメリカで成長しているクルーズだと勘違いした。

日本の海運会社が，こうした北米でのクルーズの実態を必ずしも正確に認識していなかったことは，クリスタル・クルーズの初代社長であった竹野弘之氏の著書『私はクルーズ業一年生』（海事プレス社）の中でも述懐されている。

3.7　日本のクルーズ市場

日本政府（国土交通省）は，クルーズ元年以降，毎年，クルーズ産業の実績を調査して，「クルーズ人口」を公表している。このクルーズ人口は，アメリカの業界団体 CLIA，ベルリッツ社発行の Douglas Ward 著『Cruising & Cruise Ships』，スウェーデンの ShipPax 社発行の年鑑『STATISTICS』などでも毎年公表されているが，基準が必ずしも統一されていない。一般的には，船上で 2 泊以上するクルーズに乗船した人の数としている場合が多いが，日本政府の調査では，1 泊以上のクルーズに乗船した人の数をクルーズ人口としている。また，日本からのパックツアーで，バルト海の 1 泊のクルーズフェリーに乗船した人の数も含まれているなど，他の統計とは若干違った性格をもっている。

クルーズ元年に 15 万人であった日本のクルーズ人口は，翌 1990 年には 17.5 万人となり，その後の相次ぐクルーズ客船の投入で 1995 年には 22 万人に達したが，バブルの崩壊，新規投入船の業績悪化に伴う撤退から，急速に減少して，1999 年には 16 万人にまで減少した。

2000 年には，マレーシア資本のスタークルーズが神戸発着の定点定期釜山クルーズを開始して再び 20 万人を超えたが，2 年後には日本マーケットから撤退して，2003 年には 15 万人を切るまで減少した。その後，国内ではクルーズアドバイザー認定制度を発足させるなどの対策がとられ，また飛行機で海外に行き，クルーズ客船に乗船する人も増えて，リーマンショックなどの世界的景気後退の影

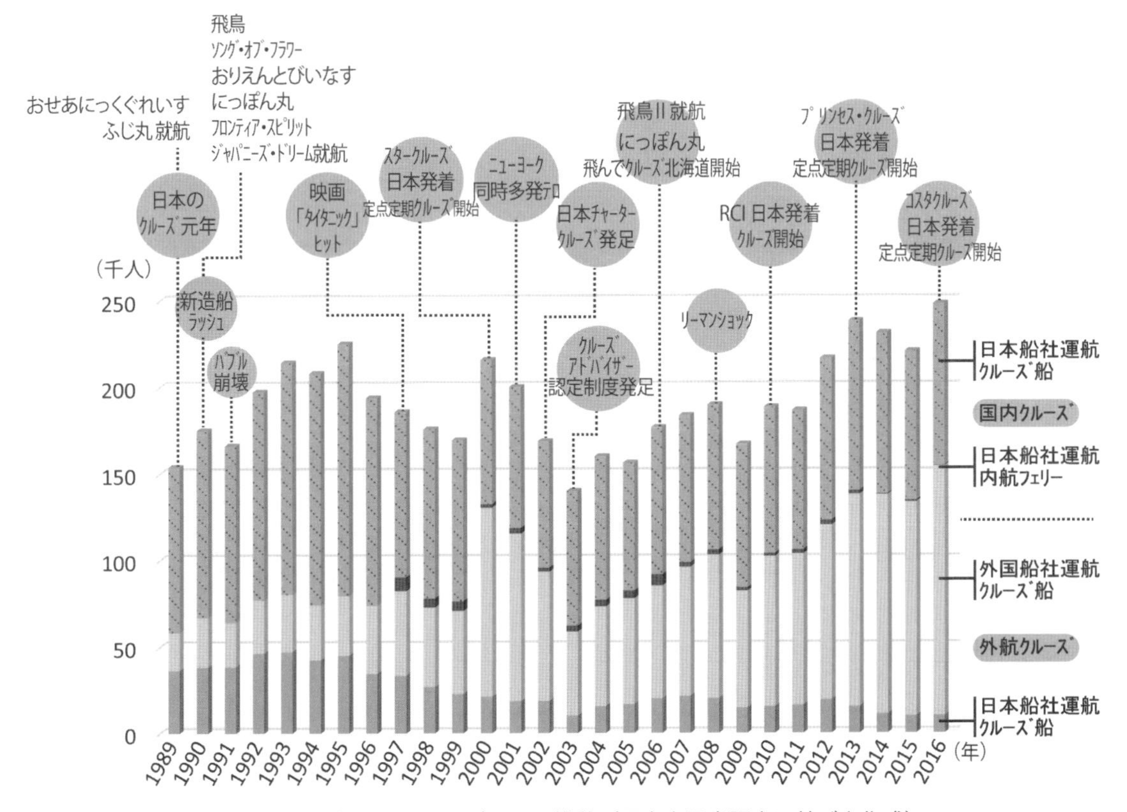

図 3-26　日本人のクルーズ人口の推移（国土交通省調査に基づき作成）

響などで成長率にはばらつきはあるものの，ほぼ順調に推移している。

2010 年からは，外国籍カジュアルクルーズ客船の日本発着のクルーズが不定期ながら始まり，2013 年からは夏季における日本発着定点定期クルーズも始まった。

図 3-27 には，図 3-26 の中の外国の船会社が行うクルーズ客船の乗船客だけを抜き出した。この図から，外国の船会社の運航するクルーズへの日本人乗客数は，クルーズ元年以来，ほぼコンスタントに増加の傾向を示し，特に 2013 年以降は成長率に増加の兆しが見える。

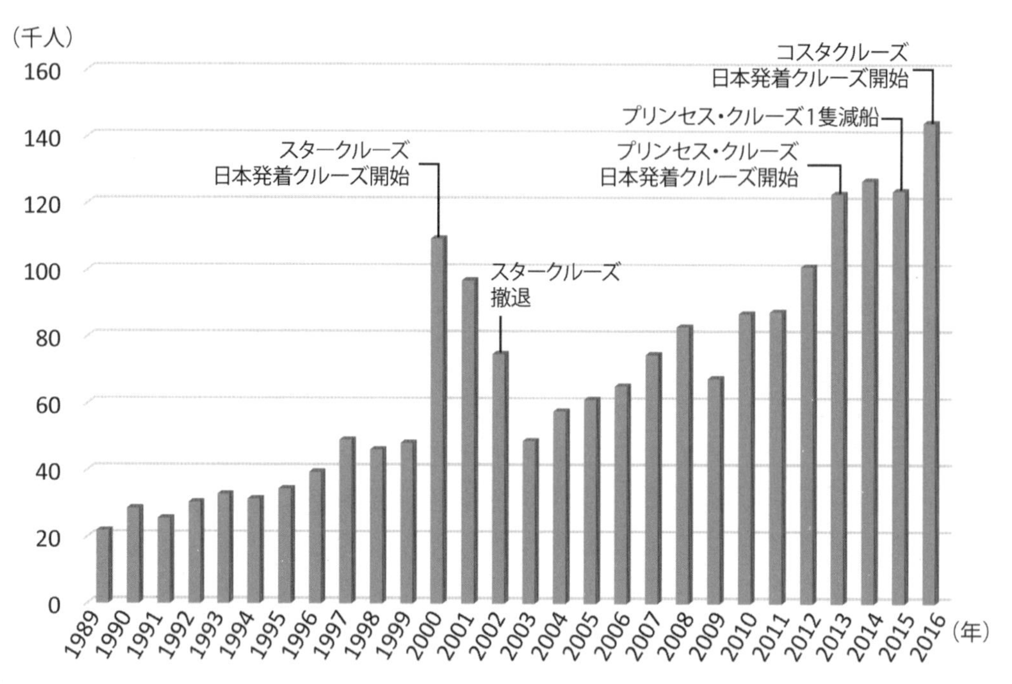

図 3-27　外国船社が運航する船に乗った日本人数（クルーズ人口）（国土交通省調査に基づき作成）

また，クルーズ産業の経営を分析するためには，クルーズ人口だけでなく，1 人の乗客が何日間乗船しているかが，非常に重要となる。図 3-28 に示す外航クルーズの泊数別乗船客数の図から，5 ～ 7 泊および 8 ～ 13 泊の乗船客が最近増えていることがわかる。一方，図 3-29 に示す内航クルーズでは，1 泊クルーズの乗船客が長年 20% 以上を占めており，4 泊までが 90% 近くを占めている。ただし，2010 年以来，5 ～ 7 泊の乗客数が増えているが，これは日本一周などのクルーズが人気となっているためである。

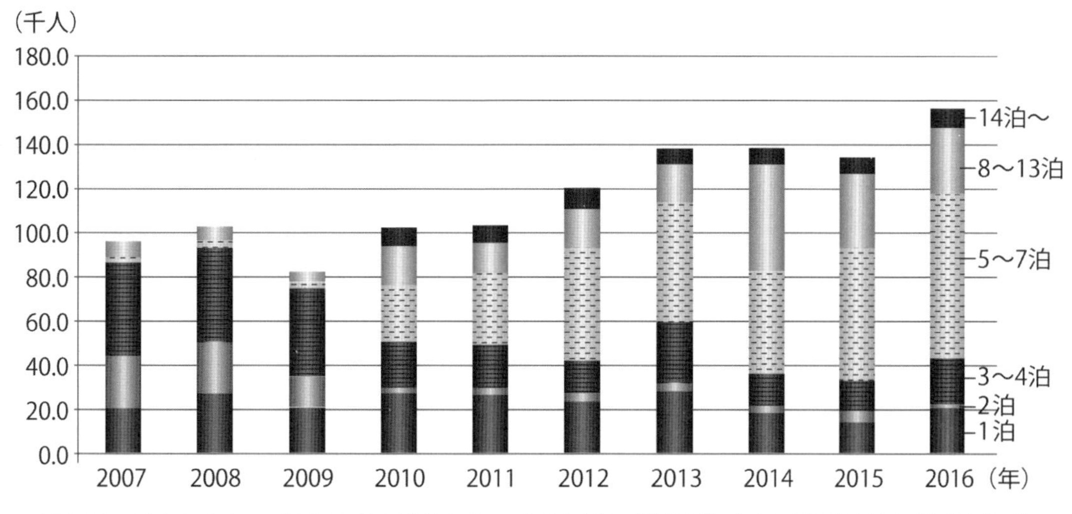

図 3-28　日本人クルーズ人口のうち外航クルーズ乗船客数（泊数別）（国土交通省調査に基づき作成）

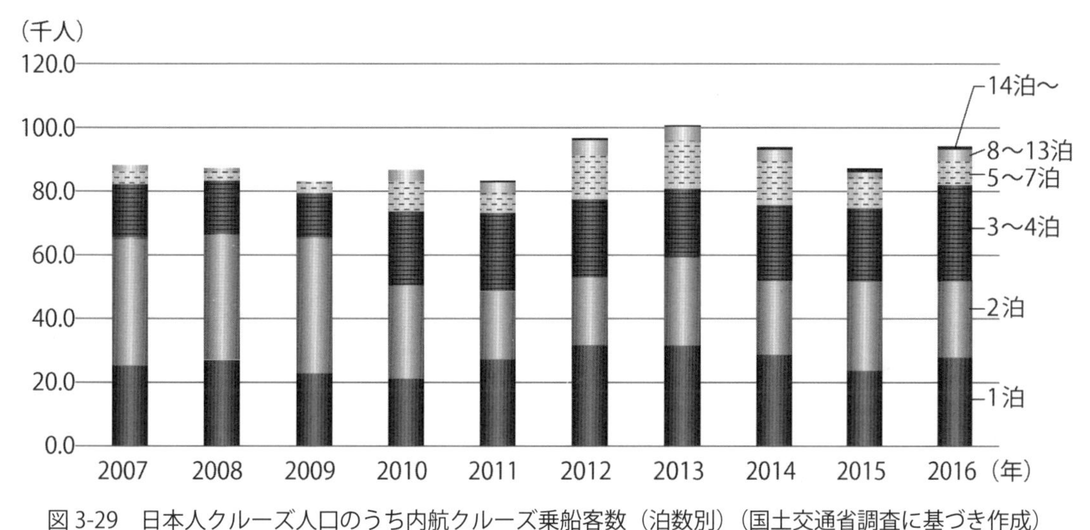

図 3-29　日本人クルーズ人口のうち内航クルーズ乗船客数（泊数別）（国土交通省調査に基づき作成）

3.8　販売システムの充実

日本でのクルーズの販売は，当初はクルーズの運航会社が直接行うことが多かったが，最近は各旅行会社がクルーズの販売を積極的に行うようになってきた。

また，旅行会社がクルーズ客船をチャーターして，自社企画のクルーズを販売する事例もしだいに多くなってきている。当初は，十分な乗客が集まらずに，多額の赤字となった事例も少なくなかったが，最近は，各社の販売網が広がったこともあって，チャーターに積極的となっている。

この旅行会社によるクルーズ販売を大きく促進したのが，第 5 章で詳しく述べるクルーズアドバイザー認定制度である。この制度によって，各旅行会社のカウンターで，クルーズの専門知識を有するクルーズ・コンサルタントおよびクルーズ・マスターの称号をもった社員が対応できるようになった。

3.9　日本のクルーズ・マーケット爆発への処方箋

アメリカで成長した現代クルーズのビジネスモデルと日本籍船のクルーズとを比較すると大きく異なっており，日本のクルーズは豪華客船による伝統的クルーズ（トラディショナル・クルーズ）に近い。

現在世界的なブームになっている現代クルーズのコンセプトを導入することが，日本のクルーズ・マーケットの底辺を拡大し，ひいては爆発させるためには欠かせない。こうした提案は，政府などのクルーズ振興のための委員会などで，これまでにも幾度か行われているが，残念ながら，現代クルーズの以下の基本コンセプトが日本のクルーズ産業に取り上げられることはなかった。

現代クルーズは，①定点定期，②1週間程度までの短期，③飛行機と連携したフライ＆クルーズ，④大型化によるクルーズ料金の低廉化，⑤大型化による船内の楽しみの選択肢拡大，というキーワードで，その性格を表すことができる。こうした現代クルーズの成功要因の導入が必要とされる。

ただし，上述の「現代クルーズの成功のキーワード」の一部を取り入れようという新しい動きも，最近，見られるようになった。商船三井客船がJTB北海道と組んで，2006年から小樽起点で「にっぽん丸」で実施している「飛んでクルーズ北海道」という企画は，8月末から9月中旬までの短い期間ながら，フライ＆クルーズおよび定点クルーズの日本におけるプロトタイプとして注目されており，ほぼ毎航海満船状態で，リピーター客よりも新規顧客の方がはるかに多いという，これまでの日本のクルーズでは見られなかった現象が現れている。この企画は2008年に第1回のクルーズ・オブ・ザ・イヤーを受賞し，さらに2015年，発売開始10周年の機会に再び同賞に輝いている。

また，2006年から東アジアクルーズに進出したコスタクルーズをはじめとする欧米のクルーズ運航会社は，中国マーケットにターゲットを当てたが，日本および韓国のクルーズ・マーケットも注視していた。

まず，RCIは2012年から，単発的ではあるものの日本発着のクルーズを始めた。使用したのは7万総トン型の「レジェンド・オブ・ザ・シーズ」であり，その後14万総トンの「ボイジャー・オブ・ザ・シーズ」によるクルーズも行っている。

本格的な日本発着クルーズを始めたのはプリンセスクルーズで，2013年に8万総トンの「サン・プリンセス」を投入した。さらに翌年には，日本で建造された11万総トンの「ダイヤモンド・プリンセス」を日本人客向けに改装のうえ，「サン・プリンセス」と2隻でのクルーズを始めたが，初期の集客10万人の目標が達成できずに，3年目からは「ダイヤモンド・プリンセス」の1隻体制とした。

一方，コスタクルーズは，中国発着の韓国・日本クルーズにおいて，一部をインターポーティング（9.1節参照）によって韓国および日本で集客することを2012年から始めた。また，2015年からは夏季限定の日本発着の日本海クルーズを，博多，舞鶴，金沢と釜山で集客するインターポーティングで開始した。2017年からは，冬期にも太平洋側の港を発着するクルーズを試みている。

【参考文献】

［1］池田良穂 編：さようなら　にっぽん丸，日本内航客船資料編纂会発行，1977 年 5 月
［2］竹野弘之：タイタニックから飛鳥 II へ―客船からクルーズ船への歴史―，成山堂書店，2008 年 12 月
［3］竹野弘之：私はクルーズ業一年生―クリスタル・クルージス創業の思い出―，海事プレス社，2009 年 1 月
［4］わが国における客船クルーズ振興に関する提案，フェリー・客船情報 2002，船と港編集室発行
［5］山田廸生：世間は船旅をどう見ているか，フェリー・客船情報 2002，船と港編集室発行

≪問題≫

3-1　日本で最初に本格的なクルーズを行った船の名前と，実施した年を述べよ。

3-2　移民船の廃止と，日本のクルーズのつながりについて説明せよ。

3-3　1980 年代後半に日本の海運会社が続々とクルーズ事業を始めたのはなぜか。

3-4　日本のクルーズ元年とはいつか。

3-5　日本のクルーズ元年に建造された 2 隻のクルーズ客船の名前を挙げよ。

3-6　クルーズ元年直後にたくさん就航した日本のクルーズ客船の多くが事業に失敗したのはなぜか。

3-7　日本のクルーズ人口の推移を説明しなさい。

第4章 クルーズ・マーケット

4.1 クルーズ・マーケットの分類

クルーズには，①本格的なクルーズ客船による宿泊型の洋上周遊旅行，②定期航路の旅客カーフェリーまたはRoPax（RoRo Passenger Shipの略）を利用してクルーズが楽しめるクルーズフェリー，③宿泊を伴わないデイ・クルーズなどの種類がある。

まず，①の本格的クルーズが，現在世界的にブームとなっているクルーズの主流である。船は旅客のみを扱う純客船で，宿泊，食事，各種イベントのできる施設をもつ。クルーズの期間は1泊の短いものから，世界一周のように100泊程度のものまでいろいろだが，現在，最もポピュラーなのは1週間もしくは3～4日の比較的短いクルーズとなっている。このマーケットは，2016年の時点で，世界で約2,500万人の需要があり，14兆円程度の産業規模となっている。

②のクルーズフェリーは，定期航路の旅客フェリーを利用したクルーズビジネスである。毎日，同じ航路に定期的に就航する「定期旅客カーフェリー」が，船内設備およびサービスを充実させて，移動のための需要だけではなく，多くの「船旅」を楽しむ観光需要を生み出した。1980年代にスウェーデンとフィンランドを結ぶバルト海横断航路に登場し，それまでは年間400万人程度の移動のためのマーケットだったのを，純粋に船旅を楽しむという新しいマーケットを創造したことによって，年間1,500万人以上のマーケットにまで膨らませた。このマーケット成長の原動力として，船上の免税品販売の力が大きかったが，1990年のEUの経済統合後，域内の多くの国際航路においては船上で

図4-1　デッキに広いサンデッキを有するカリブ海の本格的クルーズ客船「ソング・オブ・アメリカ」

図4-2　バルト海のクルーズフェリー「シリヤ・シンフォニー」。ストックホルムとヘルシンキを1泊で結ぶ定期旅客カーフェリーだが，2泊3日のクルーズ客需要を開拓して大きくマーケットを拡大した。

免税品が扱えなくなったことから，免税品販売に頼らないスタイルのクルーズフェリー・ビジネスに変化している。日本でも，一部の旅客カーフェリーが旅客設備・サービスを充実させてクルーズフェリー化して，船旅を楽しむ旅行客の需要を掘り起こしている。ただし，「クルーズフェリー」という言葉には確固とした定義はなく，一般的に旅客設備の充実したカーフェリーであればクルーズフェリーと呼ぶことも多い。しかし，マーケットという視点からいえば，移動需要だけでなく，かなりの比率の船旅需要を創生したフェリーを「クルーズフェリー」と呼ぶのが正しい。

デイ・クルーズは宿泊施設をもたない客船での船旅で，本格的なレストランでの食事の提供をメインとするレストラン船，海上からの観光を中心として軽食程度を提供する観光船，海上からの観光だけにターゲットをしぼった遊覧船などがある。東京港，神戸港，広島港，博多港などの都会の港や琵琶湖などに就航しているデイ・クルーズ客船が有名。また，海外でもニューヨーク，サンフランシスコなどのディナークルーズでは，タキシードなどの正装をして楽しむ高級船も運航されている。

このようにクルーズといってもさまざまな種類があるが，本書では基本的に①の本格的クルーズのみを取りあげ，②および③のクルーズフェリーとデイ・クルーズのマーケットについては取りあげない。

図 4-3　琵琶湖のデイ・クルーズ客船「ミシガン」

図 4-4　神戸港起点のデイ・クルーズ客船「コンチェルト」

コラム⑥　ピースボート

世界の平和を求めた日本の市民団体「ピースボート」が，クルーズ客船をチャーターして世界一周クルーズを長年行っている。このクルーズは，社会主義国や紛争国，また北朝鮮のように世界的に孤立した国などを訪問して，草の根の市民交流を図る目的で運航されるものだが，最近は，一般の乗客も集客するようになっている。100 日間の世界一周クルーズに 100 万円を切る価格で乗船できる。

4.2　世界のクルーズ・マーケット

現代クルーズのマーケットは，これまで好不況の影響もあまり受けずに年8～10%の成長を続けている。

一般に，マーケットの規模は「クルーズ人口」として表される場合が多い。ただし，マーケットの経済的な規模を表すには，1年間にクルーズ客船で宿泊した延べ人数を表す「年間人泊数」の方が厳密だ。例えばクルーズ人口では，100泊の世界一周をした乗客は「1人」と数えられるが，「人泊数」では「100人泊」と数えられる。年間人泊数に1泊あたりの平均クルーズ料金を掛けると，クルーズ会社の年間のクルーズ運賃売上高となる。

4.3　クルーズ・マーケットの構造

他のマーケットと同様にクルーズ・マーケットもピラミッド型をしており，頂点から順に，探検クルーズやラグジュアリー・マーケット，プレミアム・マーケット，カジュアル・マーケットとなる。このマーケットの分類には厳密な定義はないが，クルーズ料金を指標にして探検クルーズやラグジュアリーが1泊あたり4～5万円から，プレミアムが2万円程度から，カジュアルが1万円程度からというのがおおよその目安といえる。これ以外にも，乗客1人あたりの乗組員数，サービスグレード，公室やキャビンの設備などを総合的に評価したランキングから分類することもある。

ダグラス・ワード氏のクルーズランキングは世界的にも有名であり，その評価の詳細が本として毎年発行されている。

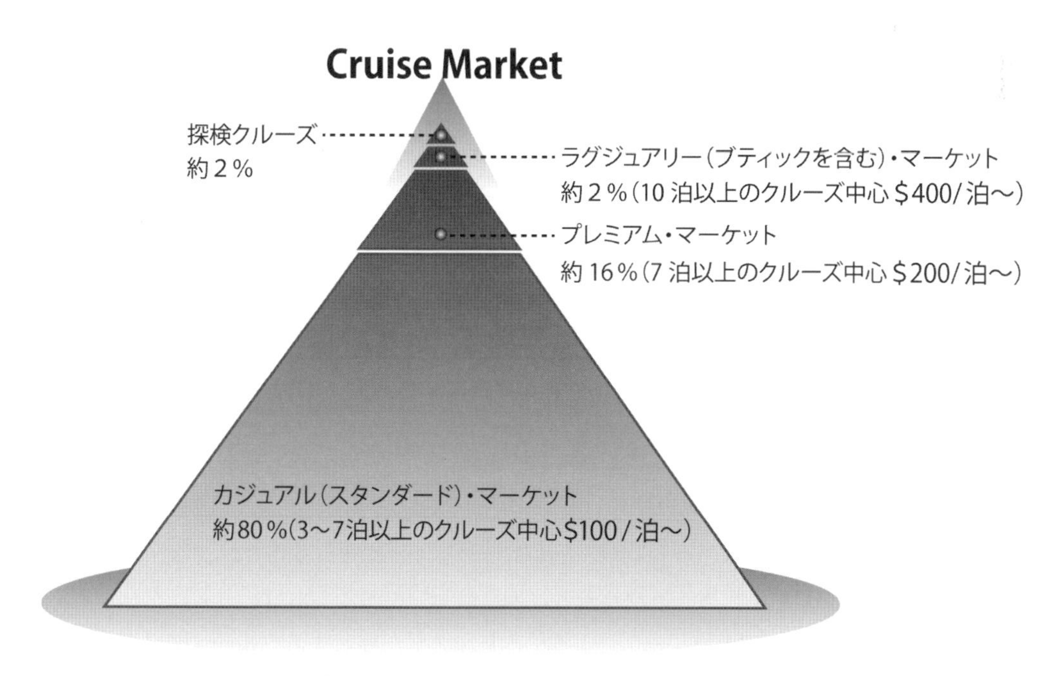

図4-5　世界のクルーズ・マーケットの構造

コラム⑦　クルーズ客船のランキング

クルーズ客船の評価にはいろいろあるが，ダグラス・ワード著の「クルーズガイド」のランキングが有名である。最終的なランキング結果については，著者の趣味嗜好が強くでているので注意をした方がよいが，評価指標の設定と評価点は参考になる点が多い。

ランキングは，
5 スタープラス
5 スター
4 スタープラス
4 スター
3 スタープラス
3 スター
2 スタープラス
2 スター
1 スタープラス
1 スター
の 10 段階評価で，船自体（25%），居室（10%），食事（20%），サービス（20%），エンターテイメント（5%），クルーズ体験（20%）という項目と重みによって，著者の評価による総合点で決定される。

船の大きさによって，大型船（定員 1,200 人以上），中型船（500 ～ 1,200 人），小型船（201 ～ 500 人），ブティック船（50 ～ 200 人）に分けられてランキングがなされている。

図 4-6　ダグラス・ワード著の『Cruising & Cruise Ships』（2018 年版）

4.3.1　ラグジュアリー・マーケット

費用に糸目をつけずハイグレードな船旅を楽しみたい人をターゲットにしたマーケットで，全体のおおよそ 2% 程度を占めている。

比較的小型船を使ったブティック・クルーズと，大型船を使った豪華客船クルーズに分かれており，前者としてはシーボーン・クルーズ・ライン，シルバー・シー・クルーズ，ラディソン・セブンシーズ・クルーズなどが，後者にはキュナード・ライン，クリスタル・クルーズなどがある。

日本では，クルーズというと，このクラスの超豪華船での船旅というイメージが定着しているが，クルーズ・マーケットの中では，すでにニッチ・マーケット（Niche market）となっており，リピーターを中心にした営業展開が一般的である。リピーター率は高く，80% 程度となっている場合が多い。

各社ともに，定点定期ではないワールドワイドなクルーズを展開しており，クルーズ期間も 2 週間以上が一般的で，3 ヶ月もの世界一周クルーズにも就航している。これは，リピーターが多いため，カジュアル・クルーズの定点定期クルーズのように同じルートを周遊するクルーズでは飽きられる可能性が高いことが大きな理由だ。

日本のクルーズ客船の多くは，料金的にはこの分類に入るが，需要の掘り起こしのためもあって1～2泊のクルーズも多く実施している。

4.3.2　プレミアム・マーケット

厳密に分類することは難しいが，高級なハードとソフトを兼ね備えて，落ち着いた大人のクルーズを提供することをメインコンセプトとしている。一部の富裕層だけでなく，若干広いレベルの乗客をある程度大量に集客する必要性があるため定点定期クルーズを基本としているが，季節によって起点港を変えて運航している場合も多い。海外では，セレブリティ・クルーズ，ホランド・アメリカ・ライン，プリンセス・クルーズ，ドリーム・クルーズなどがこのセグメントに入る。

ただし，前述したように厳格にカジュアル・マーケットと分けることは難しい。定点定期クルーズでも，一般にクルーズ期間は1～2週間程度と，カジュアル・マーケットに比べると長い場合が多いのが特徴といえる。

4.3.3　カジュアル・マーケット（スタンダード，大衆，マス・マーケット）

大定員の大型クルーズ客船を数多く運航して乗客1人あたりのコスト削減を図り，一般の人々が気軽に楽しめる「船旅」を提供することを基本コンセプトとしており，世界でブームとなっている「現代クルーズ」のビジネスモデルの中心的存在である。一般大衆をメインターゲットにして，大型船を使って一度に大量の乗客を乗せることで成り立っていることから，「マス・マーケット」と呼ばれることもある。リピーター率はラグジュアリー，プレミアム・クラスの船に比べると比較的低いが，これは常に新規客を獲得できている結果である。このカジュアル・マーケットがクルーズ・マーケットの裾野を拡大し，マーケット全体の成長の牽引車となっている。

定点定期を基本とし，クルーズ期間は1週間程度までが多い。また，さらに短い3～4日間クルーズも急速に増加している。

北米では，このカジュアル・クラスのクルーズが一般的なレジャーとして定着しつつあり，年収3

図 4-7　カジュアル・クルーズの雄カーニバル・クルーズ・ラインが運航する7万総トン型船「ファシネーション」

図 4-8　欧州企業として急成長中の MSC クルーズが運航する新鋭船「MSC オーケストラ」

万ドル（約300万円）以上の層がそのターゲットとされている。短いクルーズなので，年に何回も楽しむ人が多く，夏の長期休暇よりも，第2，第3のショートバケーションとして選択する場合が多い。

カリブ海クルーズのパイオニアであるカーニバル・クルーズ・ライン，RCI（ロイヤル・カリビアン・インターナショナル），NCL（ノルウェージャン・クルーズ・ライン），アジアの新興クルーズ企業であるスタークルーズ，欧州のMSCなどがこのクラスに入る。

4.3.4　特殊クルーズ・マーケット

規模は非常に小さいが，特殊な嗜好をもつ人にターゲットを絞ったニッチ・マーケットとして探検クルーズ，帆装クルーズ，河川クルーズなどがある。

探検クルーズは，秘境と呼ばれる地域や極地へのクルーズを主目的としており，需要がそう大きくないのと，小さな港の利用が多いため，比較的小さなクルーズ客船を使っている場合が多い。「リンドブラッド・エクスプローラ」がそのパイオニアであり，現在，約40隻が稼動している。中には，本格的な砕氷船をクルーズ客船に改造したものもあり，北極点などへのクルーズも実施されている。

図4-9　極地クルーズ客船のパイオニア「リンドブラッド・エクスプローラ」（撮影：菊池宗克）

帆装クルーズは，ヨットと同様に本格的に帆走を楽しむことを主目的とするものと，帆装を備えて，風のある時に乗客へのエンターテイメントとして帆走気分を味わってもらうことを目的とするものの2つに大別される。いずれも，世界で数隻が稼動するだけではあるが，ニッチ・マーケットして定着している。海外では，「クラブメッド2」「ウインドスター」などが有名。日本では，民間が運航する帆走クルーズ客船はないが，練習帆船「海王丸」（航海訓練所）や「みらいへ」（グローバル人材育成推進機構）が，帆走訓練を目的として一般を対象としても稼動しており，これも一種の帆装クルーズ客船といえる。

河川クルーズは，欧州ではライン川，ドナウ川，ボルガ川などに，アメリカではミシシッピ川，エジプトのナイル川，アジアでは中国の長江などに，宿泊施設を有する本格的なクルーズ客船が多数就航している。この河川クルーズは，昼間に航海し，夜間は観光地に停泊する場合が多く，河川に沿った観光地を効率よく訪問できるというメリットがある。また，揺れる心配がほとんどないので，船酔

図4-10　帆装クルーズ客船「シー・クラウド」。従来型の帆を張るクラシック型。

図4-11　コンピュータ制御の現代的セールシステムをもつ帆装クルーズ客船「クラブメッド2」。フランスの地中海クラブが運航。

いになりやすい人でも十分に楽しめるという特性もある。

例えばライン川クルーズでは，スイスのバーゼルから，フランス，ドイツを通り，オランダのロッテルダムまたはアムステルダムまで5日間をかけて航海する。途中には，川岸に立つ古城やワイン畑，河川沿いの町々，狭い急流で有名なローレライなどをゆっくりと観光することができ，流れを制御する堤ではロック（閘門（こうもん））を使って船が上下するのを体験することができる[3]。

図4-12　ライン川クルーズに就航する河川用クルーズ客船

図4-13　ロシアの河川用クルーズ客船

コラム⑧　洋上大学船

客船を教育の場として活用する試みもいくつかあった。日本のクルーズ客船が手がけた洋上研修クルーズもその1つといえるが，専用の洋上研修客船もこれまで幾隻か登場している。主に，中古客船を改造した船が多かったが，採算をとるのはなかなか難しく，今ではほとんど姿を消した。（3.6節参照）

4.4　ソースマーケット

クルーズに乗船する乗客の居住地域あるいは国籍に基づくマーケットを，ソースマーケットと呼ぶ。2016 年の SHIPPAX 社の統計によると，

- 北米　　54.7%（アメリカ＋カナダ）
- 欧州　　24.1%
- アジア　12.4%
- オセアニア　4.5%（オーストラリア＋ニュージーランド）
- 南米　　1.8%
- その他　2.5%

となり，現代クルーズが生まれ育った北米が大きなシェアを占めており，欧州，アジアがそれに続いている。

さらに国別にすると，2016 年現在で下記のとおりである。

- アメリカ　51.16%
- 中国　8.76%
- ドイツ　8.34%
- イギリス＋アイルランド　7.80%
- オーストラリア＋ニュージーランド　4.55%
- カナダ　3.51%
- イタリア　3.29%
- フランス　2.56%
- スペイン　2.01%
- ブラジル　1.86%
- 日本　0.99%

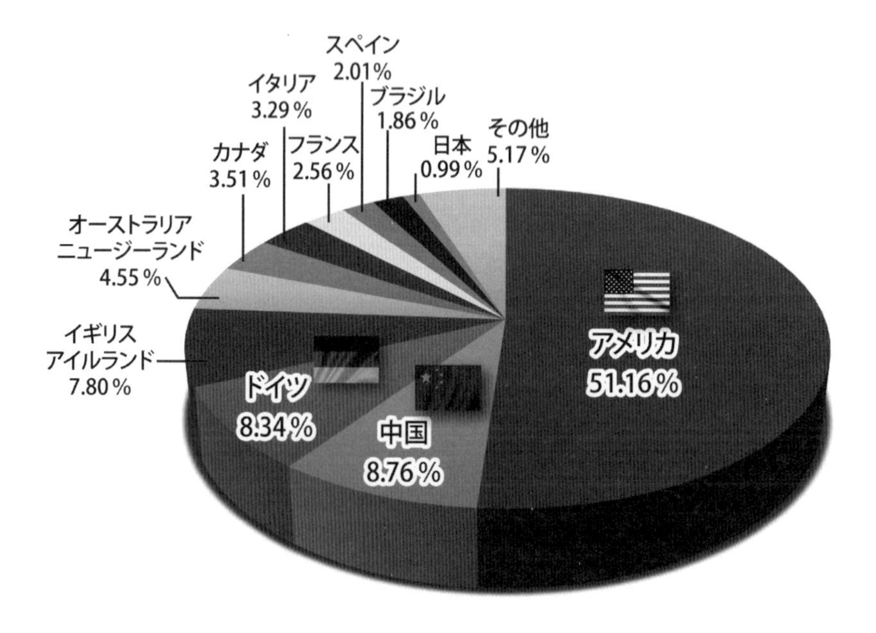

図 4-14　各国のクルーズ人口の比率

このソースマーケットのシェアの2016年前後の変化をみると，北米，ドイツ，イギリスがほぼ停滞状態にあるのに対し，中国は急増している。また，イタリア，フランスは漸減，ブラジルは急減している。

各国の人口に対するクルーズ人口の比率は，それぞれのソースマーケットでの観光の中にクルーズが占める浸透率を示す重要な指標となっている。以下，2016年時点での地域または国別のクルーズ浸透率を示す。

- オーストラリア＋ニュージーランド　3.98%
- アメリカ　3.80%
- イギリス＋アイルランド　2.60%
- ドイツ　2.50%
- カナダ　2.40%
- イタリア　1.30%
- スカンジナビア　1.21%
- スペイン　1.05%
- ブラジル　0.22%
- 日本　0.19%
- 中国　0.15%

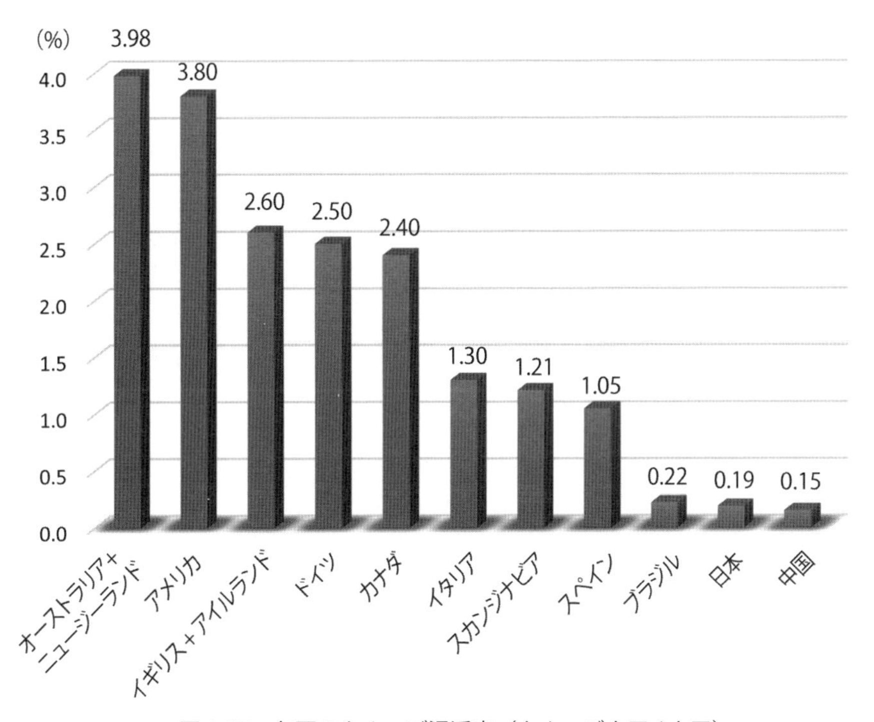

図4-15　各国のクルーズ浸透率（クルーズ人口/人口）

このクルーズの浸透率の経年変化は図4-15のようになる。アメリカがほぼ一定の割合で上昇しているのに対し，イギリスは1990年代後半，ドイツ，イタリア，オーストラリアでは2000年代後半から急増している。これは，それぞれのソースマーケットに現代クルーズが進出した年代とおおよそ一致している。

日本マーケットでは，2013年から海外の現代クルーズが日本発着クルーズを開始し，クルーズ人口が増え始めており，浸透率にも上昇の気配が見え始めている。

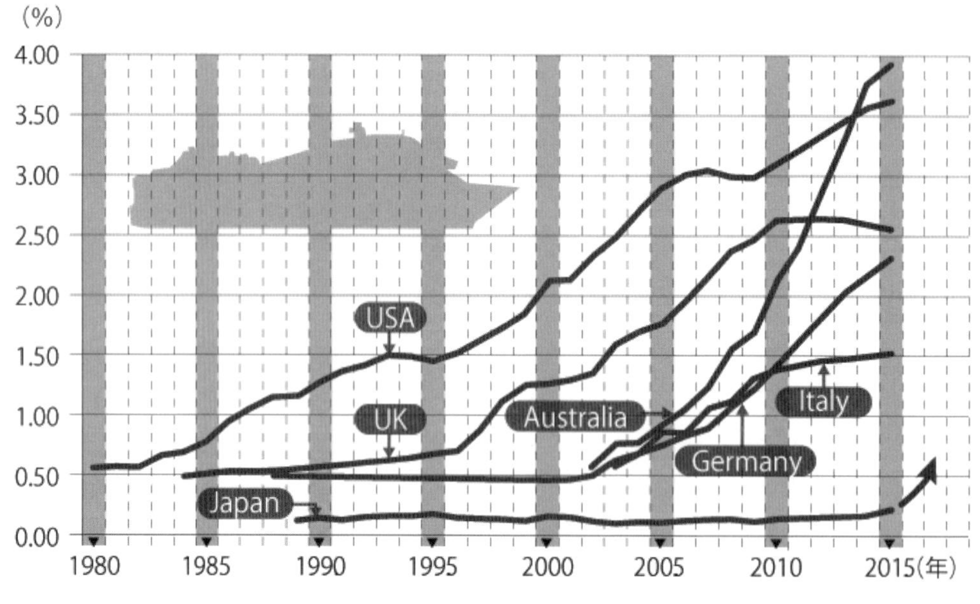

図 4-16　各国のクルーズ浸透率の推移（各国のクルーズ人口 / 人口）

コラム⑨　分譲型クルーズ客船

クルーズがブームになって，いろいろなクルーズのビジネスモデルが登場しているが，船内キャビンをマンションのように分譲するモデルとして，レジデンシーが企画・建造した「ザ・ワールド」（43,188 総トン，最大定員 699 名）が 2002 年に竣工した。しかし，分譲が進まず，空室を一般乗客に提供して営業しているが，経営は思わしくなく，2010 年からは所有者からなる管理団体による運航が続けられている。

図 4-17　那覇港に寄港した分譲型クルーズ客船「ザ・ワールド」

【参考文献】

[1] 世界のクルーズ客船 2009–2010，世界の艦船別冊，海人社，2009 年 5 月

[2] 世界のクルーズ客船 2013–2014，世界の艦船別冊，海人社，2013 年 11 月

[3] 池田良穂：静かな海と楽しい航海，朔北社

演習

⑴ アメリカのカーニバルグループは，グループ内でラグジュアリー・マーケットからカジュアル・マーケットまでをターゲットとした多彩なクルーズ客船を，運航会社を分けて営業している。このようにブランドを分けてビジネス展開をするメリットとデメリットについて論じなさい。

⑵ ラグジュアリー・クラスのクルーズ客船の料金と，陸上で高級ホテルに宿泊して食事やショーなどのイベントを楽しみながら旅をする場合の費用を比較しなさい。

≪問題≫

4-1　クルーズフェリーとはなにか。

4-2　デイ・クルーズとはなにか。

4-3　クルーズ人口とはなにか。

4-4　クルーズ・マーケットのラグジュアリー，プレミアム，カジュアル・クラスとは，それぞれどのようなものか。

4-5　クルーズのソースマーケットとはなにか。

4-6　クルーズのソースマーケットが大きい地域を順に 4 つ挙げよ。

4-7　イギリスのソースマーケットが 1990 年代後半から急増したのはなぜか。

第5章 クルーズビジネスの構成

5.1　クルーズ運航会社

クルーズ客船を使ったクルーズ商品を企画・販売するのが，いわゆるクルーズ運航会社である。日本では，商船三井客船，日本クルーズ客船，郵船クルーズの 3 社がある。この 3 社が運航するクルーズ客船はいずれも「日本籍船」で，基本的に日本の法律が適用される（旗国主義）。

また，日本国内だけを回るクルーズは，日本籍船だけしか行うことができない。これは国内輸送に携われるのは日本籍船だけに限られるという，「カボタージュ」という規則によっている。すなわち，日本国内だけを巡るクルーズを実施することの多い日本のクルーズ運航会社は，日本籍のクルーズ客船を使う必要があるわけである。

クルーズ運航会社は，自社でクルーズ客船を所有しているとは限らず，船を所有する船主（せんしゅ）が別にあり，そこから船を借りている場合も多い。これを「用船」「チャーター」と呼ぶ。商船三井客船が運航する「にっぽん丸」は親会社である商船三井の所有，郵船クルーズが運航する「飛鳥Ⅱ」は親会社の日本郵船が所有している。

海外のクルーズ運航会社では，カーニバル・クルーズ・ライン，ロイヤル・カリビアン・インターナショナル，プリンセス・クルーズ，ホーランド・アメリカ・ライン，MSC クルーズ，キュナード・ラインなどがある。

海外のクルーズ運航会社では，クルーズ自体の販売だけでなく，各都市から起点港までの航空機チケット，クルーズ前後の起点港周辺での宿泊なども一括して手配をしていることが多い。

海外のクルーズ運航会社では，リベリアなど船にかかる税金の安い国にペーパーカンパニーを設立して，その会社が船を所有している場合も多い。これを便宜置籍船（べんぎちせきせん）という。便宜置籍船とすると，船員の国籍も原則自由となるから，相対的に人件費の安い国の船員を雇うことができ，船員費を削減することが可能となる。

コラム⑩　もっともカボタージュの影響が大きいアメリカ

アメリカにはジョーンズ法と呼ばれる内航海運に関する法律があり，アメリカ籍の船舶は，アメリカ国内で建造され，アメリカ船員によって運航されなければならない。アメリカの造船業は極めて高価な軍艦の建造に特化していて，他の造船国に負けない価格での商船の建造は不可能な状態にある。このため，アメリカ国内のクルーズを行うためには，極めて高価なアメリカ製客船を使って，高給のアメリカ人船員を使わねばならず，極めてコスト高となる。このためクルーズがもっとも盛んな北米水域には，アメリカ籍のクルーズ客船はほとんどいない。北米の港から出港するクルーズは，必ずアメリカ以外の国に寄港することによって外航航路として，便宜置籍船を使用できるようにしている。アメリカのアラスカクルーズがカナダのバンクーバーを起点として発展したのもカボタージュ規制を避けるためであり，アメリカ本土からのハワイクルーズが少ないのもカボタージュのためである。ハワイ諸島内のクルーズでもアメリカ籍船を使用する必要があり，これまでいくつかのクルーズ運航会社が試みているが，コスト高が原因で撤退する事例が多い。

5.2　マネジメント会社とマンニング会社

船の運航・操船には特殊な技能が必要で，海技免状が必要となる。最近は，クルーズ運航会社が直接，船員を雇うのではなく，船員の雇用から運航までを専門に行うマネジメント会社を活用することが多くなっている。

また，こうした船員を世界各国で集めて，船会社やマネジメント会社に供給する会社も多くあり，マンニング会社と呼ばれている。特にフィリピンやインドは世界の船員の一大供給源となっており，そこに拠点を置くマンニング会社は多い。

海外のクルーズ客船では，サービス部門の船員の多くがチップ収入を主収入としている場合も多い。そのため，クルーズのパンフレットに標準的なチップ料金が表示され，下船の 1 日前にはチップ用の封筒が各キャビンに配られる船も多い。標準的なチップ額は，ウェイター，キャビンスチュワードで 1 日あたり 5 ドル程度，バスボーイ，チーフスチュワードなどはその半額程度とする場合が多い。一般的に，月 20 万円程度の収入が確保されるように，担当するダイニングルームのテーブルの数やキャビンの数が決められている。また，飲物のサービスをするバー・ウェイターなどについては，飲物代にチップが含まれるようになってきている。こうしたチップを主な収入とする船員のレベルを確保するために，クルーズ客船では厳格な乗客アンケートを実施しており，各個人のサービスを評価したアンケート表（コメントカード）を下船時に集め，常にサービスの状況を把握している。

ラグジュアリー・クラスのクルーズ客船では，チップ制を廃止している船もある。その分，船員の給料が増加し，クルーズ料金が高額になることとなる。

また，アジアではチップ制になじみがないことから，クルーズ料金の支払い時に，別途，チップを徴収する船会社もある。

5.3　販売代理店（GSA）

海外のクルーズ運航会社は，日本に代理店を置いて，マーケティングおよびチケットの販売などを委託しており，国内に1社だけの場合を総代理店，複数社の場合を販売代理店と呼んでいる。

クルーズのみを扱う代理店がある一方で，大規模旅行業者がクルーズ運航会社の代理店となっている場合もある。また，最近は日本支社を置く海外のクルーズ運航会社も増えている。

主な販売代理店は以下のとおり（2016年現在）。

- MSCクルーズジャパン：MSCクルーズ
- カーニバル・ジャパン：プリンセス・クルーズ，キュナード・ライン，シーボーン・クルーズ
- コスタクルーズ日本支社：コスタ・クルーズ
- スタークルーズ日本オフィス：スタークルーズ，ドリームクルーズ
- ノルウェージャンクルーズライン日本支社：ノルウェージャン・クルーズ・ライン
- クルーズバケーション：シーボーン・クルーズ，ハパグロイド・クルーズ
- オーバーシーズ・トラベル：ホーランド・アメリカ・ライン，シーボーン・クルーズ，ディズニー・クルーズライン
- ミキ・ツーリスト：ロイヤル・カリビアン・インターナショナル，セレブリティ・クルーズ，アザマラ・クラブ・クルーズ
- アンフィトリオン・ジャパン：カーニバル・クルーズ・ライン，シークラウド・クルーズ，アヴァロン・ウォーターウェイ，セレスティアル・クルーズ
- メリディアン・ジャパン：スター・クリッパーズ，セレスティアル・クルーズ
- 郵船クルーズ：クリスタル・クルーズ
- 郵船トラベル：ディズニー・クルーズライン
- パシフィックリゾート：シーボーン・クルーズ・ライン
- インターナショナル・クルーズ・マーケティング：シルバーシー・クルーズ，ポール・ゴーギャン・クルーズ，ルフトナー・クルーズなど
- クルーズプラネット：ホーランド・アメリカ・ライン，シーボーン・クルーズ
- マーキュリートラベル：ポナン，サガ・クルーズ，シードリーム・ヨットクラブなど

5.4　クルーズ専門旅行会社

アメリカで発祥した現代クルーズの成長過程において，クルーズ専門の旅行会社が大きな役割を果たしたことは前述したが，現在でも，北米マーケットではこうした旅行会社がマーケット拡大の最前線で機能している。

また，ドイツ語マーケットだけに特化したドイツのクルーズ・マーケットにおいても，古くから，クルーズ専門の大規模旅行業者が，クルーズ客船を年間チャーターして独自のクルーズ企画を販売することでクルーズ・マーケットを拡大した。

日本においても，クルーズ商品を中心に扱う旅行会社がしだいに数を増やしている。これは，クルーズは他の旅行商品とは違った特性をもっており，クルーズに関する十分な知識をもった旅行会社に顧客が信頼をおいているからに他ならない。例えばクルーズ客船には，高齢者向き，若者向き，特殊な嗜好者向きなどのさまざまなタイプがあり，それぞれの顧客に合ったクルーズを的確に紹介することができるのは，こうしたクルーズ専門旅行会社ということとなる。

日本の主なクルーズ専門旅行会社は以下のとおり。

- クルーズのゆたか倶楽部：03-5294-6261
- 郵船トラベルクルーズセンター：03-5213-9987
- JTB PTS クルーズラウンジ横浜：045-285-9620

5.5　クルーズアドバイザー認定制度

日本でのクルーズ振興を目的として，日本外航客船協会を中心として「クルーズアドバイザー認定制度」が設けられている。旅行業に携わる人々にクルーズの専門知識をもってもらい，販売の最前線でクルーズを顧客に積極的に勧めてもらうことが設立の目的である。

毎年，試験が行われており，合格すると「クルーズ・コンサルタント」「クルーズ・マスター」の称号を得ることができる。クルーズ・コンサルタントは，毎年，受験者が増加し，2017 年の時点で合格者の累計は 7,000 名を超えた。また，さらにクルーズセミナー講師やパネリストを務め，クルーズに関する豊富な知識と経験を有していると認定されているクルーズ・マスターも 60 名を超えている。

5.6　業界団体

クルーズ運航会社は業界団体を設立して，マーケットの拡大や共通の利害に関する対策をとっている。代表的な業界団体は以下のとおり。

- CLIA（Cruise Lines International Association）（アメリカ）
  1975 年に設立された，北米を中心とするクルーズ運航会社 60 社が加盟する業界団体。クルーズに関する紹介，各種の統計などをホームページに公表している。
  https：//www.cruising.org/cruise-vacationer
- 日本外航客船協会（JOPA）
  1990 年に設立された，日本の外航客船を運航する会社で組織する業界団体で，クルーズおよび船旅の振興に関するさまざまな事業を行っている。クルーズアドバイザー認定制度もその 1 つ。また，クルーズの商品企画などを表彰するクルーズ・オブ・ザ・イヤーも行っている。
  http：//www.jopa.or.jp/index.html

5.7　クルーズの専門雑誌

クルーズの専門雑誌の主なものを以下に挙げる。

- クルーズ（隔月，海事プレス社発行）：一般ユーザー向け
- クルーズトラベラー（季刊，クルーズトラベラーカンパニー）：一般ユーザー向け
- SHIPPAX INFO（月刊，ShipPax Information，スウェーデン）：プロ向け
- Seatrade Cruise Review（季刊，Seatrade Communications Ltd.，イギリス）：プロ向け
- Passenger Ship Technology（Riviera Maritime，イギリス）：技術者向け

図 5-1　SHIPPAX INFO

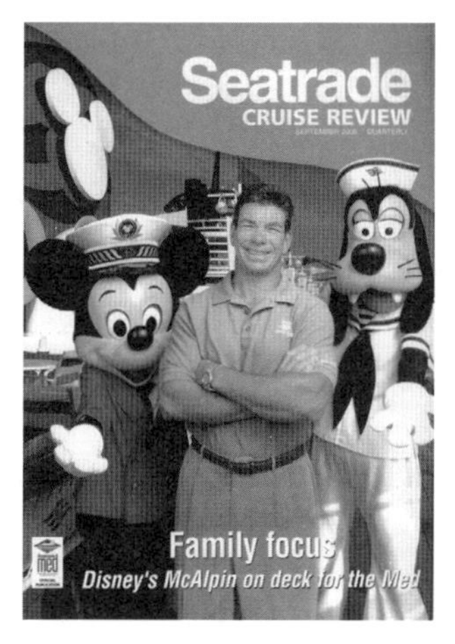

図 5-2　Seatrade Cruise Review

図 5-3　Passenger Ship Technology

5.8　クルーズの国際会議

Seatrade 社の主催するクルーズ会議をはじめ，各種の国際会議が定期的に開催されている。

- FERRY SHIPPING CONFERENCE ONBOARD（Cruise & Ferry Info 主催）：フェリー船上で毎年開催されるユニークな国際会議
- Cruise Shipping Miami（Seatrade 主催）
- All Asia Cruise Convention（Seatrade 主催）
- Med Cruise & Superyacht Convention（Seatrade 主催）

演習

(1) リベリア籍のクルーズ客船で，東京起点の 1 泊クリスマスクルーズを実施することを企画したいが，その問題点を挙げよ。

(2) プリンセス・クルーズの運航する「ダイヤモンド・プリンセス」が，クルーズの途中，室蘭，東京，鹿児島に寄港することとなった。日本のお客が，室蘭から鹿児島までの乗船を希望している。

どのような対応が可能か。

⑶ CLIA のホームページから最新の CRUISE MARKET OVERVIEW を入手し，北米のクルーズ業界の最新の動静をまとめなさい。

≪問題≫

5-1　カボタージュとはなにか。

5-2　各国にカボタージュ規制があるのはなぜか。

5-3　便宜置籍船とはなにか。

5-4　マンニング会社とはなにか。

5-5　クルーズアドバイザー認定制度とはなにか。

5-6　CLIA とはなにか。

5-7　JOPA とはなにか。

5-8　日本のクルーズ雑誌にはどのようなものがあるか。

第6章 クルーズ客船のハード

6.1　クルーズ客船の造船所

クルーズ客船は，世界各地の造船所で建造される。貨物船とは異なり，客船は安全面でのレベルが高く，また，乗客が快適に過ごすための船内の内装などにも特別な配慮が必要となる。したがって，クルーズ客船の建造ができる造船所は，世界的にも限られるのが現実である。

世界のクルーズ客船の90%以上は，欧州の造船所で建造されている。特にフィンランド，イタリア，フランス，ドイツの造船所がその中心になっている。

特に現代クルーズのクルーズ客船では，フィンランドのバルチラ造船所が1960年代末から常に新しいアイディアに基づく船を造り続けている。その後，フランスのアトランティック造船所，ドイツのマイヤーベルフト造船所，イタリアのフィンカンティエリ造船所がクルーズ客船の建造を手がけ，現在ではフィンカンティエリ造船所の建造量がもっとも多くなっている。2008年に，フィンランドのアーケル・フィンヤード（元バルチラ造船所，元マーサ造船所）とフランスのアーケル・フランス造船所（元アトランティック造船所）は，韓国の新興造船所STXの傘下に入り，STXヨーロッパとなったが，STXの経営破綻からフィンランドの造船所はマイヤーベルフト造船所，フランスの造船所はフィンカンティエリ造船所の傘下に入った。

日本では，三菱重工業が，海外向けのクルーズ客船「クリスタル・ハーモニー」「ダイヤモンド・プリンセス」「サファイア・プリンセス」をはじめ，「ふじ丸」「にっぽん丸」「飛鳥」など数隻の日本籍クルーズ客船を建造している。また，石川島播磨重工業（現JMU）が日本籍クルーズ客船「おり

図6-1　ドック内で建造中の世界最大（当時）のクルーズ客船「オアシス・オブ・ザ・シーズ」

えんとびいなす」「ぱしふぃくびいなす」，日本鋼管（現 JMU）も小型クルーズ客船「おせあにっくぐれいす」の建造実績がある。

もちろんクルーズ客船の建造実績のない造船所でも，クルーズ客船の建造ができないわけではないが，クルーズ客船の建造には極めて手間がかかるため，船価自体は一般貨物船に比べると極めて高いものの，客船建造に慣れた造船所での連続建造体制をもってはじめて利益が出せるといわれている。

図 6-2　屋内ドックで建造中の大型クルーズ客船。ドイツのマイヤーベルフト造船所。

図 6-3　神戸の三菱重工業の船台で建造されて進水するクルーズ客船「にっぽん丸」（商船三井客船）。船の誕生を祝う進水式は，とても感動的。

6.2　コンセプト設計

クルーズ客船のコンセプトを構築することをコンセプト設計もしくはコンセプトデザインと呼び，これには，クルーズ運航会社，船主，造船所の担当者が一体になってあたる必要がある。また，クルーズ客船のインテリア（内装）設計については，専門のデザイナーやデザイン会社も存在している [3]。

コンセプトの設計にあたっては，ターゲットとするマーケットセグメント（ラグジュアリー，プレミアム，カジュアルなど）を決めて，それに合わせることが重要となる。クルーズ客船には，一般大衆をメインターゲットとする現代クルーズのための客船，富裕層をターゲットとするラグジュアリー・クラスの豪華客船，秘境や極地をまわる探検クルーズ客船，ヨット型の帆装クルーズ客船など，さまざまなタイプがある。それぞれのタイプによって，客室，公室をはじめ各種施設が異なる。それぞれのマーケットに適したコンセプト設計をする必要がある。

6.3　安全性

客船の安全性は，国際的な規則で規定されている。これは，世界を回る客船の場合，どこにおいても統一された規則でその安全性を確保しておく必要があるためである。

国際連合の専門機関として国際海事機関（IMO：International Maritime Organization，本部ロンドン）が，その国際規則を決めている。

その中心となるのはSOLAS条約（International Convention for the Safety of Life at Sea：海上における人命の安全のための国際条約）で，客船「タイタニック」の海難を契機にして国際規則として作られ，毎年，必要な改定が行われている。この法律が適用されるのは，基本的に国際航路に就航する船舶であるが，2009年から日本政府は内航船についてもSOLAS条約を原則適用することを決定して，安全性に関するレベルの統一を図っている。

客船の建造にあたっては，このSOLAS条約に適合するように設計，建造を行う必要があり，各国政府機関またはその委託を受けた船級協会（Class）が検査を実施する。

安全性に関する項目は非常に広い範囲にわたり，主要なものとして，転覆を防ぐ復原性規則，火事の場合の安全性を確保する防火規則，衝突時の沈没や転覆を防ぐ損傷時復原性規則，非常時に船を放棄して乗客乗員を海上に避難させるための救命施設にかかわる規則などが定められている[4][5]。

6.4　建造過程

船の建造に携わる技術者を「造船技術者」と呼び，英語ではNaval Architect，すなわち「海の建築家」という意味となる。

造船所では，船主からクルーズ客船建造の「引き合い」があると，その基本計画を行って，その建造費用を見積もる。船主との間で合意に達すれば，いよいよ建造が始まる。

船は，船台やドックで建造されて，海に浮かぶ。船体は，主に鋼鉄（スチール）で造られる。製鉄所で製造された鋼材を切断，加工して，1隻の船へと組み上げていく。最近は，ブロック建造法と呼ばれる建造法が使われる。これは船台やドック以外の工場内で高精度のブロック（船体構造を構成するかたまり）を製造し，それを巨大なクレーンで船台やドックに移動・設置して，溶接でつなぎ合わせるもので，船体を比較的短期間で造ることができる。

客船は船体だけでなく，内部の装置や施設がたいへん大事になる。こうした内部を造る作業を「艤装工事（ぎそうこうじ）」と呼ぶ。船を動かすエンジンなどの機械類，電気設備と配電，船内の配管，船室や公室の内装など，艤装工事は多岐にわたる。ブロックを造る時点で，できるだけ艤装工事を行って，効率的に行う手法を「先行艤装（せんこうぎそう）」という。

船台やドックで完成して海上に浮かんだ船体内で，最終的な艤装工事が行われた後，各種の検査と海上試運転を経て船は完成する。1隻の客船が完成するまでには実際の作業に約1～2年，引き合いの時点からすると3～4年程度かかる。

船主は，造船所に，契約時，起工時，完成時の3回に分けて，建造費を分割して支払うのが一般的となっている。

6.5 乗り心地

クルーズ客船の場合には「船酔い」の問題が大きい。船酔いは「乗り物酔い」の一種で，船が波の影響で揺れることによって発症する。症状としては，むかつき，嘔吐などが典型的で，揺れがおさまったり，船を下りたりするとすぐに症状はなくなる。

この船酔いは，主に船上での上下加速度（速度の変化率）の大きさとその周期に強く関係することがわかっている。特に，周期が 5 ～ 6 秒の時に発症率が高くなる。主要な原因は，内耳器官（特に耳石）からの運動情報と，目などからの運動情報とが異なると，脳の中で情報混乱が起こるためとされており，目を閉じて目からの情報を遮断したり，デッキに出て水平線近くを見ることによって目からも「動いている」という情報を積極的に入れたりすることで，症状が緩和される。逆に，船内に閉じこもったり，本などの文字を読んだりして，目から「止まっている」という情報を脳に送ると，船酔いをしやすくなる。この他，体調や自己暗示などの心理的要因も船酔いには大きく影響する。

船酔いには慣れの効果が顕著に現れ，一般には 2 ～ 3 日間船に揺られると，脳が情報混乱を解消できるようになり，船酔いの程度は大きく減少することがわかっている。また，脳に入る情報に混乱があっても適正に処理ができるように，脳を積極的に教育することも効果がある。

船酔いの主因となる「船の運動」は，波の高さ（波高：波の谷から山までの鉛直距離）と周期に強く依存する。波の高さは，台風などの強い低気圧の中では 6 ～ 10 m 近くにまで発達し，波の周期は 6 ～ 10 秒程度に集中している。波の長さは「周期の 2 乗の約 1.5 倍」なので，周期 10 秒の波の長さは 150 m 程度となる。したがって，船の長さが 150 m 以上あれば，船の揺れは小さくなり，船酔い

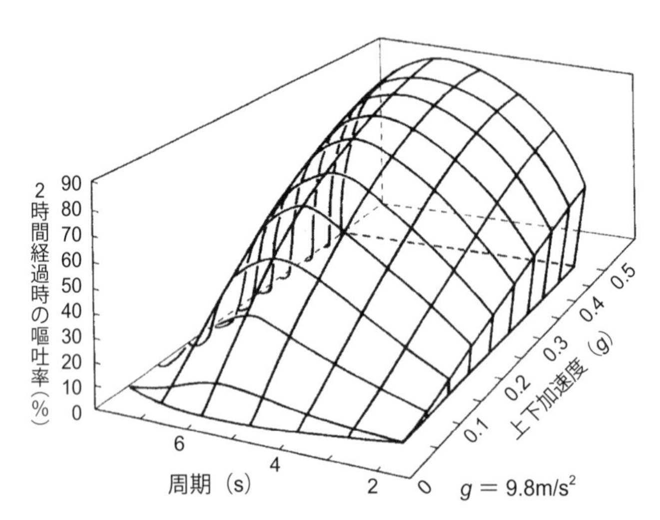

図 6-4　ハンロンが 11 m の振幅で上下する部屋に被験者を乗せて，2 時間経過した時の嘔吐者の割合を求めた図表。MSI を計算するときには，この実験結果が用いられる。（池田良穂著：新しい船の科学，講談社 ブルーバックス，1994 年）

する確率が減少する。このように船が大型化すると，船酔いの問題はしだいに少なくなる。長さが300m近い7万総トン以上の大型船では，かなりの荒天下においても船酔いをする心配はない。

今では船の建造時に，船酔いを評価することが可能となっている。一般に船酔いはMSI（Motion Sickness Incidence）という指標で表される。これは，乗客の何％が嘔吐にまで至るかを示すもので，一般的に2時間にわたってその運動に晒された場合の値を用いる。船酔いは，船首方向から波を受ける向かい波中においてもっともきつく，特に船首，船尾部においてきつい。船体中央から若干船尾寄りの場所に，もっとも船酔い発生率が低いところがある。

運動を減少させる装置としては，フィンスタビライザーなどのライドコントロールシステムがある。ライドコントロールシステムを使うと，高速船の場合には縦揺れ（pitching）も減らすことが可能だが，一般的なスピードの船では横揺れのみが軽減できる。したがって，船酔いの心配を解消する手段としては，船の大型化がもっとも効果があるといえる。

6.6　キャビン

船の客室をキャビンという。クルーズ客船のキャビンは，基本的にホテルの部屋と同じ機能をもつ。ベッド，机，椅子，収納用家具，トイレ，バス（またはシャワー）などが装備され，もっとも小さいもので15m^{2}程度からで，標準的なものは18～20m^{2}程度，大きい部屋では100m^{2}を超えるものもある。また，最近はバルコニー付のキャビンも多くなっている。クルーズ客船の船内は，等級差のないモノクラスが一般的であり，キャビンの大きさとグレードの違いがクルーズ料金の差となっている。家族用には，壁もしくは天井に収納できるプルマンベッドと呼ばれる2段ベッドを有する部屋もあり，このベッドは昼間は収納され，夜になるとベッドメイキングされる場合が多い。

外側のキャビンで窓のあるものをアウトサイド・キャビン，中側で窓のないキャビンをインサイド・キャビンという。最近の超大型船では，船体中心線に船首尾に長く延びる吹き抜けのアーケードを作って，そのアーケード側に窓をつけたインサイド・キャビンを設けた船もある。

6.7　公室設備

乗客がだれでも利用できるスペースを公室またはパブリックスペースと呼ぶ。基本的な公室としては，ダイニングルーム，レストラン，ショーラウンジ，展望ラウンジ，サンデッキ，ライブラリーなどがある。基本的にはパブリックスペースの利用およびショーやイベントの鑑賞は無料であり，飲み物やサービスを頼まない限り，支払いの必要はない。

大型船になると複数のレストランやラウンジを有し，好みに応じて選べるように配慮されている。

最近は，健康管理のため各種のスポーツ施設が配置されるようになっている。ジム，プール，バスケットボールのコート，ジョギング用コース，ミニゴルフをはじめ，ロッククライミング，アイススケート，サーフィンプール，ボクシングジムまで揃えている船も出現している。これらは，顧客ターゲットのニーズに合わせて装備されるのが一般的であるが，最近の10万総トンを超える超大型クルーズ客船では老若男女あらゆる世代階層の人々のニーズに合わせた多彩な施設を有している。例えば「三世代船」を売りにしているクルーズ客船では，祖父母から孫までが一緒にクルーズをしながらも，それぞれの年代に合わせて楽しめる種々のサービスを提供している。

6.8　ダイニングルーム

一般的にクルーズ客船は，かつての定期客船時代の 1 等客用の高級なダイニングサービスを売り物にしている。すなわち，乗客全員に食事のテーブルを指定し，決まったボーイによるサービスを提供している。テーブルには 6 ～ 10 人程度の座席があり，一緒になった人々をテーブルメイトと呼ぶ。一般に，夕食は 2 回に分けて提供され，最初をファーストシッティングまたはメインシッティング，2 回目をセカンドシッティングまたはレートシッティングと呼ぶ。最近は，予約時にテーブルが自動的に指定される場合も多いが，最初の夕食後に申し出るとテーブルを変わることができる場合もある。

また，高級船では乗客全員が一度に食事ができるだけの席を設けている船もある。

一方，最近は，こうした伝統的なダイニングサービスをやめて，他の乗客との合い席をできるだけ避けて，夫婦，家族，友人だけのテーブルとする船や，一切テーブルを決めずに自由に空いているテーブルを選べるシステムに移行する船も多くなりつつある。

テーブルサービスのあるダイニングルーム以外に，カフェテリア式のレストラン，さらにフレンチ，イタリアン，ステーキ，中華，寿司などの専門レストランを有する船も増えており，多少の追加料金が必要な場合もある。

キュナード・ラインの運航するクルーズ客船「クイーン・メリー 2」「クイーン・ビクトリア」「クイーン・エリザベス」では，高級キャビンの乗客とスタンダードキャビンの乗客の利用するレストランを厳密に分けてきた。これは，かつての定期客船時代の等級を残したものともいえる。同様のクラス分けをする大型クルーズ客船も増えている。これは数千人の乗客を乗せる大型船では，カジュアルからラグジュアリーまで広い層の集客が必要となり，船内でレストランを分けることで高級キャビン利用者に満足感を与えることを意図している。

6.9　ギャレー

船の厨房のことをギャレーと呼ぶ。数百人～数千人の食事を効率よく調理し，配膳するための工夫がもられている。また，揺れる船の上での調理や配膳作業なので，陸上のレストランの厨房とは違った工夫が随所に取り入れられている [6]。

食材倉庫についても，航海日数に応じた食材を積み込み，肉，魚，野菜，果物などを新鮮な状態で保存する工夫がされている。

6.10　船上での水

船上で使う水は，港で清水タンクに給水するほか，造水機によって海水から真水が造られている。造水方法としては，海水を熱して蒸気にして抽出する方法と，水分だけを通す膜を使って濾過する方法の 2 つがある。前者は蒸発式，後者は逆浸透式と呼ばれる。

一般的に，乗客・乗員が 1 人あたり 1 日に使用する水は 220 リットルが目安であるが，その他にプールや清掃などにも水は使われるので，2 倍程度の造水能力を備えている船が多い。

16 万総トンのクルーズ客船「フリーダム・オブ・ザ・シーズ」の場合，4 基の造水機で 1 日 2,700 トンの水を造る能力があり，さらに 5,800 トンの清水タンクをもっている。

6.11　空調

快適な船上生活には空調が必須であり，暖房，冷房，換気の役割をもつ。船内で使われる電力の，1/3 が推進などの機器類に，1/3 が空調に，残りの 1/3 がホテル部門の種々の電力として使われている。

空調には，外部から新鮮な空気を取り入れるとともに，船内の温度調整された空気を循環して用いることで，省エネを図っている。一般には，キャビンでは外部空気量が 30%，公室などでは 50%，ギャレーなどでは衛生と臭気対策で 100% となっている。船内温度は，夏季で 25℃，冬季で 22℃程度に設定されているが，キャビンについてはホテルなどと同じく，乗客の好みに応じて自由に設定できるようになっている場合が多い。このために，セントラル方式によって各空調室からダクトを通して送られた空気が，キャビン天井内に設置された吹出ユニット内の風量調整用のダンパーとサイリスタ式の電気再熱器によって温度調整されている [7] [8]。また，火災時の煙や火が空調ダクトを伝って広がらないようにするためのダンパーやファンなどの設置も必要となり，空調システムは客船建造において非常に高度な技術が必要な項目となっている。

6.12　船の推進

船は一般的に，油を燃料とするディーゼルエンジンによって，船尾の水面下に配置したスクリュープロペラを回して進む。スクリュープロペラは，扇風機とほぼ同じ原理で，回転する羽根でまわりの流体を後方に加速することによって生ずる反力によって船を推し進めている。

最近は，ディーゼル機関で直接スクリュープロペラを回すのではなく，ディーゼル発電機で発電し，電気モーターでスクリュープロペラを回すタイプの推進システムも増えて，これをディーゼル・電気推進と呼んでいる。

また最近の大型クルーズ客船では，船底に吊るしたポッド（繭型の容器）内に交流モーターを入れてプロペラを駆動するポッド推進も多く採用されるようになってきた。このポッド推進器は，360 度回転させて任意の方向に推力を発生させることが可能であり，舵のいらない推進器として注目を集めている。

図 6-5 「ボイジャー・オブ・ザ・シーズ」の巨大なポッド推進器（提供：RCI）

【参考文献】

[1] 谷井建三，池田良穂：精密イラスト 船ができるまで―豪華客船＜ふじ丸＞，偕成社，1989 年

[2] 池田良穂：船の最新知識，ソフトバンククリエイティブ，2008 年

[3] 清家健：客船・フェリーのインテリアデザイン，KANRIN，日本船舶海洋工学会発行，17 号，2008 年 3 月

[4] 佐藤功，小佐古修士：客船の安全性について，KANRIN，日本船舶海洋工学会発行，27 号，2009 年 11 月

[5] 太田進，梅田直哉：クルーズ時代に対応する新しい客船安全基準，KANRIN，日本船舶海洋工学会発行，17 号，2008 年 3 月

[6] 北出隆久：フェリーのギャレー機器，KANRIN，日本船舶海洋工学会発行，17 号，2008 年 3 月

[7] 越智仁一，山田一俊：客船とその空調装置の進化，KANRIN，日本船舶海洋工学会発行，17 号，2008 年 3 月

[8] 小佐古修士，椎山邦昭：大型客船の空調システム設計についての紹介，KANRIN，日本船舶海洋工学会発行，17 号，2008 年 3 月

演習

⑴ 船酔いをしやすいというお客がクルーズの相談にやってきた。どのようなクルーズを勧めればよいか考えなさい。

⑵ 船酔いのしやすい乗客へ渡す注意事項をまとめなさい。

≪問題≫

6-1　大型クルーズ客船は主にどこで建造されているか。

6-2　日本で大型クルーズ客船を建造した実績のある造船所はどこか。

6-3　客船の安全性はどのように担保されているか。

6-4　人はなぜ船酔いをするのか。

6-5　船による乗り物酔いがもっとも激しい原因はなにか。

6-6　船のどの部分にいるともっとも船酔いしやすいか。

6-7　キャビンとはなにか。

6-8　クルーズ客船の公室にはどのようなものがあるか。

6-9　クルーズ客船の水はどのように確保されているか。

6-10　ポッド推進器とはどのような推進器か。

資料　デッキプランと客室・レストラン

船内の配置図をデッキプランといい，船内で生活するための大事な情報となる。船内の各階ごとの客室や公室の位置が示されており，船内新聞でのスケジュール表とデッキプランが必需品となる。

「にっぽん丸」デッキプラン

B1 階～ 8 階まであり，キャビンはすべてアウトサイド（外側で窓付き）。

B1：エンジンなどの機械類

1 階：船員居室，乗客キャビン

2 階：メインダイニングルーム，ギャレイ（調理配膳室），フロントなど，乗客キャビン

3 ～ 6 階：乗客キャビンと各種公室

7 階：プール，スポーツデッキ，展望ラウンジ

8 階：オープンデッキ

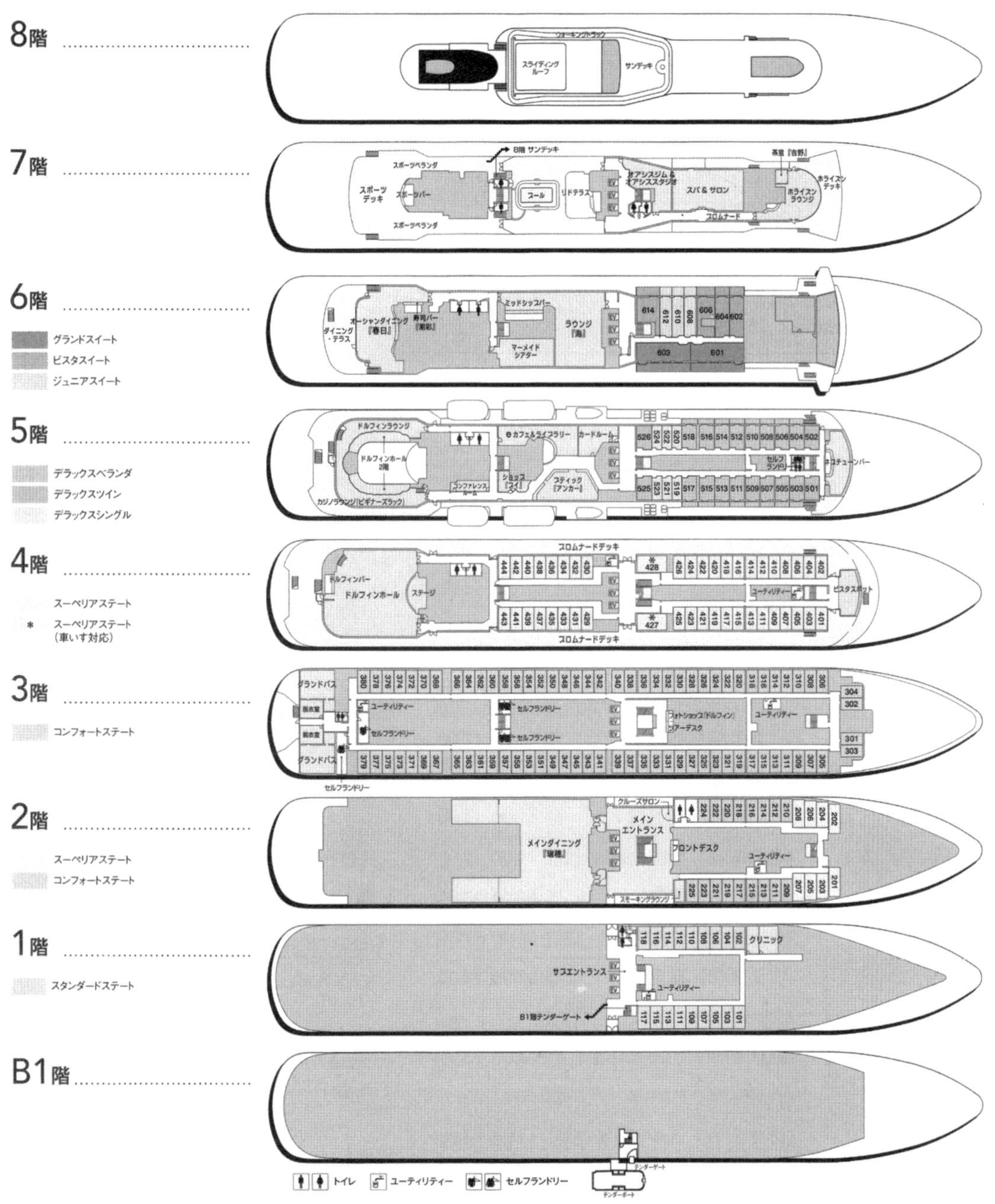

●にっぽん丸主要目　総トン数：22,472トン　船客定員：202室524名(最大)398名(ツインベース)　主機関：ディーゼル 10,450馬力×2　全長：166.6m　全幅：24.0m　喫水：6.6m　最高速力：21ノット　船内電圧：100ボルト／60Hz

（「にっぽん丸」のクルーズパンフレットより）

世界最大のクルーズ客船「ハーモニー・オブ・ザ・シーズ」デッキプラン

【就航年】オアシス[2009年12月]/アリュール[2010年12月]
ハーモニー[2016年5月]/シンフォニー[2018年4月]
【総トン数】225,282トン/226,963トン(ハーモニー)/230,000トン(シンフォニー)
【乗客定員】5,400名/5,494名(ハーモニー/シンフォニー)
【乗組員数】2,175~2,394名
【全長】361m 【全幅】65m 【全高】65m
【喫水】9.1m 【巡航速度】22.0ノット

※下記はハーモニー・オブ・ザ・シーズのデッキプランです。※オアシス・オブ・ザ・シーズ、アリュール・オブ・ザ・シーズ、シンフォニー・オブ・ザ・シーズはカテゴリー分けおよび施設などが一部異なります。※ロイヤル・カリビアン日本語ホームページ等でご確認ください。

客室カテゴリー一覧 ※客室+バルコニーの広さ(広さはハーモニーのものです。)

スイート客室

RL ロイヤルロフトスイート/141.5㎡+78.3㎡
TL スターロフトスイート/67.0㎡+38.0㎡
L1 L2 クラウンロフトスイート/50.6㎡+10.5㎡
VS ヴィラスイート/106.0㎡+44.2㎡
A1 A2 アクアシアタースイート/
A3 A4 56.1-76.4㎡+54.7-71.7㎡
OS オーナーズスイート/51.6㎡+22.5㎡
GT グランドスイート(2ベッドルーム)/53.8㎡+22.1㎡
GS グランドスイート/34.4㎡+9.7㎡
J3 J4 ジュニアスイート/26.6㎡+7.4㎡

バルコニー客室

1A ファミリーバルコニー/25.1㎡+7.6㎡
1C 2C スーペリアラージバルコニー/16.9㎡+7.4㎡
1D 2D 4D 5D 6D 7D 8D スーペリアバルコニー/16.9㎡+4.6㎡
1I 2I 4I ボードウォークバルコニー/16.9㎡+4.8㎡
1J 2J セントラルパークバルコニー/16.9㎡+4.8㎡

海側客室

1K ファミリー海側/25.1㎡
1N 2N 6N スタンダード海側/16.6㎡
2O ステュディオ海側/8.9㎡

内側客室

1R ファミリー内側/24.1㎡
1S セントラルパークビュー/18.4㎡
1T プロムナード/18.0㎡
4U スタンダード内側(バーチャルバルコニー)/15.9㎡
1V 2V 3V 4V 6V スタンダード内側/13.8㎡
2W ステュディオ内側(バーチャルビュー)/8.9㎡

デッキ3
1N 2N
1V 2V

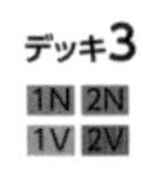

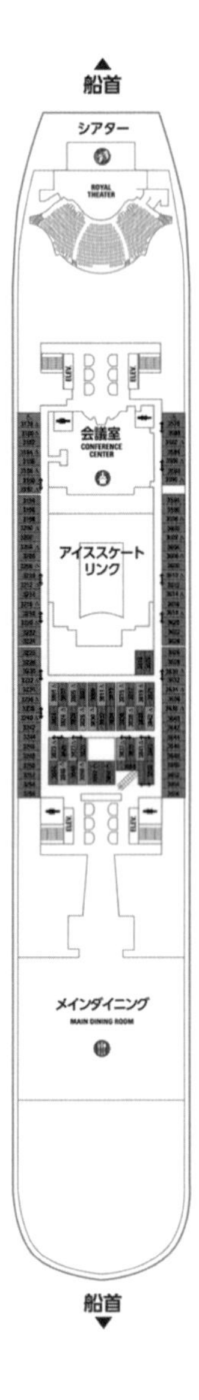

デッキ4
2O
2W

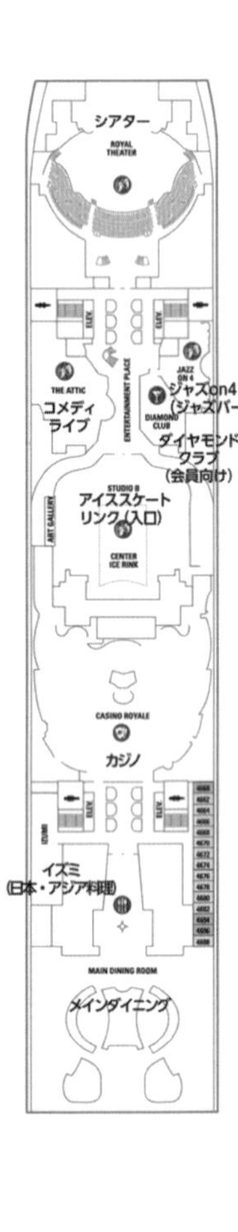

23万総トンの世界最大（2017年現在）のクルーズ客船で，全18階で構成されている。9階～14階まで，船体の中心線上に吹き抜けのプロムナードがあり，そこに面したキャビンにも窓がある。6,000人以上の乗客を乗せるため，一部，窓のないインサイドキャビンも配置されている。

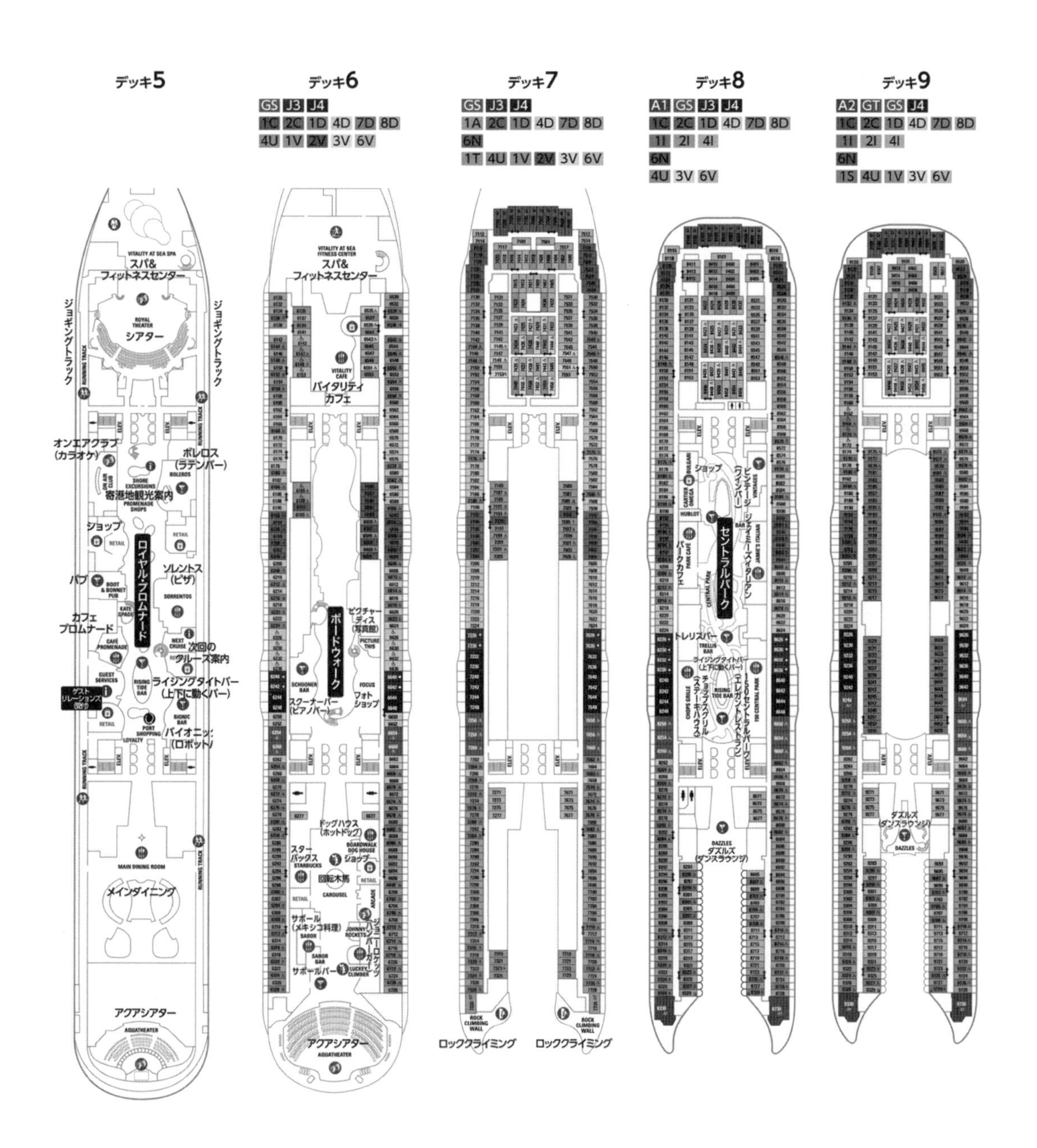

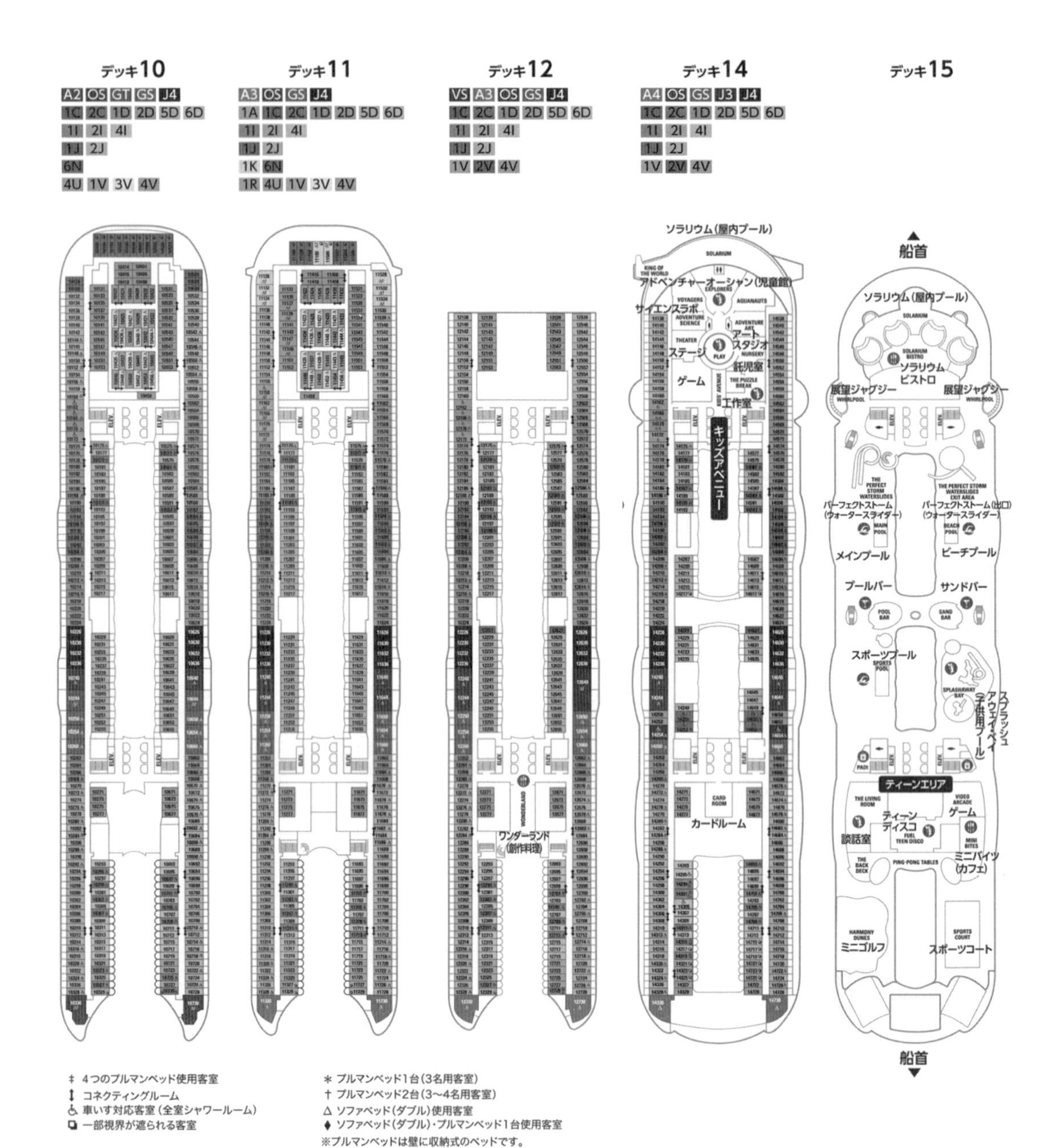
デッキ10
A2 OS GT GS J4
1C 2C 1D 2D 5D 6D
1I 2I 4I
1J 2J
6N
4U 1V 3V 4V
デッキ11
A3 OS GS J4
1A 1C 2C 1D 2D 5D 6D
1I 2I 4I
1J 2J
1K 6N
1R 4U 1V 3V 4V
デッキ12
VS A3 OS GS J4
1C 2C 1D 2D 5D 6D
1I 2I 4I
1J 2J
1V 2V 4V
デッキ14
A4 OS GS J3 J4
1C 2C 1D 2D 5D 6D
1I 2I 4I
1J 2J
1V 2V 4V
デッキ15
ワンダーランド（創作料理）
ソラリウム（屋内プール）
アドベンチャーオーシャン（児童館）
サイエンスラボ
アート・スタジオ
ステージ
託児室
ゲーム
工作室
キッズアベニュー
カードルーム
船首
ソラリウム（屋内プール）
ソラリウム
ビストロ
展望ジャグジー
展望ジャグジー
パーフェクトストーム（ウォータースライダー）
パーフェクトストーム（出口）（ウォータースライダー）
メインプール
ビーチプール
プールバー
サンドバー
スポーツプール
スプラッシュ・アウェイ・ベイ（子供用プール）
ティーンエリア
ティーン・ディスコ
ゲーム
談話室
ミニバイツ（カフェ）
ミニゴルフ
スポーツコート
船首
‡ 4つのプルマンベッド使用客室
↕ コネクティングルーム
♿ 車いす対応客室（全室シャワールーム）
❏ 一部視界が遮られる客室
＊ プルマンベッド1台（3名用客室）
† プルマンベッド2台（3〜4名用客室）
△ ソファベッド（ダブル）使用客室
♦ ソファベッド（ダブル）・プルマンベッド1台使用客室
※プルマンベッドは壁に収納式のベッドです。

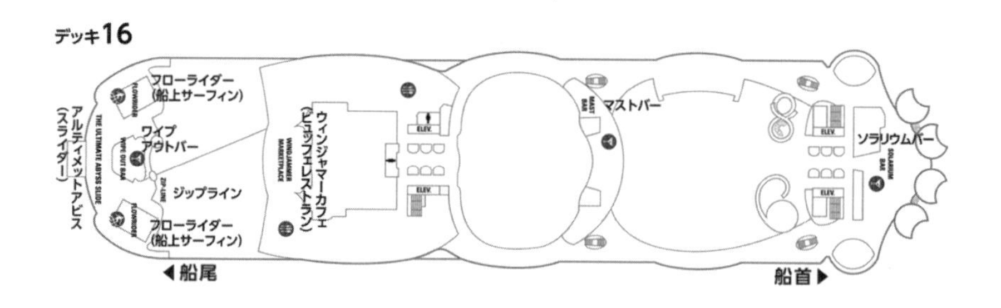

（デッキ17・18はデッキ16／ビュッフェレストランの上方に位置）

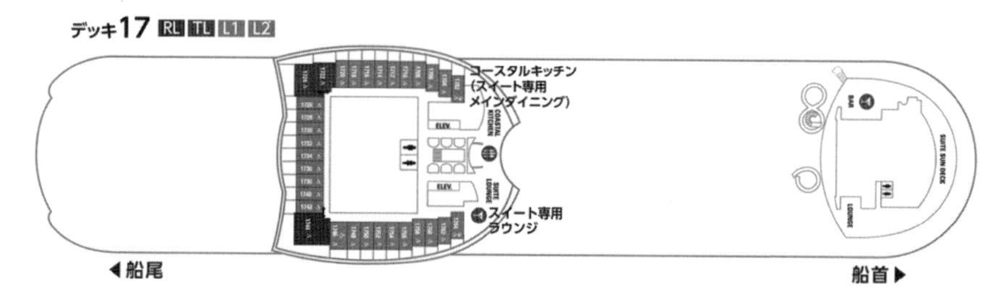

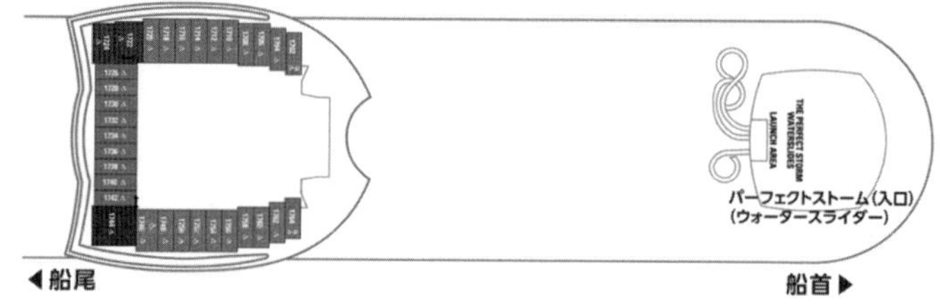

（提供：ミキツーリスト）

以下，74 ～ 75 ページに，客室とレストラン，バーの種類を示す。

客室のご紹介

Stateroom

開放感あふれる2階建ての「ロフト客室」や、大海原を映し出す「バーチャルバルコニー」など、客室にも驚きのアイデアが盛りだくさん！お一人様から三世代ファミリーまで、あらゆるニーズに対応した心地よい空間をご用意しています。

Royal Suite CLASS

スイート客室の特典

オアシスクラスのスイート客室のお客様は"ロイヤル・スイートクラス"サービスをご利用いただけます。

クラス	スタークラス	スカイクラス	シークラス
対象客室カテゴリー	RL、OP GP、TL、VS A1、A2	L1、L2 A3、A4、OS GT、GS	J3、J4
バトラーサービス"ロイヤル・ジニー"	●		
チップの別途支払不要	●		
スペシャリティレストラン無料	●		
ビバレッジパッケージ無料	●		
インターネット無料	●	●	
スイート専用ラウンジ	●	●	
専用ダイニング「コースタル・キッチン」	●	●	●※1
ワンランク上のバスアメニティ、ピロートップ、マットレス	●	●	●

※1 空きがある場合、ディナーのみご利用いただけます。

※2018年2月現在の情報です。各特典は予告なく変更する場合があります。　※スイート客室の種類によって、特典は異なります。

スイート客室

贅沢な空間が個性豊かにラインナップ。

ロイヤルロフトスイート

ロイヤルロフトスイート

RL

・客室141.5㎡ ・バルコニー78.3㎡
・2ベッドルーム ・ツインベッド（ダブルに変更可）
・バスタブ付きバスルーム ・定員6名

ヴィラスイート

VS（ハーモニー／シンフォニーのみ）

・客室106.0㎡ ・バルコニー44.2㎡
・4ベッドルーム ・ツインベッド（ダブルに変更可）
・バスタブ付きバスルーム ・定員9名

クラウンロフトスイート

L1 L2

・客室50.6㎡ ・バルコニー10.5㎡
・ダブルベッド（ツインに変更不可）
・シャワーのみ ・定員4名

スターロフトスイート

TL

・客室67.0㎡ ・バルコニー38.0㎡
・2ベッドルーム ・ツインベッド（ダブルに変更可）
・シャワーのみ ・定員4名

［オアシス／アリュールにはこちらも］

オーナーズパノラマスイート

OP

・客室99.8㎡ ・バルコニー14.9㎡
・ツインベッド（ダブルに変更可）
・バスタブ付きバスルーム ・定員4名

グランドパノラマスイート

GP

・客室84.9㎡ ・バルコニー14.9㎡
・ツインベッド（ダブルに変更可）
・バスタブ付きバスルーム ・定員4名

アクアシアタースイート

アクアシアタースイート

A1 A2
・客室62.5㎡～76.4㎡
・バルコニー56.6㎡～71.7㎡
・2ベッドルーム・定員8名
・ツインベッド（ダブルに変更可）
・バスタブ付きバスルーム

A3 A4
・客室56.1㎡～60.5㎡
・バルコニー54.7㎡～58.6㎡ ・定員4名
・ツインベッド（ダブルに変更可）
・バスタブ付きバスルーム

オーナーズスイート OS

・客室51.6㎡
・バルコニー22.5㎡
・ツインベッド（ダブルに変更可）
・バスタブ付きバスルーム
・定員4名

グランドスイート GT

（2ベッドルーム）

客室53.8㎡
・バルコニー22.1㎡
・2ベッドルーム
・ツインベッド（ダブルに変更可）
・バスタブ付きバスルーム
・定員8名

グランドスイート GS

・客室34.4㎡
・バルコニー9.7㎡
・ツインベッド（ダブルに変更可）
・バスタブ付きバスルーム
・定員5名

ジュニアスイート J3 J4

・客室26.6㎡
・バルコニー7.4㎡
・ツインベッド（ダブルに変更可）
・バスタブ付きバスルーム
・定員2～5名

バルコニー客室

潮風を感じるバルコニーは船旅の醍醐味。

ファミリーバルコニー

スーペリアバルコニー

1D 2D 4D 5D 6D 7D 8D

・客室16.9㎡ ・バルコニー4.6㎡（※1D、2Dの一部は7.4㎡）
・定員2～4名

ファミリーバルコニー

1A

・客室25.1㎡ ・バルコニー7.6㎡
・定員6名

スーペリアラージバルコニー

1C 2C

・客室16.9㎡ ・バルコニー7.4㎡
・定員2～4名

ボードウォークバルコニー

ボードウォークバルコニー

1I 2I 4I

・客室16.9㎡ ・バルコニー4.8㎡ ・定員2～4名

セントラルパークバルコニー

1J 2J

・客室16.9㎡ ・バルコニー4.8㎡ ・定員2～4名

海側客室

陽が差し込む、明るい窓付き客室。

スタンダード海側

スタンダード海側

1N 2N 6N

・客室16.6㎡ ・定員2～4名

ファミリー海側

1K

・客室25.1㎡ ・定員6名

ステュディオ海側

おひとり様OK

2O（ハーモニーのみ）

・客室8.9㎡ ・定員1名

内側客室

コンパクトで機能的。バーチャルバルコニーも登場！

プロムナード

スタンダード内側

プロムナード

1T

・客室18.0㎡ ・定員2～4名

スタンダード内側

1V 2V 3V 4V 6V

・客室13.8㎡ ・定員2名

セントラルパークビュー

1S

・客室18.4㎡ ・定員2～4名

ファミリー内側

1R

・客室24.1㎡ ・定員6名

［ハーモニー／シンフォニーのみ］

バーチャルバルコニー

バーチャルバルコニー

4U

・客室15.9㎡ ・定員2名

ステュディオ内側

おひとり様OK

2W

・客室8.9㎡ ・定員1名

客室設備 ■標準設備／テレビ、電話（目覚まし機能付）、金庫、ラジオ、化粧台、クローゼット、ミニバー（保冷庫）、室内温度調節器、アメニティ（タオル、石鹸、リンスインシャンプー）、ヘアドライヤー、110V（日本同様A型プラグ）及び220Vの電源 ■全室にツインベッド（ダブルに変更可）。※クラウンロフトスイート、ステュディオ客室はダブルベッド（ツインに変更不可）。3/4人目以降はソファベッド（ダブルサイズ）またはプルマンベットです。全室にミニバー（グランドスイート以上の客室には冷蔵庫）があります。朝夕は清掃のサービスがあります。■バスタブ／スターロフトスイート、クラウンロフトスイートを除くスイート客室にあります。スイート以外の客室およびスターロフトスイート、クラウンロフトスイートはシャワーのみとなります。

※客室の広さは、おおよその目安です。　※ここに記載した客室・バルコニーの広さはハーモニーのものです。オアシス／アリュール／シンフォニーは若干異なる場合があります。

レストラン&バー

Restaurant & Bar

「オアシスクラス」は、40店以上のレストランやバーがひしめく"グルメシップ"。美しい夜景を望めるロマンチックなダイニングや緑豊かな庭園カフェ、あっと驚く仕掛けのあるバーなど、多彩なグルメをご用意。美味しくてワクワクする、食の時間をご提供します。

OA オアシス　AL アリュール　HM ハーモニー　SY シンフォニー

基本のお食事

豪華なフルコースから、スイーツまで無料でお好きなだけ召し上がれます。

コースタルキッチン

※スイート客室専用ダイニング。(ジュニアスイートは夕食のみ空きがあれば利用可)

メインダイニング

ウィンジャマーカフェ(ビュッフェ)

ルームサービス

※ルームサービスは、朝食(コンチネンタルブレックファースト)を除き有料となります。

スペシャリティレストラン

ハイクオリティなお料理を召し上がれる特別なレストラン。

メニュー表や料理に魔法の仕掛けが!

150セントラルパーク

チョップスグリル

イズミ

ワンダーランド

スペシャリティレストラン	内容	カバーチャージ	OA	AL	HM	SY
150セントラルパーク	エレガントなレストラン【夕】	$45	●	●	●	●
チョップスグリル	ステーキハウス【昼・夕】	昼$19～$20.9 夜$49～$53.9	●	●	●	●
ジョバンニズテーブル	カジュアルイタリアン【夕】	$30	●	●		
ジェイミーズイタリアン	カジュアルイタリアン【昼・夕】	$35～$38.5			●	●
ワンダーランド	創作料理【夕】	$49～$53.9			●	●
サボール	モダンメキシコ【昼・夕】	$19	●	●	●	●
サンバグリル	ブラジル料理【夕】	$30		●		
イズミ	日本&アジア料理【昼・夕】	アラカルト	●	●	●	●
ソラリウムビストロ	ヘルシーフード【夕】	一部有料	●	●	●	●
シェフズテーブル	貸切フォーマルダイニング【夕】	$95	●	●	●	●

※料金は変更になる場合があります。有料レストランの料金には18%のサービスチャージが追加されます。5歳以下無料、6～12歳は一律$8です。　※アラカルト料金の店舗では、メニューごとに料金が表示されています。
※いずれも飲み物は別料金です。

カフェ&軽食

気軽に楽しめるカジュアルなお食事。
無料のお店もたくさん!

ジョニーロケッツ

カップ&スクープ

ソラリウムビストロ

パークカフェ

カフェ&軽食	内容	カバーチャージ	OA	AL	HM	SY
カフェプロムナード	24時間営業のカフェ【24時間】	無料	●	●	●	●
ソレントス	ピザ【昼・夕】	無料	●	●	●	●
スターバックス	カフェ【朝・昼・夕】	アラカルト	●	●	●	
ジョニーロケッツ	ハンバーガー【昼・夕】	$6.95～$8.95	●	●	●	●
カップ&スクープ	カップケーキ／アイスクリーム【朝・昼・夕】	アラカルト	●	●		
ドッグハウス	ホットドッグ【昼・夕】	無料	●	●	●	●
ボードウォークドーナッツ	ドーナツ【朝・昼・夕】	無料		●		
パークカフェ	デリスタイルのカフェ【朝・昼】	無料	●	●	●	●
バイタリティカフェ	ヘルシーな軽食【朝・昼・夕】	一部有料			●	●
ソラリウムビストロ	ヘルシー料理【朝・昼】	無料	●	●	●	●
ミニバイツ	カフェ【朝・昼・夕】	無料			●	●

※オープン時間は船内新聞にてご確認ください。　※セルフサービスを除き、いずれも飲み物は有料です。

バー&ラウンジ

クルーズライフを華やかに彩る、
個性豊かなバーが勢ぞろい。

プールバー

ビンテージ

ロボットバーテンダーがおもてなし!

バイオニックバー

4デッキ間を上下に移動! 洋上初の動くバー!

ライジングタイドバー

バー&ラウンジ	内容	OA	AL	HM	SY
バイオニックバー	洋上初のロボットバー			●	●
シャンパンバー	シャンパンとカクテル	●	●		
ライジングタイドバー	上下に動くバー	●	●	●	●
ボレロス	ラテンバー/ダンスラウンジ	●	●	●	●
パブ	英国パブ	●	●	●	●
スクーナーバー	ピアノバー	●	●	●	●
ビンテージ	ワインバー	●	●	●	●
ワイプアウトバー	カフェバー	●	●	●	●
ソラリウムバー	眺めの良いバー	●	●	●	●
プールバー	プールサイドのバー	●	●	●	●
スイートラウンジ	スイート客室専用	●	●	●	●

※オープン時間は船内新聞にてご確認ください。

（提供：ミキツーリスト）

第7章 クルーズ客船のソフト

7.1　船員構成

クルーズ客船を運航するのは，基本的に船長および海員で，海員雇入契約の証明書である海員名簿に記載をされた船員となる。船員とは，その証明書となる船員手帳を有して，船での職務に従事できる人を指し，その船員手帳は監督官庁が発行する。

船の最高責任者として全権を掌握するのが船長で，長年航海士としての実績があり，1等航海士の海技免状の保持者の中から船会社が任命する。

一般に，最近の大型クルーズ客船では，船を安全に運航する「運航部門」と，乗客の接客を担当する「ホテル部門」とに分かれており，運航部門のトップは，船長が兼務するか，スタッフキャプテンと呼ばれる副船長がなり，ホテル部門のトップはホテルマネジャー（ホテルディレクターと呼ぶ船もある）となっている場合が多い。

運航部門には，甲板部，機関部，通信部（最近は甲板部が兼ねている場合もある），医療部があり，甲板部には1，2，3等航海士と部員，機関部には機関長，1，2，3等機関士と部員，通信部には通信長と通信士，医療部には医師と看護師などが配置されている。

ホテル部門では，ホテルマネジャーの下に，クルーズディレクター（旅客接遇部門長），フード＆ビバレッジマネジャー（飲食部門長），レストランマネジャー，エグゼクティブシェフ（調理部門長），客室部門マネージャー，チーフパーサー（事務部門長）が，各部門の指揮をとっている。

中でも，クルーズディレクターが率いるクルーズスタッフは，クルーズ各船独特のものである。船

図7-1　船長主催カクテルパーティ時に並んだ士官メンバー

内のショーをはじめとするさまざまなイベント，寄港地での上陸やオプショナルツアーなど，旅客の娯楽に関するあらゆることの世話をする仕事で，彼らの働きがクルーズの評判を大きく左右するといわれている。クルーズディレクターは，ショーの司会なども担当し，その多くが元エンターテイナーという経歴をもっている。

フード＆ビバレッジマネジャーの率いる部門では，船上売上の大きな部分を占めるアルコール飲料などの販売が重要となる。クルーズ料金には，食事やイベントなど，ほとんどすべてのものが含まれているが，アルコール飲料などは，船長主催パーティなどを除くと普通は入っていない。このため，船内中で飲み物の販売が行われている。

ただし，これらの部門の分け方や各責任者の呼び方は，クルーズ会社によってまちまちで，統一されたものではない。

コラム⑪　船員の肩章

船の士官の制服には，肩の部分に「肩章」が，また袖口にも同様のものが付けられていて，それを見ると，その人の船員としての役割がわかる。船長，機関長は 4 本線，1 等，2 等，3 等士官になるに従い本数が減る。線の間の色は部署によって違っていて，甲板部は黒，機関部は紫，通信部が緑，事務部は白，医療部は赤が一般的。

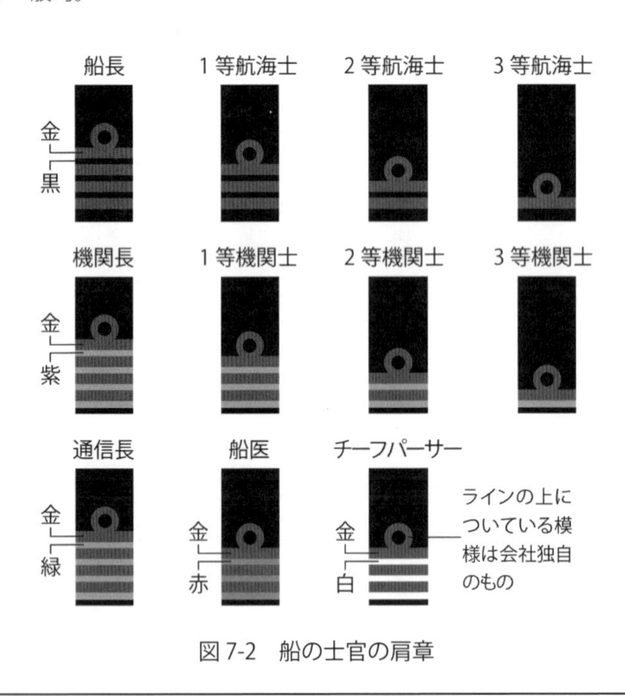

図 7-2　船の士官の肩章

7.2　各部門の守備範囲

7.2.1　運航部門

(1) 甲板部

甲板部は，船を動かす部門であり，海技士の資格をもつ士官（Officer）が常に指揮をとり，複数の部員とともにブリッジ（操舵室）において安全な運航に携わる。1，2，3等航海士が，4時間ごとに交代で当直（ワッチ）して24時間体制で船を動かし，港内や狭水道での操船には船長が自らブリッジで指揮をとる。

図7-3　ブリッジ

コラム⑫　ブリッジ

船の操舵室のことを「ブリッジ」，またはより正確に「ナビゲーション・ブリッジ」という。日本語では「船橋」と呼ぶが，これは英語のブリッジをそのまま訳したためである。

操舵室をブリッジ（橋）と呼ぶのは，昔，外輪船の時代には，両側の外輪の上に渡した橋状の構造の上で操船したからだといわれている。

(2) 機関部

機関部は，船の心臓ともいえる主機関（船を推進するためのエンジン）や補機（発電機など）をはじめ，あらゆる機関の運転とメンテナンスを行う。機関長のもとに，1，2，3等機関士が，甲板部と同様に4時間ごとに交代して常にすべての機器が正常に稼動することに責任をもつ。機関室（エンジンルーム）内のエンジンコントロール室で監視をするが，ブリッジ内のエンジン監視ボードによって監視を行うこともある。

図 7-4　エンジンコントロール室

(3) 通信部

かつては，船舶通信士の資格をもつ船員が，ブリッジ近くの通信室に常時待機して，船と陸上や他船との間の通信をするのが義務付けられており，緊急時に発信するモールス信号などには特殊な技能が必要であった。現在は音声通信手段の発達によって，当直航海士が直接交信することが可能となり，通信部は一般船舶では縮小の方向にあるが，多くのクルーズ客船では，船内通信や外部との通信のメンテナンスに高度な電気・通信技能が必要とされるため，通信部として機能している。

(4) 医療部

クルーズ客船には，医師と看護師が乗船しており，急病の乗客・乗員への処置が行える体制が整っており，長期間航海をする大型クルーズ客船では手術室も完備されている。

7.2.2　ホテル部門

(1) エンターテイメント部門

陸上のリゾートホテルなどに比べて大きく異なるのが，クルーズディレクターを責任者とするエンターテイメント部門で，この部門の能力がクルーズ客の満足度を大きく左右するといわれている。

この部門の充実は，現代クルーズのもつ大きな特徴といえ，あらゆる乗客の満足度を最大化するために，クルーズ全体をエンターテイメントとして捉えてその構成を構築している。乗船した瞬間の「楽しいクルーズへの期待」を演出するファーストインプレッション，クルーズ中の各乗客のニーズに合わせたさまざまなイベント，そして下船前夜における「楽しかった」「また乗りたい」という気持ちを最大化するためのラストインプレッションの与え方などに，細心の注意をはらったものとなっている。

クルーズディレクターの下に，クルーズスタッフ，プロのエンターテイナーが配置されて，クルーズ期間中のエンターテイメントの計画から実行までをこなしている。また，子供用プログラムなども企画・実行する。

⑵ 飲食部門（フード＆ビバレッジ）

クルーズ客の多くが，食事の質とサービス，そして量に大きな期待をもっているので，この飲食部門はエンターテイメント部門とともに顧客満足度を向上させる上で重要な部門となっている。またクルーズ会社によっては，次の⑶のレストラン部門と一体化している場合もある。

特に飲物部門の中核的な存在であるバー部門は，クルーズ運航会社にとっては，クルーズ料金に含まれない収入になるので，この部門の売上を増やすことは経営において重要となる。船内の各ラウンジやサンデッキなどに設置されたバーカウンターからのサービスが中心である。

⑶ レストラン部門

メインダイニングなどでの飲食のサービスを行う部門で，メートル・ディ（Maitre d'Hotel）の下に複数のヘッドウェイター（Head waiter）が配置されている。ヘッドウェイターは，各テーブルで直接サービスを行うウェイターおよびその補助員（アシスタントウェイター，バスボーイなどと呼ばれる）を監督する。ウェイターは，一般的に1航海をとおして特定のテーブルを担当し，メニューの紹介，注文，料理の給仕を行う。一般には5～8人掛けのテーブルを2～3卓担当する。一方，補助員は，水のサービス，空いた皿の片付けなどを行いながら，将来のウェイターとしての仕事を実地で学ぶシステムをとる船が多い。

⑷ 調理部門

非常に大量の食事を一度に供給するため，効率のよい調理，配膳が必要となる。特に，海外のクルーズ客船では，前菜，スープ，サラダ，メインディッシュ，デザートごとに複数のチョイス（選択肢）があるのが一般的であり，その数を予測しながらの下ごしらえが必要となる。

複数のシェフの下でコックが実際の調理にあたる。また，パンやデザートは別系統となっている場合が多い。

例えばカリブ海の最大旅客定員2,600名の7万総トン型船で，シェフ5人，コック35人，ペイストリー37名となっている。

図7-5　調理部門のギャレー

(5) 客室部門

客室のクリーニングをするスチュワードは，一般に固定した 7 ～ 10 室のキャビンを担当して，毎日 2 回，一般的には朝食時間帯と夕食時間帯に，部屋のクリーニング，ベッドメーキングを行う。

(6) 事務部（パーサーズ・オフィス）

船の事務長をパーサーという。このパーサーが管理するパーサーズ・オフィスは，全旅客の事務を所轄するほか，船内でのクレジットカード類の取り扱い，両替などの乗客サービス，各地での下船・乗船手続き，船内新聞の発行などを行う。

海外のクルーズ客船でも，最近はパーサーズ・オフィスに日本人パーサーを配置して，日本人旅客に対応している場合が多い。この場合，海外の旅客への対応もしなくてはならないので，日本語だけでなく，英語および第 2 外国語での応対ができる程度の語学力が必要となる。

7.3　緊急時の船員の役割

船員は全員，緊急時の乗客の避難などの誘導を行わねばならない。船長が船の放棄を決めた場合には，全船員は各乗客をそれぞれに指定された場所（マスター・ステーション）に集合させて，救命ボートまたは救命いかだ（ライフラフト）に乗せ，船から安全に脱出させる義務がある。

安全な避難を確実に行うために，各航海のたびに乗客・乗組員全員参加の避難訓練（ボートドリル）がある。乗客は，各キャビンのドアおよびクルーズカード（多機能ドアキー）に記載されているマスター・ステーションまで，迅速に避難する最短の経路を確認し，乗組員は乗客のマスター・ステーションまでの誘導，点呼，救命ボートへの乗船などの確認を行う。従来はキャビンに備え付けてあるライフジャケットを着て避難訓練を行うのが一般的であったが，「コスタ・コンコルディア」の海難事故を契機に，大型船ではマスター・ステーションまたは救命ボート周辺に人数分のライフジャケットが備え付けられるようになり，キャビンのライフジャケットを着て集合することを義務付けない船も多くなった。

図 7-6　最近の避難訓練の様子。ライフジャケット着用の必要はなく，指定のマスター・ステーションに集合して，点呼を受け，避難の仕方，ライフジャケット着用方法を学ぶ。出港前に行い，全員参加の義務がある。

7.4　陸上との連携

陸上には運航管理者が置かれ，常に船長と連絡を取りながら安全な航海を担保している。かつては，船の安全はすべて船長の責任だったが，現在は，通信手段が発達して常に陸上とのコミュニケーションが保てるようになり，船と陸上が一体となって安全運航に責任をもつことを義務化した国際安全管理コード（ISM Code：International Safety Management Code）が制定されている。

7.5　CIQ

外国籍船が日本に寄港する時には国の CIQ が重要となる。CIQ とは，税関（Customs），出入国管理（Immigration），検疫（Quarantine）の英語の頭文字をとったもので，国境を越える時の荷物への関税，パスポートによる人の検査，病原菌や害虫の持ち込みを防ぐための検査を国が行っている。

出入国管理については，パスポート検査と同時に，顔写真の撮影，指紋の登録が義務付けられており，空港では長蛇の列ができて通過するのに時間を要することも珍しくない。

クルーズ客船の場合には，朝に入港して，夕方には出港するパターンが多く，観光に使える時間が短く，しかも最近は数千人の乗客・乗員が乗っていることから，CIQ に要する手間と時間をできるだけ短縮することが求められていた。

そこで日本政府は，2012 年に顔写真などを不要として簡易手続きで上陸できる「寄港地上陸許可制度」を設けたため，乗客が上陸できるまでの時間が大幅に短縮された。さらに 2015 年に「船舶観光上陸許可制度」を設けて，特定の客船について短時間の上陸観光をする乗客のビザを不要とした。

演習

クルーズという仕事につくための就職活動の計画を立てなさい。

≪問題≫

7-1　船長の役割はなにか。

7-2　ホテルマネジャーの役割はなにか。

7-3　航海士と機関士の役割の違いはなにか。

7-4　クルーズディレクターの役割はなにか。

7-5　緊急時の船員の役割はなにか。

7-6　国際安全管理コードの求めているものはなにか。

7-7　CIQ とはなにか。

7-8　クルーズ客船の CIQ と空港での CIQ の違いはなにか。

参考資料

RCI の「フリーダム・オブ・ザ・シーズ」の船内新聞に入っていた「BEHIND THE NAME TAGS」という資料で，クルーズ客船の運航を支えるキーマンを紹介している。運航やサービスをしている船員を身近に感じてもらうための情報で，船内ボードに掲示されている船も多い。

CRUISE NEWS

BEHIND THE NAME TAGS

CAPTAIN BILL WRIGHT, MASTER

Captain Bill Wright, 52, joined Royal Caribbean Cruises Ltd. in June 1992 and was promoted in September 2001 to the position of Senior Vice President of Safety, Security & Environment for the fleets of both Celebrity Cruises and Royal Caribbean International. In January 2005 Captain Wright overtook the responsibilities of Senior Vice President of Marine Operations for Royal Caribbean International. Prior to that, Captain Wright was Master of *Radiance of the Seas, Enchantment of the Seas, Grandeur of the Seas* and *Sovereign of the Seas.* Captain Wright also sailed as Master of Celebrity Cruise Lines exclusive *Xpedition* in the Galapagos Islands. Captain Wright began his career at the age of 15, working as a deckhand on a commercial dive ship during summer vacations. He is fluent in Norwegian and received his Master Mariner's license in Norway in 1991. In 1992, Captain Wright received his Master's degree in Marine Engineering from Vestfold College in Norway. He has delivered lectures on navigational systems and advanced ship handling at maritime colleges and institutions worldwide. Captain Wright and his wife maintain their home in the small mountain village of Roros in Norway. While Captain Wright is responsible for the Marine Operations of the entire Royal Caribbean International Fleet, it is his disticnct honor to serve as your Captain on *Freedom of the Seas.*

CINDY DANGEL, LOYALTY & FUTURE CRUISE MANAGER

Your Loyalty Ambassador and Future Cruise Manager Cindy comes from beautiful British Columbia, Canada. Her passion for travel brought her to Royal Caribbean International in 1992. During Cindy's 14 years with Royal Caribbean International, she has had the opportunity not only to see the world but also to be part of a growing company. Cindy has had the good fortune to experience all the different classes of ships in our expanding fleet, starting from the days of the Nordic Prince and Song of Norway to the current Vision Class, Radiance Class, Voyager Class and Freedom Class ships. Traveling the world and meeting so many wonderful people continues to make her job very rewarding. She looks forward to meeting you and assisting you with planning and booking your next cruise vacation while onboard the Freedom of the Seas. Cindy is also very pleased to introduce Suzanne Van Der Westhuizen as her assistant Loyalty Ambassador and Future Cruise Consultant.

JOHANN PETUTSCHNIG, EXECUTIVE CHEF

Johann is originally from Klagenfurt, Austria where he started his culinary career. Prior to joining Royal Caribbean International, Johann had the honor of working for the King of Norway. Along the way, he also earned several awards from various international culinary competitions. When Johann joined Royal Caribbean International in July 2005 onboard the Navigator of the Seas, he brought with him seven years of shipboard culinary management experience working with other cruise lines. Among his many professional credentials, he holds a diploma as a Certified Chef from the American Culinary Institute. While on vacation, Johann enjoys spending his time fishing in Norway. Johann hopes that you enjoy all of the fabulous dishes in each of the various restaurants that he and his team work tirelessly to create each week. Bon appetit.

MARTIN RISSLEY, HOTEL DIRECTOR

Martin Rissley joined Royal Caribbean International in November 2000. He has served as Hotel Director on *Voyager of the Seas* and *Adventure of the Seas.* He previously spent 21 years in the hotel and resort industry. His career started in the United Kingdom before moving to Lausanne, Switzerland, Sydney, Australia and on to Vancouver, Canada. There he joined Mandarin Oriental Hotels, followed by Delta Hotels and Resorts and Coast Hotels and Resorts. His home base is Victoria, British Columbia where he lives with his wife and three children. Martin has several Hotel Management diplomas, is a Certified Hotel Administrator, a distinguished member of the Association of Tourism Professionals, SKAL and the Chaine Des Rotisseurs, Maitre d' Hotel. Martin welcomes you onboard the beautiful *Freedom of the Seas* and wishes you a wonderful vacation experience.

KEN RUSH, CRUISE DIRECTOR

Ken can boast of being from the *Courtesy Capital of the World,* La Mirada, California. After studying acting , Ken pursued a successful career as a Master of Ceremonies for many major events including charity auctions, nightclubs and cabarets. In the fall of 1984, he took a cruise aboard the Azure Seas that changed his life. Within months of the cruisevacation, he was hired as a Cruise Staff member. Because of Ken's passion and determination, he was promoted quickly through the ranks holding several positions from Shore Excursion Manager, Production Manager to Cruise Director. He has been Cruise Director on numerous Royal Caribbean International ships and is very honored to have been chosen as part of the inaugural team taking out all of our new ships. With twenty one years in the cruise industry, Ken is happy to be your host for the best vacation of your life. Future plans include a long successful career with Royal Caribbean International or possibly his own talk show appropriately named, *The Rush Hour.*

JOAO MENDONÇA, FOOD & BEVERAGE MANAGER

Born in Lisbon, Portugal, at the age of 17 he decided to enroll in the most prestigious hotel school in Portugal, the Estoril Hotel and Management school. After working in several luxury hotels in Lisbon, Joao decided that it was time to move on to something more exciting and went to work in the Ambassador Hotel in London. He came to Royal Caribbean International at the age of 20 and has never left. He started as an Assistant Waiter and then progressed to Waiter, Head Waiter, Maitre d', Windjammer Manager and now Food and Beverage Manager. He has worked on several Royal Caribbean International ships such as the *Sun Viking, Nordic Prince, Song of America, Sovereign of the Seas, Majesty of the Seas, Monarch of the Seas, Viking Serenade, Legend of the Seas, Rhapsody of the Seas, Splendour of the Seas, Brilliance of the Seas, Navigator of the Seas* and now *Freedom of the Seas.*

第8章 クルーズ事業の経済性

8.1　クルーズの運航コスト

クルーズ産業は，よく「設備（装置）産業」といわれる。クルーズ事業を行う上でのインフラとなるクルーズ客船が非常に高価で，その設備への投資が事業採算に極めて大きな影響があるためだ。

船の建造に要する費用を「船価」という。船価は，その時々の造船業の好不況によって大きく変動するが，1～2万総トンの小型クルーズ客船でも100～150億円，最近主流となっている10万総トンの大型クルーズ客船だと600億円前後，15万総トンで800～1,000億円が必要となる。2009年に完成した22万総トンのクルーズ客船「オアシス・オブ・ザ・シーズ」は1,300億円という。

この高額な船価を，クルーズ運航会社は耐用年数内に回収する必要があり，このために毎年のコストに分散して計上する費用を「減価償却費」という。客船は建造されてから20年程度は使われるので，単純にその耐用年数で割ってみると，毎年の減価償却費は，上述の小型船で5～7.5億円，10万総トン級の大型船になると30～50億円となる。すなわち，毎年この減価償却費以上の収益が出せなければ，経済的には成り立たないことになる。船の減価償却年数は，国によってさまざまで，15～30年間程度の幅があるといわれている。日本では基本的に15年なのに対し，アメリカのクルーズ会社では30年程度だという。年数が長い分，アメリカのクルーズ客船では毎年のコストが小さく抑えられ，クルーズ料金を下げることが可能となる。

クルーズ客船の運航に必要なコストは，固定費と変動費に分けられる。固定費とは，たとえクルーズ客船を運航しなくても発生する年間のコストで，上述のクルーズ客船の建造費の減価償却費，金利費，保険料，修繕費，船員費などの人件費などがある。一方，変動費はクルーズ客船の運航に伴って計上されるコストで，燃料費，食料費などがある。

また，運航費，船費，一般管理費といった分類法もある。運航費は変動費であり，船費と一般管理費は固定費となる。

一般にクルーズ客船では，以下のように各コストを見積もる。

①　固定費

- 減価償却費：船価×0.9÷15（耐用年数15年とし，15年後の簿価を船価の10%と仮定）
- 船員費　　：船員平均年収×船員数（日本人船員では年間700～1,000万円，フィンピン人など外国人船員では年間150～200万円程度）
- 金利費：借入額×金利
- 保険料：船舶の運航に伴うさまざまなリスクに対する保険の費用

- 修繕費：船舶の定期的なドック費用，日常の補修費用
- 店費　：陸上事務所費＋陸上要員人件費＋広告費など

② 変動費

- 燃料費：1 時間・1 馬力あたりの燃料消費量×エンジン馬力×燃料価格×年間航海時間
- 食料費：1 人・1 日あたりの食材費×乗員数×年間運航日数
- 港費　：（港湾使用料＋タグボート費用＋パイロット費用）×年間入港回数

事例：クルーズ運航コストの内訳 [1]

日本郵船の子会社クリスタル・クルーズの場合の運航コストの内訳は概略以下のとおり。

5 ～ 6 万総トンのラグジュアリー・クラスの客船を 3 隻運航時の概略値。

- 運航費　　：22％（食材費，港費，運河使用料，燃料費，消耗品購入費など）
- 船費　　　：54％（用船料，船員費，修繕費，保険料）
- 一般管理費：24％（店費，広告宣伝費など）

船費は 1 隻・1 日あたり大略 11 万ドル程度といわれる。

8.2　クルーズの損益分岐解析

クルーズ事業の可能性調査には，損益分岐解析が用いられることが多い。損益分岐解析とは，事業を行うことによるコスト（費用）と，その事業による売上（収入）が同じになる点を求める解析法で，利益がゼロ，すなわち利益も損失も出ない点を求めるもの。すなわち，その事業が，黒字になるか，赤字になるかの分岐点を解析する手法である。

クルーズ客船の場合には，この損益分岐点における「乗客のクルーズ料金」を求め，それがターゲットとするマーケットに対して妥当かを検討することがよく行われる。

損益分岐料金を下げるためにもっとも効果的なのは，大定員化である。同じ大きさの船で定員だけを増やすと，ほぼ定員に反比例して損益分岐料金は下がる。しかし，これは一般的には，船内が窮屈になり，サービスグレードも落ちることとなる。したがって大定員化を船体の大型化と組み合わせて，ターゲットとするマーケットに相応しい大きさと定員のバランスをとることが必要となる。大型化すると損益分岐料金が低下する傾向になるのは，大型化による船価の上昇，燃料費の上昇，船員費の増加率が，旅客定員の増加率よりも一般的には小さいことによっている。

この他，損益分岐料金を下げるための方策としては，船の速度を落とすことによる燃料費の削減，外国人船員の使用による船員費の削減がある。特に，海外のクルーズ客船では，サービス要員の多くを賃金の安い国からの人材でまかなっており，日本籍の外航クルーズ客船でも，現在では，フィリピン人などのサービス要員を使っている。

8.3　クルーズ客船の需要予測

前節で，クルーズ客船の大型化，大定員化によって，損益分岐料金を下げることができると述べたが，これはあくまで，需要が十分にあれば，という仮定のもとでの議論である。損益分岐解析では，費用と収入が一致する点を求めるわけだが，この収入は次のようにして計算される。

収入＝平均料金×乗客数＝平均料金×旅客定員×消席率

ここで消席率とは，定員に対してどれだけの割合の乗客があったかを示す指標で，カリブ海などで人気のカジュアル船では 100 ～ 110%，ラグジュアリー・クラスのクルーズ客船で 60 ～ 70% 程度といわれている。カジュアル船で 100% を超えているのは，消席率は全室を 2 人部屋とした時の定員に対するものであり，一部のキャビンを家族連れなどで 2 人以上が使用したためである。

カリブ海のクルーズ客船においては，消席率 80% 程度が損益分岐点といわれ，これを超えた消席率分が利益になり，この消席率を達成できないクルーズは撤退を余儀なくされる。すなわち，いくら大定員化をしても消席率が低ければ収入は増えず，損益分岐料金は下がらない。

したがって，十分な需要を生み出し，消席率を上げることが重要となる。需要予測には種々の方法があるが，ここでは筆者らが行った AHP 法を用いたクルーズ・マーケットの需要予測とその結果 [2] について紹介する。

まず，顧客が旅行を決定する要因として「交流」「心の安らぎ」などの 9 つをあげ，「温泉」「クルーズ」「テーマパーク」「海外旅行」の 4 つの旅行のどれを選択するかを，AHP 法を用いて分析した。その結果，クルーズの潜在需要は，年代，性別にかかわらず全体のほぼ 30% を占めており，極めて高いことがわかった。

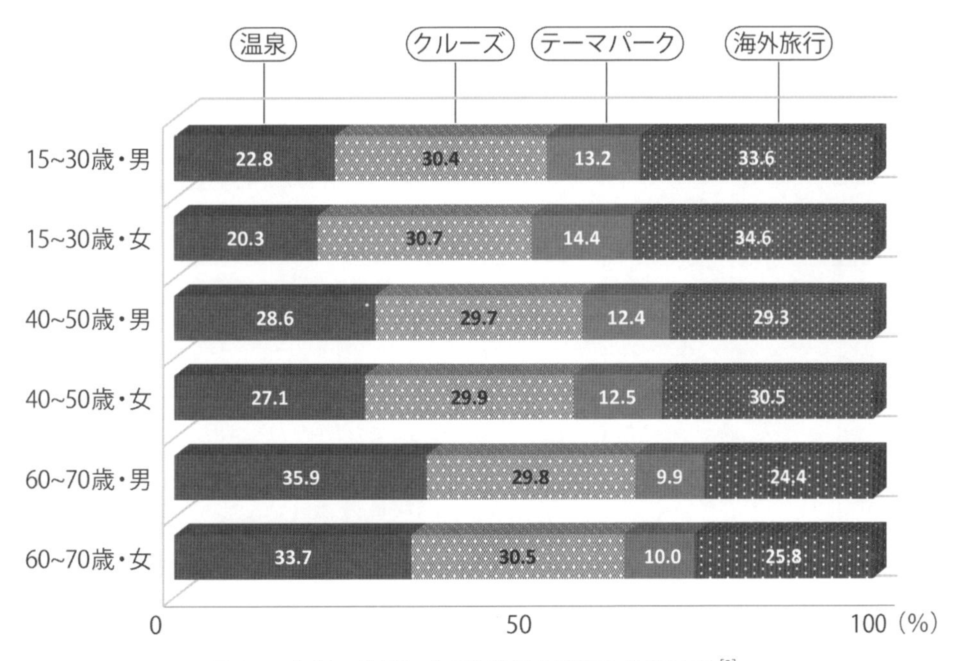

図 8-1　年齢・性別による観光選択嗜好の分析結果 [2]

次に，クルーズに出かけることの障害となっている「期間」「料金」「船酔い」「認知度」が，どの程度需要に影響を及ぼしているかを分析した結果を示す。図 8-2 には，クルーズ期間が需要に及ぼす影響を示す。クルーズ期間が長くなるほど需要は減少し，その予測値は 1999 年の実績値とよく合っている。

図 8-3 には，クルーズ料金が需要に及ぼす影響を示す。この結果から，1 泊あたりのクルーズ料金が高いほどクルーズの需要は急激に減少することがわかる。クルーズ料金を，現在の日本籍クルーズ客船の 1 泊あたり 42,000 円から 20,000 円にすると需要は約 4 倍，10,000 円にすると約 9 倍の需要が見込める。ただし，日本のクルーズとして採算がとれる料金は，1 泊 16,000 円以上であり，16,000 円の場合には 10 万総トン型大型船（旅客定員 3,164 名，乗組員 1,185 名，エンジン出力 84,761 馬力，船価 488 億円と仮定）を使う必要があるとの結論が導かれている [2]。

この研究をさらに進めた成果が，文献［3］に紹介されているので，参照してほしい。

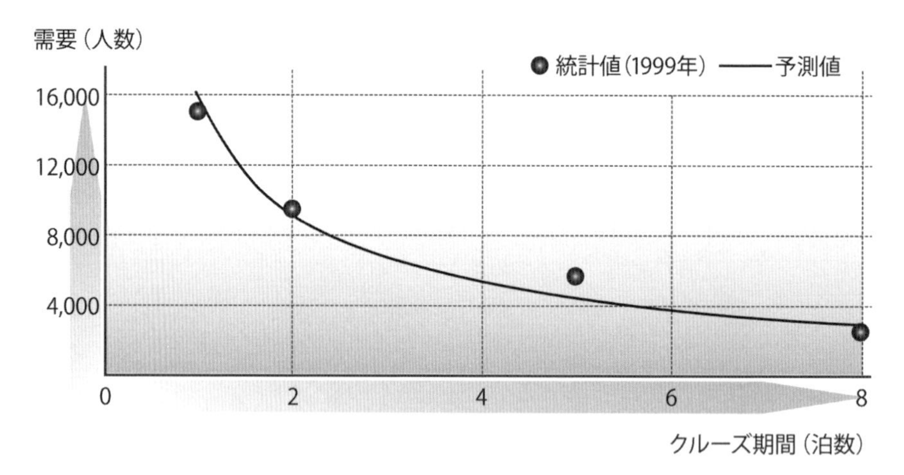

図 8-2　クルーズ期間が需要に及ぼす影響の分析結果

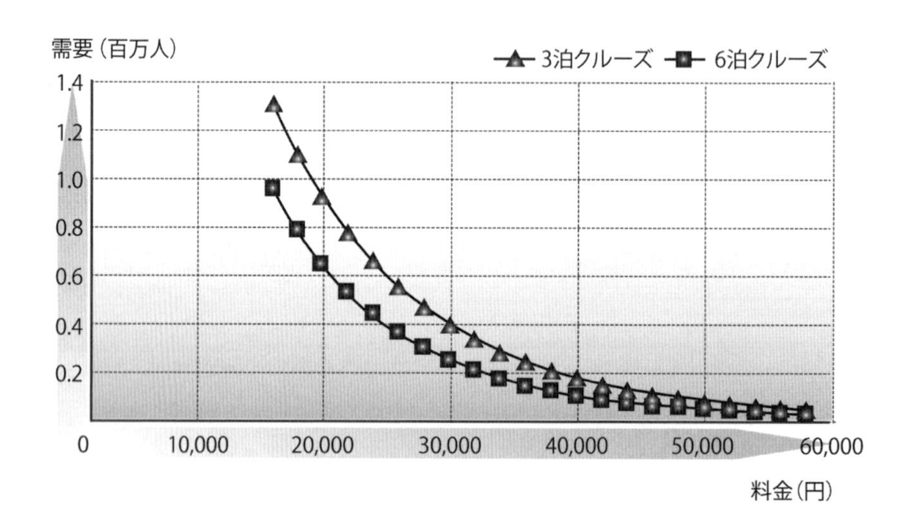

図 8-3　クルーズ料金（1 泊あたり）が需要に及ぼす影響の分析結果

コラム⑬　AHP 法

いくつもの要素が関連する意思決定の複雑な過程を科学的に分析できる手法として AHP 法（Analytic Hierarchy Process）がよく使われる。日本語では「階層分析法」と呼ばれる。意思決定に，どの要素がどのくらい寄与しているかを定量的に分析できる。したがって，AHP 法は，「深層心理を分析する」とか，「フィーリングを分析する」などといわれており，工学系の計画学やビジネススクールで広く教えられている。

意思決定の場合に，いくつもの要素が，それぞれどの程度の役割を演じているかを調べるために，1 つずつの要素を 1 対ずつ，どちらがどの程度重要か評価し，その数値をマトリックス化して計算することによって，各要素の相対的な重要度を分析する。

例えば，図 8-1 の分析では，まず旅行を決定する要因として「交流」「心の安らぎ」などの 9 つの要因について，利用者に 1 対比較で評価をしてもらい，それぞれの要因のもつ重要度を求め，次に旅行専門家に「温泉」「クルーズ」「テーマパーク」「海外旅行」の 4 つの旅について，それぞれの旅が「交流」「心の安らぎ」などの 9 つの要因をどの程度満足させるかを評価してもらって，その 2 つの分析結果を比較することにより，どの程度の利用者がどの旅行を選ぶかを分析している。

8.4　事例研究

瀬戸内海用のクルーズ客船の経済性評価の実例を挙げておこう。13,000 総トン，定員 300 名弱の架空のクルーズ客船の場合で，損益分岐解析をしてクルーズ料金を出したものである [4]。

表 8-1　クルーズ客船のデータ

基本要目	
船価	¥5,000,000,000
船員数	104 人
耐用年数	15 年
食材費（1 日 1 人あたり）	¥3,300
1 日航海時間	8 時間
馬力	8,160 ps
旅客定員	272 人
航海日数	7 日
排水量	7,552 トン
総トン数	13,000 トン

表 8-2　コスト計算内訳

船費	合計		¥1,167,000,000
	内訳	船員費	¥728,000,000
		修繕費（船価×0.01）	¥50,000,000
		保険料（船価×0.01）	¥50,000,000
		固定資産税（船価×0.007）	¥35,000,000
		減価償却費（船価×0.9÷耐用年数）	¥300,000,000
		その他	¥4,000,000
一般管理費	—		¥100,000,000
金利	年6％		¥300,000,000
年間固定費（上記合計）	—		¥1,567,000,000
1日あたりの運航費（食費含まず）	合計		¥1,786,896
	内訳	燃料費（18円/1時間1馬力あたり）	¥1,175,040
		港費	¥111,856
		その他	¥500,000

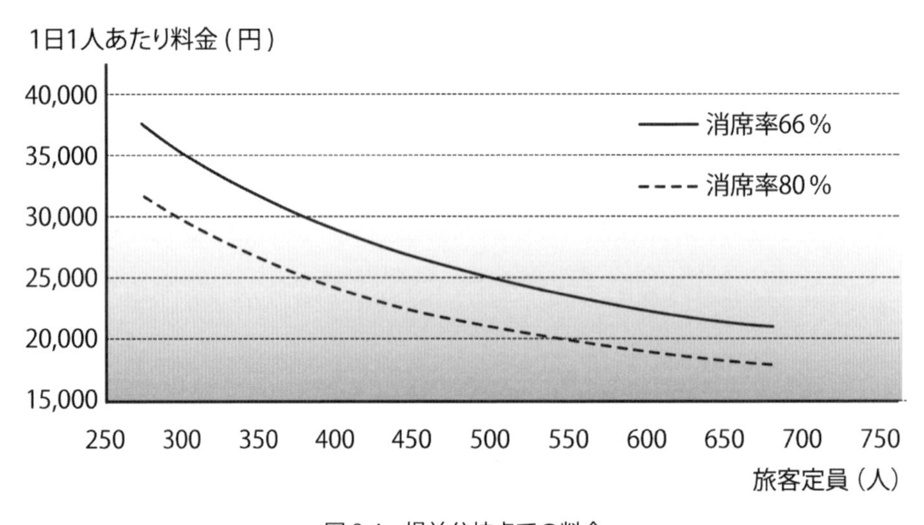

図 8-4　損益分岐点での料金

上図に示す損益分岐点での料金解析結果からは，消席率が高いほど，また旅客定員が多いほど損益分岐料金が減少することがわかる。

8.5　クルーズ産業の経済波及効果

CLIA（Cruise Lines International Association，5.6 節参照）の 2017 年発行の「Cruise Industry Outlook」によると，2015 年の世界のクルーズによる経済波及効果は 1,170 億米ドル，約 14 兆円（1 ドル＝ 122

円で換算）となる。雇用人数は 96 万人，人件費は 380 億ドルとしている。

また，アジアにおけるクルーズについては，2016 年のクルーズ人口は 310 万人に達し，2015 年比で 55% 増となった。2017 年には，アジア水域で 66 隻のクルーズ客船が稼働し，そのうち 18 隻が 10 万総トン以上のメガシップという。

2016 年のクルーズ会社の消費は 10 億米ドルで，北アジア内での造船および修繕に 3 億ドルが支払われている。

- 造船・修繕　3.084 億ドル
- 卸売業　2.153 億ドル
- 旅行業者 & 事務サービス　1.347 億ドル
- 舶用機器　0.968 億ドル
- 輸送　0.827 億ドル
- 専門科学技術サービス　0.818 億ドル
- 他のサービス　1.030 億ドル

CLIA によると，2017 年の世界クルーズ人口は 2,580 万人に達する見込み。26 隻の新造船が登場し，キャパシティは 3 万床増。

河川クルーズの成長が著しく，2015 年で 184 隻が稼働しており，2017 年には 18 隻の新造船が投入された。航洋クルーズ客船は 2016 年時点で 448 隻で，2017 年の新造船は 26 隻。

CLIA に所属する旅行代理店は，2010 年には 12,000 社だったのが 2016 年には 25,000 社に増加している。

8.6　主要クルーズ運航会社の経営状況

世界のクルーズ業界では，寡占化が進んでおり，カリブ海で現代クルーズを始めたノルウェージャン・クルーズ・ライン（NCL），ロイヤル・カリビアン・クルーズ・ライン（RCCL），カーニバル・クルーズ・ライン（Carnival）の 3 社をコアにしてグループ化が進んでいる。いずれのグループも，買収や統合によって多数のクルーズ会社を傘下に収めており，ラグジュアリーからカジュアルまで多様なクルーズを実施している。新興勢力としては，香港のゲンティン・グループ，イタリアの MSC クルーズがある。

この他に独立系と呼ばれる小さなクルーズ会社が多数存在する。

主要 3 グループの業績は，リーマンショック後の世界経済の減速などで利益率が低迷していたが，原油価格が下がって燃料費の割合が小さくなったこと，大型船の投入に伴い旅客 1 人あたりのコストの低減が進んでいることから 2016 年には急回復しており，図 8-7 からもわかるように，いずれのグループも全収入に占める純利益の比率が 10% を超えている。また，全収入に占める船上売上は 25 ～ 30% になっており，クルーズ運航会社にとっては船上販売が重要な収入源になっていることがわかる。

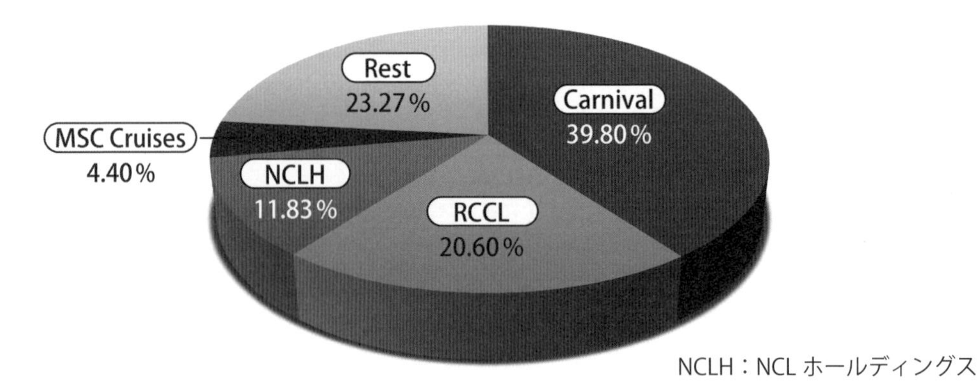

図 8-5　主要なクルーズグループの旅客数のシェア（「SHIPPAX MARKET 2016」のデータに基づいて作図）

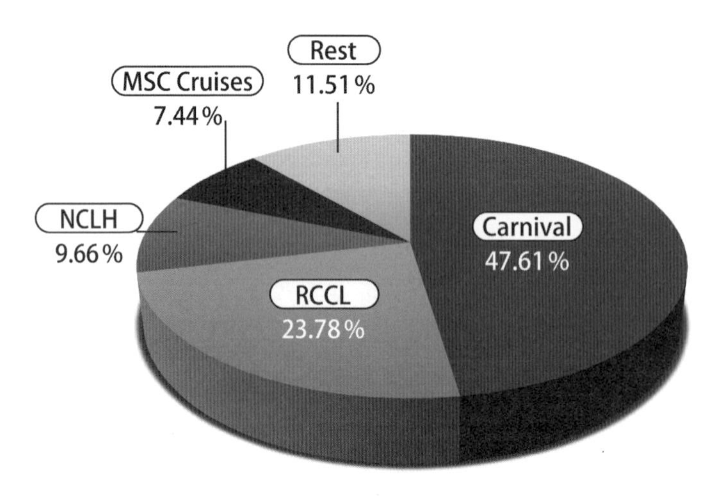

図 8-6　主要なクルーズグループの売上額のシェア（「SHIPPAX MARKET 2016」のデータに基づいて作図）

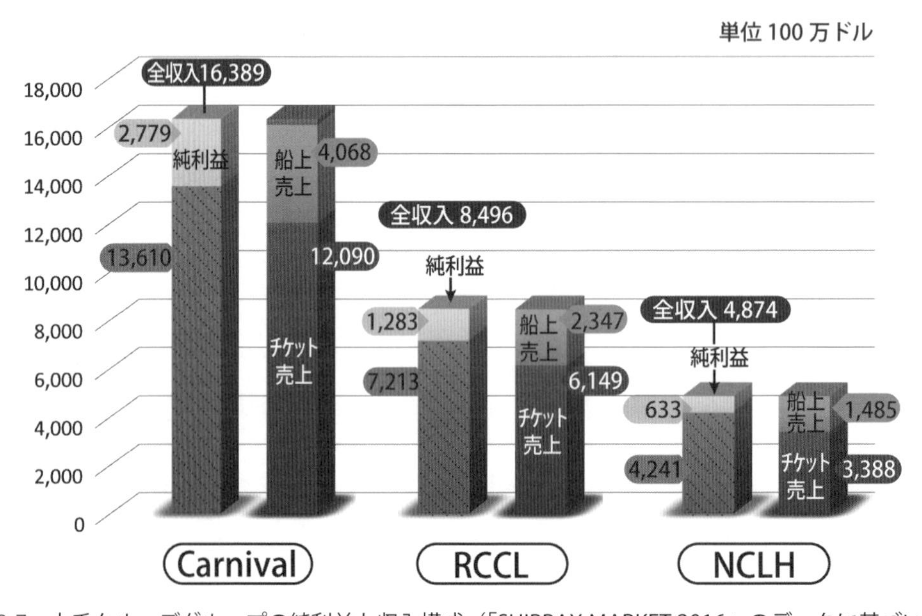

図 8-7　大手クルーズグループの純利益と収入構成（「SHIPPAX MARKET 2016」のデータに基づいて作図）

主要 3 グループのコスト構成については，下図のようになっている。

減価償却費が 16 ～ 22% 程度，人件費が 20 ～ 28%，食材費が 8 ～ 12%，燃料費が 11 ～ 16%，販売・店費が 25 ～ 27%，利子が 3 ～ 10% となっている。

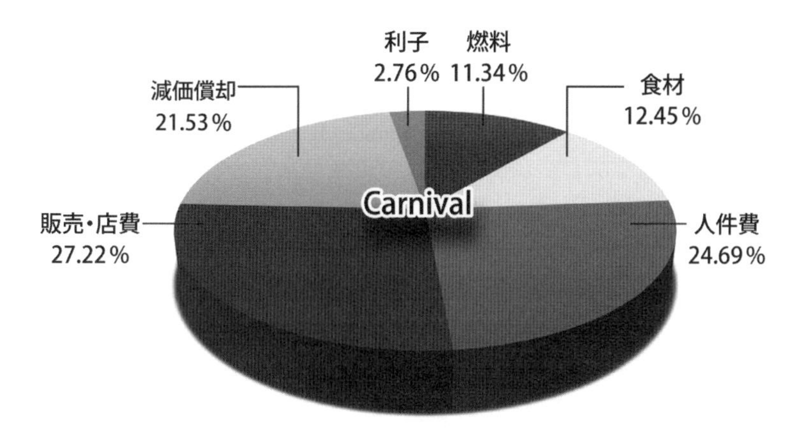

図 8-8　カーニバルグループのコスト構成（「SHIPPAX MARKET 2016」のデータに基づいて作図）

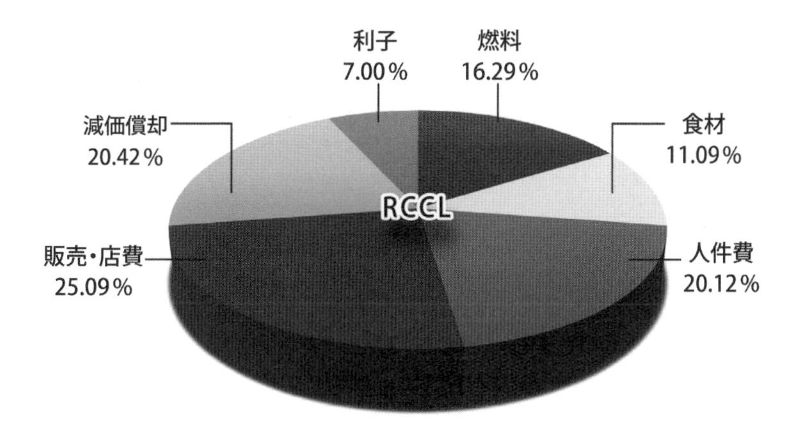

図 8-9　RCCL グループのコスト構成（「SHIPPAX MARKET 2016」のデータに基づいて作図）

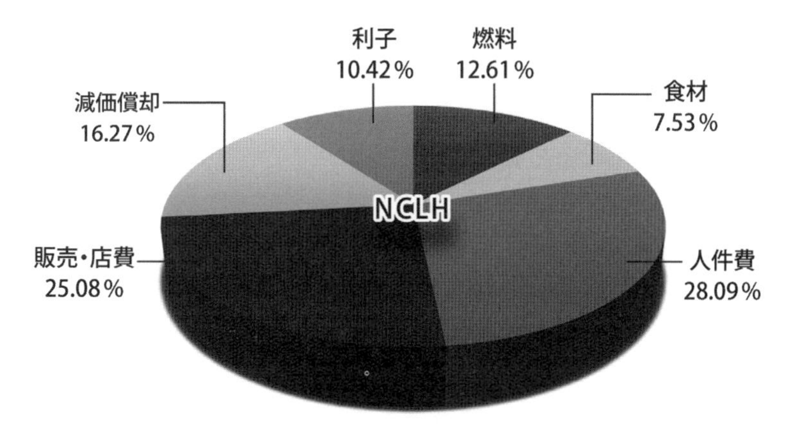

図 8-10　NCL グループのコスト構成（「SHIPPAX MARKET 2016」のデータに基づいて作図）

8.7　クルーズ産業の地域経済波及効果と観光公害

クルーズ産業が港湾都市に及ぼす経済波及効果には極めて大きなものがあり，世界各国の港はクルーズ誘致に積極的に乗り出している。現代クルーズでは，比較的短期間の定点定期クルーズが基本パターンである。

クルーズ起点港では，乗客の受付や，外航クルーズであれば CIQ（出入国審査，税関検査），乗客のトランクなどの大型荷物のハンドリングが必要となる。このため，ターミナルビルなどの施設が必要となる。一方，寄港港では大型荷物のハンドリングがないため，特別な施設を必ずしも必要としない。

クルーズ客船の入港は，港湾施設の使用料，タグボートやパイロット費用，給水や食料品，消耗品の購入，乗客・乗員の観光や買い物などの経済波及効果があり，さらに起点港の場合には，クルーズ会社の事務所費や人件費，乗客・乗員の移動費や宿泊費，各種の船用備品費や消耗品費なども加わって大きなものとなる。

2009 年に就航した当時世界最大のクルーズ客船「オアシス・オブ・ザ・シーズ」（定員 5,400 人）の場合，毎週の 1 週間クルーズの年間実施に伴う起点港であるフォート・ローダーデールへの経済効果は，試算によれば，年間 620 万ドル（約 6 億円）の港湾使用料となり，経済波及効果全体では 2.66 億ドル（約 266 億円），税収は 900 万ドル（約 9 億円）に達するという。

大阪府立大学の田口他による推定 [6] では，7 万総トン，定員 2,500 人のクルーズ客船が，大阪港を寄港港として使った場合には，近畿地方への経済波及効果は 1 回あたり 6,720 万円となり，1 週間クルーズの定点定期クルーズの寄港港になって，毎週 1 回の寄港があるとすると，年間 54 億円の経済波及効果（直接効果 28 億円）が期待される。さらに大阪が同船の 1 週間定点定期クルーズの起点港になった場合には，年間の経済波及効果は 192 億円（直接効果 160 億円）に上る。

定点定期クルーズの一大拠点として，10 隻以上の大型クルーズ客船の起点港として機能しているアメリカのマイアミ港の経済波及効果は 1 兆円を超えており，アラスカクルーズの起点港および寄港港となっているカナダのバンクーバーでは年間約 1,300 億円（2016 年度）となっている。

クルーズ客船が発着もしくは寄港することによる経済波及効果は絶大であるが，一方，クルーズ客船による負の側面，いわゆる環境公害問題も存在することを忘れてはならない。

環境公害とは，観光に伴って生ずるさまざまな弊害で，車による騒音，排気ガス，渋滞，観光客によるごみのポイ捨て，プライバシーの侵害，観光客を受け入れるための開発に伴う環境破壊などである。

クルーズが他の観光形態と根本的に異なるのは，寄港地になった場合には，数百人から数千人の観光客が一気に訪れて，7 ～ 8 時間滞在し，再び船に戻って出航するという点であろう。寄港地での滞在が大量かつ短時間というのが重要なキーワードとなる。

50 ～ 80% の乗客は，船内で販売したオプショナルツアーとしてのバスツアーに参加するので，数十台から百数十台のバスが岸壁に並び，一斉に観光地に向かうため，道路の渋滞，観光地での駐車の問題が深刻になる。1 つの観光地にたくさんのバスが集中しないように，ツアーのコースとスケ

図 8-11　クルーズハブ港として年間 1 兆円の経済波及効果があるマイアミ港
（提供：マイアミポートオーソリティ）

ジュールを全体として調整することが必要となる。

船を走らせるためのエンジンおよび電気を発生させる発電機には，ディーゼル機関が使われている場合が多い。最近は，電気モーターでスクリュープロペラを回す電気推進船も多いが，この場合にはすべてのエネルギーを発電用ディーゼル機関によって発生させる。船舶用のディーゼル機関は，熱効率が非常によく，CO_2 の排出は他の機関に比べて少ないが，窒素酸化物（NO_X：Nitrogen Oxide），硫黄酸化物（SO_X：Sulfer Oxide），黒煙などの粒子状物質（PM：Particulate Matter）などの公害物質を発生させる。クルーズ客船は，港に停泊中にも発電機を使うため，近隣の公害を引き起こすので，こうした有害物質をできるだけ出さないようにすることが求められている。このため排気ガスから NO_X，SO_X を取り除く装置を取り付けたり，硫黄分の少ない油や液化天然ガス（LNG）を燃料として SO_X の排出を抑えたりする対策がとられている。また，港に停泊時には陸上から電気を供給する陸電を使う船も出てきている。

なお，最近の新造クルーズ客船では，船内で発生するすべての汚水は浄化装置できれいにしてから海に放出されている。

【参考文献】

[1] 日本海運集会所 編：入門「海運・物流講座」（第 3 章Ⅳ 客船ビジネス），日本海運集会所発行，2004 年 12 月

[2] 池田良穂，田角宏美：日本におけるクルーズ需要推定とマーケット育成方策，関西造船協会平成 13 年春季講演会講演論文集，pp.171–172，2001 年 5 月

[3] Yoshiho Ikeda, Jaswar：A Prediction Method of Travel Demand of Cruise Ships in Japan, Journal of Kansai Society of Naval Architects, No.238, pp.215–223, Sep. 2002

[4] 辰巳貴俊：瀬戸内海用両頭型クルーズ客船の開発と性能分析，大阪府立大学工学部海洋システム工学科平成 20 年度卒業論文，2009 年 3 月

[5] 池田良穂，辰巳貴俊：瀬戸内海用両頭型クルーズ客船の開発と性能分析，日本船舶海洋工学会講演会論文集，第 8 号，2009 年 5 月
[6] 田口順他：クルーズ客船誘致の経済波及効果，第 14 回クルーズ客船 & フェリー研究会資料集，2010 年 2 月 27 日

演習

⑴ 瀬戸内海の 1 週間定点定期クルーズを 5,000 総トン（定員 150 人）と 50,000 総トン（定員 1,700 人）の船で実施する場合のそれぞれの損益分岐料金を計算し，それぞれの船の営業戦略を立てなさい。
損益分岐料金の計算については，以下のデータを使用すること。
5,000 総トン船：船価 80 億円，エンジン出力 6,000 馬力，港費：10 万円 / 港，船員数 80 人
50,000 総トン船：船価 400 億円，エンジン出力 25,000 馬力，港費：30 万円 / 港，船員数 800 人
なお，定期ドック期間（1 週間）を除いて年間フル稼働とし，消席率は 80%，速力は 16 ノット（約 30 km/h）とする。その他のデータについては，8.4 節の事例を参照のこと。

⑵ 本章で紹介されているクルーズ客船の経済波及効果の結果に基づいて，神戸港をクルーズハブ港として機能させるための戦略を立てなさい。

⑶ 「年収 300 万円程度の層にクルーズを楽しんでもらうこと」をターゲットとしてクルーズ事業を立ち上げるとして，どの程度の大きさ・旅客定員のクルーズ客船を使うべきかについて考えなさい。

≪問題≫

8-1　クルーズ産業が「設備産業」と呼ばれるのはなぜか。
8-2　クルーズ事業における固定費と変動費の内訳を挙げよ。
8-3　減価償却費とはなにか。
8-4　サービス要員のコストを下げるためにとられている方策をまとめよ。
8-5　損益分岐料金とはなにか。
8-6　クルーズの潜在需要を推定するための方法を挙げよ。
8-7　クルーズ客船が大型化するほど損益分岐料金が下がるのはなぜか。
8-8　世界のクルーズ事業の経済波及効果はどのくらいか。
8-9　クルーズにおけるチケット売上と船上売上の比率はどのくらいか。
8-10　22 万総トンの「オアシス・オブ・ザ・シーズ」の定点定期カリブ海クルーズによる起点港への経済波及効果はいくらくらいか。
8-11　クルーズによる環境公害にはどのようなものがあるか。

第9章 クルーズポート

9.1 クルーズポート

現代クルーズでは，出発するために船に乗船する出発港と，最後に下船する帰港港は同じであることが多い。こうした港を起点港または発着港と呼ぶ。英語では Home port と呼ぶことが多く，日本語に訳すと母港となるが，母港は船籍港（各船の籍のある港）を指す場合があるので，起点港または発着港と呼ぶ方が誤解が少ない。

クルーズの途中に寄る港は寄港港と呼ばれ，英語では Port of call である。寄港港では，原則として乗客は一時的な乗下船をするトランジット客である。

ただし，長期間に及ぶクルーズでは，途中の寄港港で乗船もしくは下船する区間乗船，さらに最近では定点定期クルーズでも，インターポーティングという複数の寄港港で集客をして乗客が乗下船するシステムもとられるようになった。このインターポーティングは，北米生まれの現代クルーズが欧州に進出して定点定期クルーズを実施した時に，途中の寄港港においても十分なクルーズ・マーケットがある場合が多くあるため，旅客定員の一定割合を各寄港港で集客することでクルーズ・マーケット全体が拡大できることから始まったシステムである。すなわち，このような場合，その港は，寄港港としての役割と，起点港としての役割の両方を担う必要がある。

図 9-1　カリブ海クルーズの起点港であるマイアミ港のクルーズ客船埠頭に並ぶクルーズ客船群

9.2 クルーズポートの条件

クルーズ港には，後述する港の機能だけでなく，後背地がもつべき条件がいくつかある。

まず起点港の場合，条件の 1 つは地元にクルーズのソースマーケットとしてのポテンシャルがあることであり，もう 1 つが遠隔地からのアクセス機能が整備されていて広域からの集客ができることである。前者はクルーズ運航会社にとっての重要なベースマーケットとなり，後者はマーケットを広範囲に広げるための武器となる。すなわち，遠方の各地からの乗客がスムーズに乗船できる交通網があることが，クルーズの販売地域の拡大につながる。特に，北米生まれの現代クルーズでは，飛行機とクルーズを組み合わせたフライ & クルーズがポピュラーになっていることから，空港が隣接していることが起点港の必須条件といってよい。また，欧州では鉄道を利用したレイル & クルーズなども多いので，主要鉄道駅の存在も重要となる。

また，空港や鉄道の駅など主要交通機関のターミナルから，クルーズ客船が着岸する港までのアクセスも重要なポイントとなる。さらに，自家用車で港にやってきて乗船するドライブ & クルーズの需要も大きくなっており，駐車場も必要となる。

クルーズの前後に観光ができることも，起点港としての重要な要素になっている。近くに観光資源があることは起点港としての価値を向上させるのと同時に，クルーズによる地域の経済波及効果を増加させることとなる。

一方，寄港港としての条件は，魅力的な観光資源があることに尽きる。極端な場合には，クルーズ客船が着ける港湾施設がなくても，観光の魅力だけでたくさんのクルーズ客船がやってきて，沖止めして観光客をボートで上陸させる。例えば，エーゲ海のサントリーニ島（ギリシャ），アドリア海のドゥブロヴニク（ボスニア・ヘルツェゴビナ）などである。

図 9-2　エーゲ海のサントリーニ島の沖に停泊するクルーズ客船

図 9-3　ドゥブロヴニクの沖合に停泊するクルーズ客船

9.3　港のハード機能

9.3.1　航路と水深

港内航路の必要水深は，クルーズ客船の喫水で決まる。また，港内で船が回頭するための広さは，ポッド推進器もしくは大型のサイドスラスターを複数機装備している最近のクルーズ客船では，船長の 1.2 倍あれば可能である。ボイジャー級船では約 430 m となる。また同船では航路の水深が 11 m，航路幅は 122 m，着岸壁の水深は 10 m が必要である。

なお，岸壁の近くで回頭スペースがとれない場合には，港外で回頭して後進で数 km 入港して着岸する場合（チビタベッキア港），最奥部の回頭水域で反転して出船着岸する場合（マイアミ港），切り返しによって船首方向を変える場合（ベニス港，チビタベッキア港）などがあり，高性能な操船能力を活用して狭い港内での操船を工夫している。また，風速についても 15 m/s 程度の風であれば，自身の推進器およびスラスターだけで定点保持が可能な船もある。ただし，客船の操船性能は船によって大きな違いがあるので，各船の能力に合わせて出入港の可能性評価をする必要がある。

図 9-4　大型客船が入港する狭いニース港（フランス）（グーグルアースより）

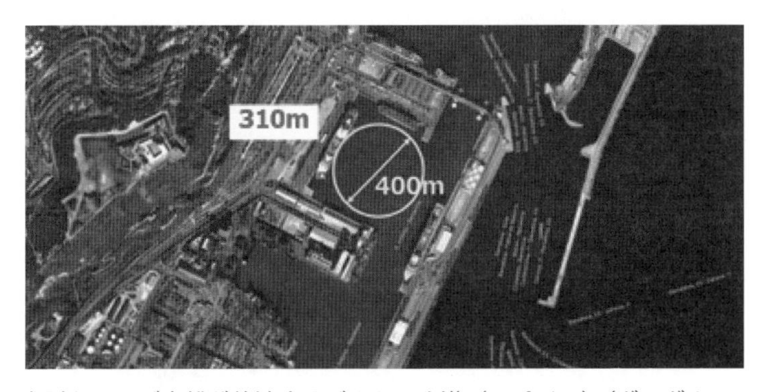

図 9-5　大型クルーズ客船が停泊するバルセロナ港（スペイン）（グーグルアースより）

9.3.2　岸壁とターミナル

起点港は，まずクルーズ客船が着岸するための岸壁が必要となる。岸壁の水深は，停泊するクルーズ客船の喫水以上の深さが必要となる。現在世界最大の 22 万総トン型オアシス級船（RCI）で喫水は 9.1 m，もっとも喫水の深いのが 15 万総トンの「クイーン・メリー 2」（キュナード・ライン）で 10.4 m である。一般の貨物船に比べると，幅が広くかつ喫水が浅い船型なので，大きさの割には喫水が浅いという特性がある。

岸壁で船は流されないようにロープで係船柱（ボラード）に係留される。この係船綱を前後にも張るために，岸壁の長さは船長より長い必要がある。また，最近のクルーズ客船では水面上の船体部分が大きいために非常に大きな風圧力が働くので，それに耐えるだけの係船綱を止めることのできる係船柱が岸壁にいる。岸壁の長さは，船の長さ＋係船綱の展開長さが必要で，大よそ船長の 1.15 ～ 1.2 倍程度の長さとなる。ただし，岸壁自体は乗客の乗下船および車などの待機スペースが確保できればよく，場合によっては係船綱をつなぐ係船柱は岸壁の外でもかまわない。例えば，14 万総トン級の「ボイジャー・オブ・ザ・シーズ」（RCI）の場合には，船首に 6 本，船尾に 6 本，さらに 4 本の係船綱で係船することが必要であり，許容荷重が 150 トン（約 1,500 ニュートン）の係船柱がいる。

船の全長は，22 万総トンのオアシス級で 362 m，15 万総トンの「クイーン・メリー 2」で 345 m，11 万総トンの「ダイヤモンド・プリンセス」で 288m である。

また，港口に橋がある場合には，水面上高さが問題となり，これをエアードラフト（ドラフトは喫水の意味）という。横浜港のベイブリッジや東京港のレインボーブリッジは，建設当時最大級だった 7 万総トン級の「クイーン・エリザベス 2」がくぐれるだけの高さを確保したが，その後のクルーズ客船の急速な大型化に伴い通過できない船が増えてきた。エアードラフトは 22 万総トンのオアシス級で 65 m，14 万総トンのボイジャー級で 63 m であり，7 万総トン級以下では 50 m 未満となる。なお，各橋の通過可能エアードラフトは，一般的には満潮時の水面から橋げた下面までの高さを基準としており，干潮時を利用した通過という条件での出入港も考慮される。

寄港港でも以上の岸壁および航路の制限は同じである。ただし，寄港港の場合には，岸壁に着岸で

図 9-6　横浜のベイブリッジの下を通過するクルーズ客船

きなくても，静穏な水域があればブイ係留または錨泊（錨による停泊）して，乗客を小型ボートで陸まで送迎することもあり，この場合には小型船用の桟橋があればよい。

起点港の場合には，大人数の乗客が大きな荷物をもって到着して，乗船手続き，外国航路の場合には出入国審査や税関検査を行う必要があるため，十分な広さのターミナルがいる。大きな荷物は乗船時にはターミナルビルの玄関で預かり，各キャビンまで配送するのが普通である。

一方，下船時は，事前に荷物を船に預けて，船側がターミナルビルの所定の場所に置いておき，外国航路の場合には入国審査後に，各自が荷物を受け取り，税関検査を受けることとなる。下船時には荷物（トランクケース）が数千個に及ぶため，それを並べるスペースがいる。

船の大きさが急激に大きくなり，旅客数が増加したため，従来の規模のターミナルビルではスペース的に狭すぎる場合も多く，状況に応じて大型化がしやすいように簡易テント型またはプレハブ式の施設を造る場合もある。

クルーズ客船の寄港数の少ない港では，貨物船埠頭でクルーズ客船を受け入れる場合も多い。比較的エプロンが広くとられているコンテナ埠頭や自動車専用船用埠頭などが運営上は使いやすい。

図 9-7　チビタベッキア港のコンテナ埠頭に建設されたテント型のクルーズ客船ターミナル

図 9-8　テント型ターミナル内は，受付，CIQ（7.5節参照），X 線検査場に分かれている。

9.4　クルーズ誘致活動と受入体制

港湾へのクルーズの誘致活動は，港湾を管理する地方自治体の港湾当局が中心となって，ポートセールスの一環として行われる場合が多い。日本では，国土交通省の港湾局がクルーズ誘致に積極的となっており，クルーズ運航会社へ日本各地の港湾情報を提供するとともに，クルーズ用港湾施設の整備にも力を入れている。

また，初入港クルーズ客船への港湾経費の免除や，入港や出港時に岸壁でイベントなどを行う港湾も多い。ただし，海外のクルーズ先進国では，特別なイベントを除くと，こうした歓迎イベントを行っている事例はほとんどなく，日本独特のものといってよい。

クルーズ客船の場合には，大量の乗客が比較的短時間の観光をすることから，港周辺だけでなく後背地の観光スポットとの連携も重要となる。このため，港湾当局だけでなく，観光当局や商工会議所，バスやタクシーなどの2次交通事業者などの綿密な連携が必要となり，各地にクルーズ客船受け入れのためのDMO（Destination Management Organization）が構成されている。

表9-1　世界の港：クルーズ乗客数トップ10（2016年）（「SHIPPAX MARKET 17」に基づく）

1位	マイアミ（アメリカ）	498万人	起点港
2位	ナッソー（バハマ）	469万人	寄港港
3位	ポート・カナベラル（アメリカ）	395万人	起点港
4位	ポート・エバグレーズ（アメリカ）	370万人	起点港
5位	コスメル（メキシコ）	364万人	寄港港
6位	バルセロナ（スペイン）	268万人	起点港＋寄港港
7位	チビタベッキア（イタリア）	234万人	起点港＋寄港港
8位	ケイマン島（イギリス自治領）	171万人	寄港港
9位	セント・トーマス（アメリカ自治領）	169万人	寄港港
10位	パルマ（スペイン）	163万人	起点港＋寄港港

表9-2　日本の港：クルーズ客船入港数トップ10（2016年）（国土交通省調査に基づく）

1位	博多（福岡）	328回（外国籍船312回）
2位	長崎	197回（外国籍船190回）
3位	那覇	193回（外国籍船183回）
4位	横浜	128回（外国籍船　41回）
5位	神戸	104回（外国籍船　32回）
6位	石垣	95回（外国籍船　91回）
7位	平良（宮古島）	86回（外国籍船　84回）
8位	鹿児島	83回（外国籍船　80回）
9位	佐世保	64回（外国籍船　62回）
10位	広島	47回（外国籍船　34回）

≪問題≫

9-1　クルーズの起点港とはなにか。

9-2　クルーズの寄港港とはなにか。

9-3　クルーズの起点港と寄港港に求められる要件の違いを挙げよ。

9-4　20万総トン級のクルーズ客船が入港する港の水深はどの程度あればよいか。

9-5　日本の港のクルーズ客船寄港回数が，九州および沖縄の港で多い理由を挙げよ。

第10章 写真で見るクルーズ客船の系譜

筆者が撮りためたクルーズ客船，乗船した船たちを紹介しながら，筆者の目から見た，この40年あまりのクルーズの歴史を振り返ってみたい。

10.1　1970年代

筆者がまだ学生だった頃，神戸港にはよく客船が入ってきた。その中には，太平洋航路やオーストラリア航路の定期客船も一部残っていたものの，定期客船からクルーズへと転用され，世界一周の途中で寄港する船もあった。神戸にあった代理店スワイヤ＆サンズに日参しては，外国客船の見学許可をもらって船内を見せてもらった。

図10-1　インドネシア航路の「チルワ」

図10-2　太平洋横断航路に就航していたOCLの「オリエンタル・パール」

図10-3　ブラジル航路の「ぶらじる丸」。南米移民船として「あるぜんちな丸」とともに活躍。引退後は，鳥羽でフローティング博物館となったが，その後撤去され，中国に売却された。

図 10-4　イギリス～オーストラリア航路からクルーズに転用されたP&Oの「オリアナ」

図 10-5　イギリス～オーストラリア航路からクルーズに転用されたP&Oの「キャンベラ」

図 10-6　クルーズの途上，日本に寄港したP&Oの「アーカディア」

図 10-7　クルーズ客船に転用されたギリシャのチャンドリス・ラインの「ラーリン」

図 10-8　世界一周クルーズの途中，神戸に向かって友が島水道を北上するスウェディッシュ・アメリカ・ラインの「クングスホルム」。時間を見計らって大阪発徳島行きのフェリーの船上から撮影した会心の1枚。

図 10-9　筆者が外航クルーズに初めて乗船したのは「コーラル・プリンセス」の釜山クルーズだった。

図 10-10　初代「にっぽん丸」には，同船が団体チャータークルーズの合間に，たまに実施する自主クルーズを見つけてはクルーズを楽しんだ。

図 10-11　三菱重工業神戸造船所でクルーズ客船に改造された「新さくら丸」には 2 回乗船した。

図 10-12　関西汽船が「さんふらわあ 7」でクルーズに進出。自主クルーズを見つけては家族，船仲間で何度も乗船した。

10.2　1980 年代

ドイツに 1 年半留学したとき，欧州水域でのクルーズを体験した。ドイツの高級クルーズ客船「オイローパ」は料金が高くてとても乗れなかったので，ドイツの旅行会社がチャーターしたソ連客船での地中海クルーズやノルウェーのフィヨルドクルーズを楽しんだが，それまで見聞きしていた船旅そのままのトラディショナル・スタイルのクルーズだった。

1987 年にカリブ海で「ソング・オブ・アメリカ」に乗船した機会に，マイアミの港湾局や RCCL の本社でクルーズに関する調査を行った。そして 1 週間のカリブ海クルーズに乗船して初めて，トラディショナル・クルーズとはまったくビジネスモデルの違った現代クルーズが急速に成長していることを知った。まさに目からうろこであった。筆者のクルーズに対する考え方が一新された瞬間であった。

図 10-13　ジェノア起点の地中海クルーズで乗船した「アゼルバイジャン」。マルタ島への寄港時。

図 10-14　ノルウェーのフィヨルドクルーズで，ソグネフィヨルド内に停泊するロシアのクルーズ客船「オデッサ」

図 10-15　カリブ海の現代クルーズを最初に体験した「ソング・オブ・アメリカ」

図 10-16　次々とマイアミ港を出航するクルーズ客船の姿に新しい現代クルーズの実態を見た。

10.3　1990 年代

いろいろな現代クルーズの客船に乗船してみた。カーニバルの「トロピカル」のメキシコクルーズ，RCCL の「サンバイキング」のアラスカクルーズ。そして 7 万総トンのメガクルーズ客船の新造第 1 船「ソブリン・オブ・ザ・シーズ」は，筆者に強い印象を与えた。

図 10-17　斬新な大型クルーズ客船が新しいマーケットを開拓することを実感した 7 万総トンの新鋭船「ソブリン・オブ・ザ・シーズ」

10.4　2000年代

その後，筆者の予想どおり，クルーズ客船は一気に大型化し，オーバーパナマックス型のクルーズ客船が続々と登場した。14万総トンの「ボイジャー・オブ・ザ・シーズ」，16万総トンの「フリーダム・オブ・ザ・シーズ」が登場するたびに，カリブ海へ飛んで斬新で巨大なクルーズ客船に乗船した。

大型化することで価格は下がり，船内での楽しみの選択肢が大幅に広がり，クルーズ・マーケットの裾野がさらに広がることを確信した。

一方，ラクジュアリー・クラスのクルーズ客船においても大型化傾向が鮮明となり，15万総トン級の「クイーン・メリー2」が登場した。

図10-18　10万総トン型の第1船はプリンセス・クルーズの「グランド・プリンセス」。就航直後に，造船所や海運会社の方々と一緒に視察に出かけた。

図10-19　14万総トン型の第1船「ボイジャー・オブ・ザ・シーズ」には，家族とともに乗船。超大型船の船内での楽しみの選択肢の広さに感動した。

図10-20　老舗のキュナード・ラインは，経営難の末にカーニバルグループに入って再生。15万総トンの「クイーン・メリー2」を登場させた。

図10-21　16万総トン型の第1船「フリーダム・オブ・ザ・シーズ」には，2006年に3世代で乗船した。小さな子供まで楽しむことのできる新しいスタイルのクルーズが確立され，クルーズは「3世代船」へと進化した。

2009年には，ついに22万総トンの「オアシス・オブ・ザ・シーズ」が登場。早速この船にも乗船した。1,300億円の建造費をかけ，6,000人の乗客を楽しませる船の登場である。メインショー以外にアイスショーやアクアショーがあり，食事のバラエティも大きく広がった。普通の人が普通に楽しめるバケーションとして，クルーズはますます進化するに違いない，そんな思いを強くした。

図10-22　2009年末にカリブ海1週間クルーズに登場した22万総トンの「オアシス・オブ・ザ・シーズ」。1960年代に現代クルーズが生まれた頃の船と比較すると，1隻で10隻分の仕事をこなす。

図10-23　「オアシス・オブ・ザ・シーズ」

図10-24　「オアシス・オブ・ザ・シーズ」の船尾側プロムナード

図10-25　「オアシス・オブ・ザ・シーズ」の屋外プロムナード

10.5　2010 年代

2010 年代になると，現代クルーズ客船が日本近海にもやってくるようになった。海外のクルーズ客船が中国の上海などを起点とするクルーズにたくさん就航して日本の港に姿を現すようになり，そして日本発着のクルーズも行われるようになった。カボタージュ規制があるため，クルーズの途中で海外の港に寄港する必要があるが，釜山以外はすべて日本の港に寄港という実質日本クルーズも多い。価格も 1 週間クルーズが 10 万円台からとリーズナブルなため，年に何回かこうしたクルーズを楽しむことができるようになった。

図 10-26　2017 年の春に乗船した「セレブリティ・ミレニアム」

図 10-27　夏季に舞鶴起点の日本海クルーズに就航する「コスタ・ネオロマンチカ」

図 10-28　通年の日本発着クルーズには「ダイヤモンド・プリンセス」が就航している。

第11章 世界のクルーズ水域

ここでは世界のクルーズ事例について，地域別に紹介する。

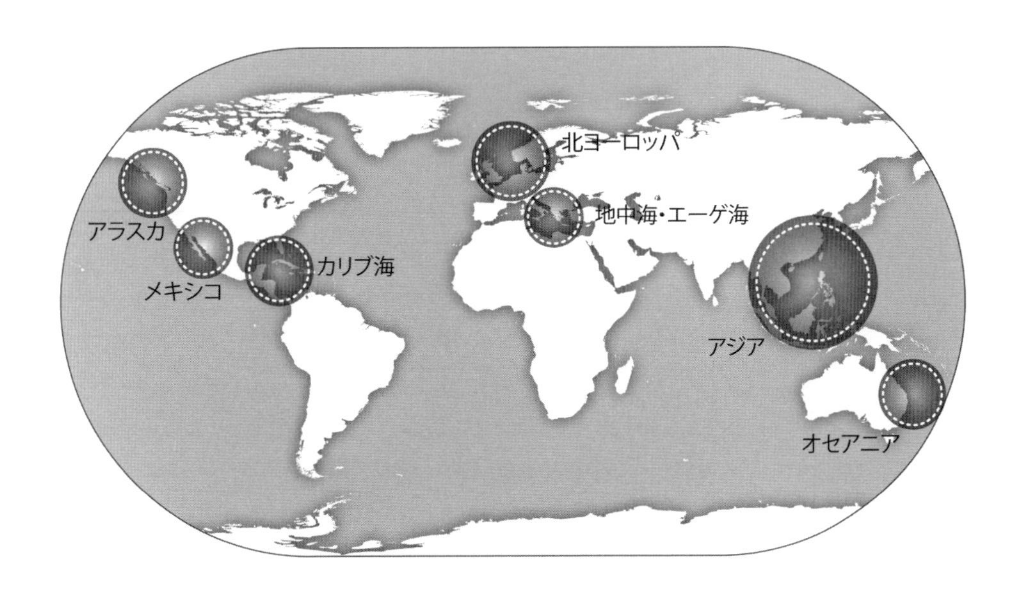

図 11-1　世界の主要クルーズ水域

11.1　カリブ海クルーズ

カリブ海クルーズは，フロリダ半島の南に位置するマイアミ港を起点とする定点定期クルーズが中心だが，最近は，ポート・エバグレーズ（フォート・ローダーデール），タンパ，ポート・カナベラルなどのフロリダ半島の港湾，さらにはサンファンなどのカリブ海に浮かぶ島を起点とする定点定期クルーズも運航されるようになっている。

カリブ海に浮かぶ島々を回遊する 1 週間クルーズは，主にメキシコにワンタッチする西カリブ海クルーズと，バージン諸島まで行く東カリブ海クルーズに大別され，さらに南に続く小アンティル諸島（ドミニカ，グレナダなど）を訪れるクルーズではフロリダ半島を起点とすると 10 日～2 週間の長期クルーズとなる。

カリブ海は常夏であるため，冬季には北アメリカ全土から避寒客が集まる。クルーズも冬季がシーズンで，世界中からクルーズ客船が集まり，隻数は倍増する。これは，アラスカや北欧が冬季にはク

図 11-2　カリブ海クルーズの主要クルーズポート

ルーズに向かなくなるために，年間を通したクルーズ適地であり，かつマーケットも大きいカリブ海にこの期間に移動するためである。

ただし夏季は，ハリケーンの発生が問題となる。このため，波の中でも船酔いの心配のない大型船が用いられるようになっており，ハリケーン発生時には，当初のクルーズルートを柔軟に変更して欠航を防いでいる。

年間を通じてカリブ海クルーズを実施している会社では，プライベート島や半島を所有して，クルーズ客だけのプライベート空間を確保して，クルーズ期間中の一日をここで過ごすように企画している場合が多い。これは，元々はセキュリティの面で運営がしやすいためであったが，乗客の満足度も高まっているという。RCI のラバディ半島（ハイチ），NCL のココケイ島（バハマ）などがある。

カリブ海の寄港地は，それなりに魅力的ではあるが，いずれもスペイン，ポルトガルなどの植民地的な雰囲気を漂わせるところがほとんどである。このため，各クルーズ会社は船上でのサービスの充実を図り，これが現代クルーズがバケーション産業として広く受け入れられるようになった大きな要因と考えられる。

11.2　地中海クルーズ

ここでは，主にイタリアより西の地中海水域について述べて，東側のエーゲ海クルーズについては次節で紹介することとする。

地中海は古いクルーズの歴史をもつが，その中心となっていたイタリアやフランスでのクルーズ産業には長い間大きな成長は見られなかった。

この地中海クルーズが本格的な成長を始めたのは，現代クルーズのビジネスモデルが，スペインのマジョルカ島を起点とする定点定期クルーズとしてイギリスの旅行会社エアーツアー社によって開発

図 11-3　地中海クルーズの主要クルーズポート

されたことによっている。同社は，チャーター機を使って，発着費用の安い地方空港からイギリス人客を運び，1 週間以内の短期クルーズを格安で楽しませた。

また，カリブ海で成長したクルーズ会社が，相次いで，カリブ海クルーズを行っていた大型クルーズ客船をヨーロッパ水域に転配して，アメリカ人マーケットだけでなく，ヨーロッパ人マーケットの開拓を行った。

現在，地中海クルーズの起点港としては，イタリアのジェノア，チビタベッキア，ナポリ，スペインのマジョルカ島とバルセロナ，フランスのマルセイユなどがある。

寄港地としては，起点港として機能している上述のスペイン，フランス，イタリア沿岸の各港の他，フランスのコルシカ島，イタリアのサルジニア島とシチリア島，イギリス領のマルタ島などの島々や，アフリカ大陸のチュニスやアルジェ，地中海の西の入り口にあるスペインのジブラルタルなどの港町が選ばれている。

船内サービスについては，英語だけでなく，フランス語，イタリア語，ドイツ語などの多言語によるアナウンスやショーの司会などがあり，また南ヨーロッパの伝統である遅い夕食時間などがカリブ海クルーズとの大きな違いといえる。

11.3　エーゲ海クルーズ

日本ではクルーズのメッカとして名高いエーゲ海は，地中海の東の端にある水域で，ギリシャとトルコに囲まれた多島海である。このエーゲ海クルーズは，かつてはアテネの外港であるピレウスを起点としてギリシャ船籍の老朽化した小型クルーズ客船が多く就航していた。

エーゲ海の島々のほとんどがギリシャ領であるため，ピレウス起点の定点定期クルーズの場合にはカボタージュ規制によりギリシャ籍客船でなければならなかったため，老朽化した小型客船でも顧

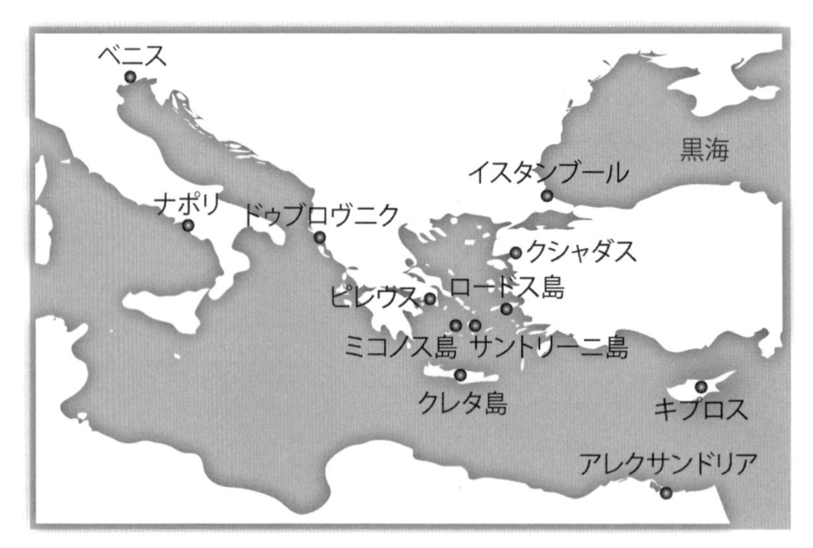

図 11-4　エーゲ海クルーズの主要クルーズポート

客をつなぎとめてこれたが，1990 年代からアメリカ型の現代クルーズの進出に伴って競争が激化し，時代の波に乗り遅れたギリシャのクルーズ会社は，企業統合をして，新造クルーズ客船の建造にも乗り出したものの成功できず，姿を消した。これに代わってエーゲ海クルーズを行うようになったのが，アメリカのクルーズ会社や，ヨーロッパの新興クルーズ会社である。

これらのクルーズ会社はカボタージュ規制にかからないようにするために，イタリアのベニスもしくはナポリを起点港として，ピレウス，エーゲ海に浮かぶサントリーニ島，ミコノス島，ロードス島，クレタ島などのギリシャの島々を回るクルーズを行っており，さらに足を延ばしてトルコのクシャダスなどに寄港する船も多い。また，最近は，ベニス起点の場合にアドリア海内にあるクロアチアのドゥブロヴニクなども寄港地として脚光を浴びている。さらにトルコのイスタンブール，エジプトのアレクサンドリア，キプロスなどの地中海東端の沿岸各港を訪れるクルーズもあるが，こうした寄港地が入ると 10 日〜 2 週間と長めのクルーズとなる。

11.4　北ヨーロッパクルーズ

北ヨーロッパのクルーズとしては，ノルウェー沿岸のフィヨルドが昔からもっとも人気のあるクルーズディスティネーションとして有名である。氷河が溶けた後に残された渓谷が海岸線から 100 km 以上も内陸に入り込み，静かな水面と切り立った断崖が続くフィヨルドがノルウェーの沿岸に数多くあり，そこを訪れるクルーズ客船は多い。主な寄港地としては，ベルゲン，オスロ，オーレスン，トロムソなどがある。起点港としては，コペンハーゲン，ブレーマーハーフェン，アムステルダム，サウサンプトンなどが選ばれている場合が多い。

夏のバルト海もクルーズ水域として人気がある。スウェーデンのストックホルムやゴトランド島，フィンランドのヘルシンキやツルク，バルト三国，ロシアのサンクト・ペテルブルグなどが寄港港と

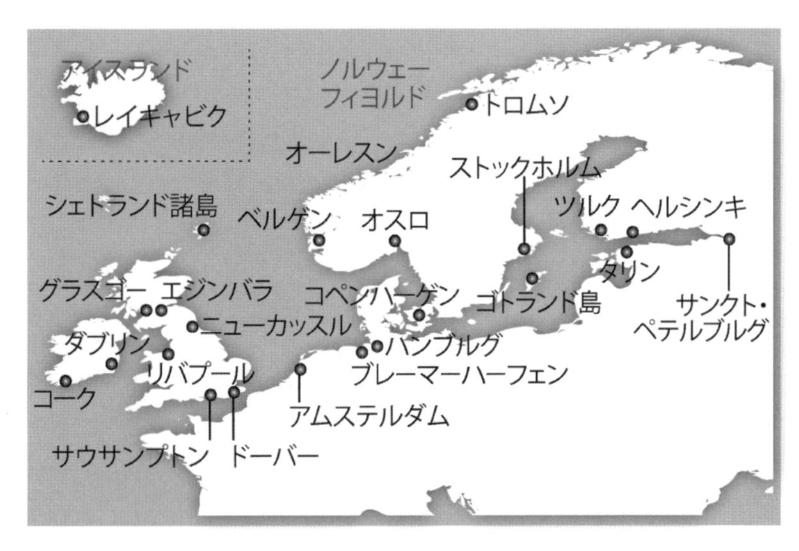

図 11-5　北ヨーロッパクルーズの主要クルーズポート

して選ばれている。寄港地各港は，最近，クルーズの受入施設の整備を積極的に行い，単なる寄港地から起点港への脱皮を図るところも少なくない。

3 つ目はイギリス周辺のクルーズで，イギリス，アイルランド，アイスランドなどを回るクルーズに人気がある。サウサンプトンもしくはドーバーを起点とし，ニューカッスル，エジンバラ，シェトランド諸島，リバプールなどのイギリスの各港，ダブリン，コークなどのアイルランドの各港，アイスランドのレイキャビクなどが寄港地となっている。

11.5　アラスカクルーズ

アラスカクルーズは，カリブ海に次いで現代クルーズが発達した水域で，当初はカナダのバンクーバーが一大拠点港となり，1 ～ 2 週間のクルーズが行われていたが，アメリカのシアトルがクルーズハブ港としての整備を積極的に行った結果，現在では二大拠点となっている。大型クルーズ客船がたくさん就航しているが，アラスカクルーズ自体は夏季だけの運航のため，冬季におけるクルーズ客船の効率的な運航ができる水域が必要となる。このため冬季には，パナマ運河クルーズを行いながらカリブ海に移動してクルーズを行う船，北アメリカ西海岸を南下してメキシコクルーズを行う船，アジア水域もしくはオセアニア水域で稼動する船などがある。

アラスカクルーズは氷河観光が中心となっているが，波のないインサイドパッセージ（たくさんの島によって太平洋の外海から遮蔽された内海）を航海するため船酔いの心配があまりないこと，自然を堪能できることなどが人気の理由であり，カリブ海クルーズに比べると費用はかなり高い。

日本からは，バンクーバーもしくはシアトルに，シンガポールと同様に 7 時間ほどの飛行で着けることから，アクセスが便利であり，今後の日本人のクルーズ適地として注目を集めるようになると考えられる。

図 11-6　アラスカクルーズの主要クルーズポート

11.6　バミューダ・北アメリカ東海岸クルーズ

ニューヨークなどを起点とするクルーズは，定期客船がクルーズに転進した頃に数多く実施されていた。その後は，比較的小型のクルーズ客船（2～3万総トン）での，ほそぼそとした運航が続けられてきたが，アメリカのテロなどの影響で飛行機に乗ることに不安をもつ層が増加した結果，クルーズ需要の大きい大都市圏から直接発着するクルーズがしだいに多くなっている。

その結果として，ニューヨークを起点とするバミューダクルーズが大きく成長し，定期客船時代に栄えたマンハッタン島の客船ターミナルが復活した。バミューダ諸島はイギリス領で，ニューヨーク

図 11-7　バミューダ・北アメリカ東海岸クルーズの主要クルーズポート

から南東 1,000 km の太平洋に浮かぶ 150 あまりの珊瑚礁の島からなるリゾート地。20 ノット船で 27 時間，すなわち前後に 2 泊 3 日の航海が必要となるが，現代クルーズの典型である 1 週間クルーズが十分構成できる水域といえる。

また，ニューヨーク，ボストンなど東海岸の主要港湾都市を起点とする北米東海岸の各地を巡るクルーズも人気を呼ぶようになり，特にカナダのニューイングランド地方へのクルーズが多く実施されるようになった。季節的には紅葉シーズンが人気であるために，秋に実施されるクルーズが多い。

11.7　アジアクルーズ

アジアでは長い間，日本を除くとクルーズ産業自体の発展はほとんどみられず，世界一周などの長期クルーズに就航する船がたまに各港に寄港するという状況であったが，1990 年代になってシンガポールにおいて最初に短期で定点定期の現代クルーズが生まれ，成長を始めた。マレーシアのカジノ，ホテル企業がスタークルーズを設立し，シンガポールおよび香港を基点とする地元マーケットに根付いたクルーズビジネスを展開した。

シンガポールでは，3 日～ 1 週間でマレー半島の沿岸を回るクルーズが華僑を中心として人気となって，今では年間を通して行われている。マレーシアの首都クアラルンプールの外港であるポートケラン，リゾート地として有名なプーケット島が寄港地として選ばれている。

また香港を起点としたクルーズでは，ワンナイトクルーズから，海南島や，世界遺産にも登録されているベトナムのハロン湾に寄港するショートクルーズなどが実施されている。

東アジアでは，2006 年から北アメリカのクルーズ会社が進出し，上海などを起点とした中国人客のための短期の定点定期クルーズが行われるようになった。1 週間以内のクルーズが多く，韓国の釜山，済州島，日本の博多，長崎，八代などが主な寄港港となっている。2016 年に中国のクルーズ人

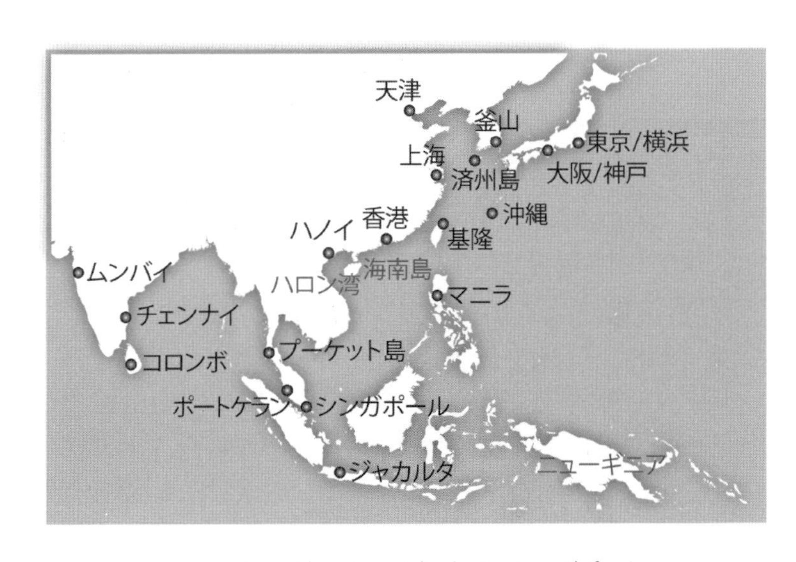

図 11-8　アジアクルーズの主要クルーズポート

口は 200 万人を超えた。また，2013 年からは外国籍船による日本発着クルーズも始まった。

11.8　日本の定番クルーズ

日本国内のクルーズは，カボタージュ規制によって日本籍のクルーズ客船だけが行うことができる。したがって，現在，国内クルーズを実施できるのは，「ぱしふぃっくびいなす」「にっぽん丸」「飛鳥Ⅱ」「ガンツウ」の 4 隻である。

これらの船の行うクルーズはさまざまだが，中には毎年必ずといっていいほど実施される「定番クルーズ」と呼ばれる企画もある。そのいくつかを紹介しよう。

11.8.1　日本一周クルーズ

10 日間程度の日程で日本を一周するクルーズで，どの船も毎年 1 ～ 2 回は実施する定番クルーズとなっている。

2018 年の事例では，「ぱしふぃっくびいなす」が 10 月 11 日に神戸を出航し，横浜，大船渡，釧路，青森，輪島，長門，福江，宿毛，堺に寄港して，横浜に戻る 10 泊クルーズを行い，料金は 41 ～ 180 万円。

「にっぽん丸」は，4 月 26 日に横浜を出航し，天草，伊万里，城崎，金沢，男鹿，江差，宮古に寄港して，横浜に戻る 10 泊クルーズを行い，料金は 50 ～ 210 万円。

「飛鳥Ⅱ」は，9 月 17 日に横浜を出航し，名古屋，瀬戸内海，広島，宮津，金沢，大船渡，仙台に寄港して，9 月 28 日に横浜に戻る 11 泊の行程で，料金は 57 ～ 290 万円。

いずれの船も，途中区間だけの乗船も受け付けている。

11.8.2　夏祭り・花火クルーズ

日本各地の夏祭りや港での花火大会を見るクルーズはたいへん人気があり，定番になっており，日本のクルーズ客船が大集合する。

2018 年の事例では，「飛鳥Ⅱ」は，「竿灯・ねぶた祭りクルーズ」（8 月 3 日～ 8 月 9 日），「阿波踊り・高松花火クルーズ」（8 月 11 日～ 8 月 16 日），「伊東花火クルーズ」（8 月 9 日～ 8 月 11 日），「鳥羽・熊野大花火クルーズ」（8 月 16 日～ 8 月 19 日）を行う。

「にっぽん丸」は，「夏休み館山花火クルーズ」（8 月 7 日～ 8 月 9 日），「東北夏祭りクルーズ」（8 月 2 日～ 8 月 7 日：秋田竿灯祭り，青森ねぶた祭り），「済州島と海峡花火祭り」（8 月 9 日～ 8 月 16 日：関門海峡花火大会，徳島阿波踊り）を行う。

「ぱしふぃっくびいなす」は，「竿灯・ねぶた東北二大祭りクルーズ」（8 月 1 日～ 8 月 6 日），「夏休み阿波踊りと関門海峡花火クルーズ」（8 月 12 日～ 8 月 16 日），「熊野大花火と南紀クルーズ」（8 月 16 日～ 8 月 19 日）を行う。

図 11-9　日本の主なクルーズ寄港地

11.9　日本の離島クルーズ

交通の便が悪く，なかなか効率的に旅行するのが難しい「離島」へのクルーズも人気だ。なかでも屋久島の人気は高く，クルーズ会社が企画に困った場合の定番寄港地の1つとなっており，「困ったときの屋久島頼み」とまで言われている。

利尻島，礼文島，八丈島，奄美大島，与論島，沖縄本島，石垣島，小笠原，種子島，屋久島などがクルーズ寄港地としての実績を挙げているが，この他にも日本には数多くの離島があり，さまざまな魅力を有する離島を訪れるクルーズを企画することが可能となっている。

離島では，岸壁施設が大型船に対応できない場合も多く，その場合には沖止めしてテンダーボートでの上陸となる。テンダーボートは船に搭載しているボートを使う場合と，港のボートをチャーターする場合があり，最近は，テンダーボートへの乗下船用の収納型プラットフォームをもつクルーズ客船も現れている。専用プラットフォームがない船では，舷側にポンツーン（箱型の台船）を係留して乗り降りのために使う場合もある。

離島では，宿泊施設が限られている場合も多いが，クルーズ客船では宿泊施設を備えているので，何人でも受け入れることができる。ただし，観光バスなどの島内の足の手配が難しいことが多く，悩ましいところである。

11.10　日本のショートクルーズ

日本のクルーズ客船では，はじめての顧客を獲得するためもあって，1～2泊の短いクルーズ企画もたくさん行われるようになった。特に1泊のものはワンナイトクルーズと呼ばれ，大都市圏発着で各船ともに積極的に実施している。なかでも，12月に行われる「クリスマスクルーズ」は定番クルーズの1つとなっている。基本的には，どこにも寄港せずに，船上でのクルーズライフだけを楽しむという究極の船旅で，根強い人気がある。

また，2港間を移動する片道だけのクルーズが数多くあるのも特徴の1つといえる。例えば，横浜～名古屋，神戸～横浜，博多～神戸，横浜～小樽などの間の1～2泊クルーズが頻繁に企画されている。

クルーズ運航会社が企画・運航する自主クルーズだけでなく，旅行会社がクルーズ客船を借りて，集客した旅客にクルーズを楽しませる事例も多くなっている。特に「ふじ丸」はこうしたチャータークルーズを中心に運航されており，「飛鳥Ⅱ」「ぱしふぃっくびいなす」「にっぽん丸」も自主クルーズだけでなくチャーターにも使用されている。

11.11　世界一周クルーズ

90～100日をかけて世界を一周するクルーズは，数は少なくなったが，キュナード・ライン，ホランド・アメリカ・ラインなどが古くから毎年実施し，日本のクルーズ客船でも定番クルーズとなっている。また，日本のピースボートは，クルーズ客船をチャーターして，年に複数回の世界一周クルーズを行っている。

季節的には，アラスカや北欧などの人気のある水域がシーズンオフとなる，北半球が冬の時期に行われることが多く，海外の世界一周のクルーズ客船は 2 ～ 3 月に日本に寄港することが多い。

一方，日本のクルーズ客船は梅雨の時期を避けるため 4 ～ 7 月に世界一周クルーズを行う場合が多い。例えば，2018 年に「飛鳥Ⅱ」が実施する世界一周クルーズは，3 月 25 日に横浜を出航し，神戸，シンガポール，プーケット，マーレ，サラーラ，ミコノス，バレッタ，チビタベッキア，バレンシア，マラガ，ジブラルタル，リスボン，ビルバオ，アムステルダム，ハンブルク，ロサイス，ダブリン，ハリファクス，セントジョン，ボストン，ニューヨーク，ボルチモア，ナッソー，カルタヘナ，プエルトケッツァル，サンディエゴ，ホノルル，横浜に寄港し，7 月 5 日に神戸に戻る 101 泊の行程。

世界一周の常連クルーズ客船としては，キュナード・ラインの「クイーン・メリー 2」「クイーン・エリザベス」，ホランド・アメリカ・ラインの「アムステルダム」がある。

世界一周のクルーズ料金は，日本船では約 300 万円から，大型の外国船では約 150 万円からとなっている。

≪問題≫

11-1　世界の主要なクルーズ水域を挙げなさい。

11-2　カリブ海クルーズの主要起点港を挙げなさい。

11-3　地中海の主要起点港を挙げなさい。

11-4　アラスカクルーズの主要起点港を挙げなさい。

11-5　日本発着クルーズの主要起点港を挙げなさい。

11-6　世界一周クルーズに要する日数はどのくらいか。

11-7　世界一周クルーズで通過する 2 つの運河を挙げなさい。

資料　102 日かけて巡る世界一周クルーズ

2018 年に久々に世界一周クルーズを再開した「飛鳥Ⅱ」。約 3 ヶ月半をかけて，世界を一周する。料金は，330 万円～ 2,625 万円。

移動も，3 食以上の食事も，ショーやイベントの費用も含んでいるので，意外にお買い得かも。

月日	寄港地	国名
3/25 日	横浜	日本
3/26 月	神戸	日本
4/2 月	シンガポール	シンガポール
4/4 水	プーケット	タイ
4/8 日	マーレ(錨泊)	モルジブ
4/12 木	サラーラ	オマーン
4/18 水	スエズ運河通航	
4/21 土	ミコノス(錨泊)	ギリシャ
4/23 月	バレッタ	マルタ
4/25 水 4/26 木	初寄港 チビタベッキア	イタリア
4/28 土	初寄港 バレンシア	スペイン
4/30 月	マラガ	スペイン
5/1 火	初寄港 ジブラルタル	イギリス領
5/2 水 5/3 木	リスボン	ポルトガル
5/5 土	初寄港 ビルバオ	スペイン
5/8 火	アムステルダム	オランダ
5/10 木	初寄港 ハンブルク	ドイツ
5/12 土	ロサイス	イギリス
5/15 火	ダブリン	アイルランド
5/22 火	初寄港 ハリファックス	カナダ
5/24 木	初寄港 セントジョン	カナダ
5/26 土 5/27 日	ボストン	アメリカ
5/29 火	ニューヨーク	アメリカ
5/31 木	初寄港 ボルチモア	アメリカ
6/3 日	ナッソー	バハマ
6/7 木	カルタヘナ	コロンビア
6/8 金	パナマ運河通航	
6/11 月	プエルトケッツァル	グアテマラ
6/17 日	サンディエゴ	アメリカ
6/23 土 6/24 日	ホノルル	アメリカ
7/4 水	横浜	日本
7/5 木	神戸	日本

食事　A・Bコース共通朝食・昼食・夕食 各100回

初寄港　飛鳥Ⅱ初寄港地

※記載されたスケジュールは、天候その他の事情により変更となる場合がございます。※掲載の写真は全てイメージです。

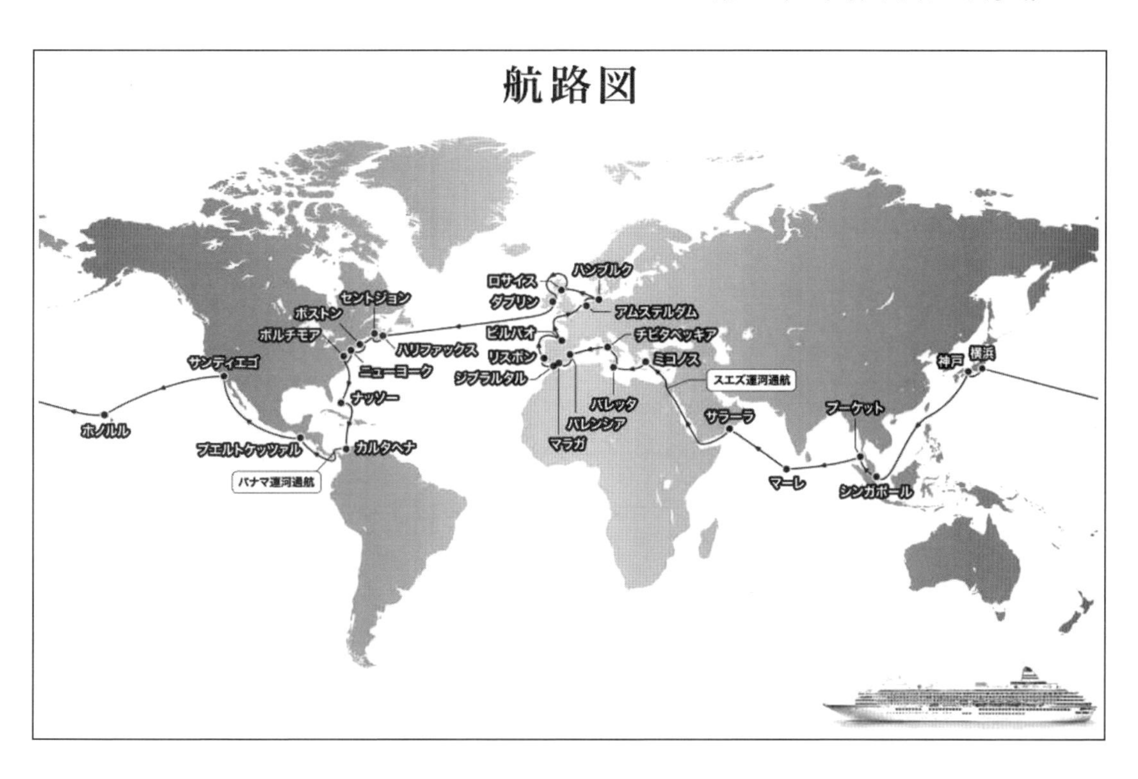

2018年 世界一周クルーズ 旅行代金〈A・Bコース共通〉

（お一人様　単位:円）

客室タイプ	K:ステート	F:ステート	E:バルコニー	D:バルコニー/D3:ティートリプル	C:スイート	A:アスカスイート	S:ロイヤルスイート
ワールド特別割引適用後の旅行代金	3,300,000	3,900,000	5,000,000	5,200,000	8,000,000	11,000,000	21,000,000
早期全額支払割引適用後の旅行代金	3,506,000	4,143,000	5,312,000	5,525,000	8,500,000	11,687,000	22,312,000
早期申込割引適用後の旅行代金	3,671,000	4,338,000	5,562,000	5,785,000	8,900,000	12,237,000	23,362,000
通常旅行代金	4,125,000	4,875,000	6,250,000	6,500,000	10,000,000	13,750,000	26,250,000

●客室をお2人様でご利用の場合のお1人様あたりの代金です。●客室をお1人様でご利用の場合、客室K・Cは上記旅行代金の160%、客室F・Eは上記旅行代金の135%、客室Dは上記旅行代金の140%、客室Aは上記旅行代金の180%、客室Sは上記旅行代金の200%となります。ただし、お1人様でご利用いただける客室には限りがございますのであらかじめご了承ください。●お申し込み時に申込金として旅行代金の10%を、2017年7月31日（月）までに中間金として旅行代金の20%を、残金は2017年11月30日（木）までにお支払いください。※2017年8月1日（火）以降にお申し込みの場合は、申込金と中間金を合計し、旅行代金の30%相当額をお支払いください。※ワールド特別割引、早期全額支払割引をご利用の場合は、各期日までに全額お支払いください。●2018年3月25日横浜～7月4日横浜、2018年3月26日神戸～7月5日神戸、2018年3月25日横浜～7月5日神戸、2018年3月26日神戸～7月4日横浜は同一旅行代金です。

（郵船クルーズホームページより）

事例研究（日本籍船）　「にっぽん丸」のクルーズ

この巻末の「事例研究」では，第1章〜第11章を読んだ上で，読者が各自で自学自習するための資料を紹介している。実際にクルーズ客船に乗船したつもりで資料に目を通し，クルーズ全体を理解してもらいたい。そのため，あまり解説文を付けていない。

この事例研究では，日本籍のクルーズ客船として，「にっぽん丸」の事例を紹介する。

「にっぽん丸」（商船三井客船）

レストランでの食事

ラウンジ

「にっぽん丸」のクルーズパンフレットより

クルーズスケジュール

月	出発日	日数	クルーズ
4月	4月6日(木)	4日間	春の喜界島・沖縄クルーズ
	4月6日(木)	19日間	飛んでクルーズ沖縄 グランドプラン
	4月9日(日)	4日間	飛んでクルーズ沖縄 Aコース ～西表・宮古～
	4月12日(水)	4日間	飛んでクルーズ沖縄 Bコース ～与那国・座間味～
	4月15日(土)	4日間	飛んでクルーズ沖縄 Cコース ～南大東島・久米島～
	4月18日(火)	4日間	飛んでクルーズ沖縄 Dコース ～石垣・台湾～
	4月21日(金)	4日間	春の那覇・奄美大島クルーズ
5月	5月7日(日)	3日間	横浜／熊本 軍艦島周遊クルーズ
	5月19日(金)	3日間	広島発着 初夏の奄美大島クルーズ
	5月22日(月)	3日間	徳島発着 初夏の八丈島クルーズ
	5月30日(火)	14日間	ぐるりにっぽん 輪島花火とロシアクルーズ ～スペシャルエンターテイメント～
6月	6月12日(月)	2日間	横浜／名古屋クルーズ
	6月19日(月)	5日間	大洗発着 小笠原クルーズ
	6月24日(土)	2日間	宮古／函館クルーズ
	6月26日(月)	4日間	新潟発着 夏の利尻・礼文クルーズ
	6月30日(金)	5日間	金沢発着 利尻・奥尻クルーズ
7月	7月13日(木)	3日間	熊野古道・新宮クルーズ
	7月21日(金)	4日間	仙台発着 夏休み 八丈島・熱海花火クルーズ
	7月28日(金)	3日間	夏休み 大洗海上花火大会クルーズ
8月	8月2日(水)	6日間	東北夏祭りクルーズ
	8月8日(火)	2日間	夏休み 館山花火クルーズ
	8月9日(水)	8日間	済州島と海峡花火・阿波おどりクルーズ
	8月17日(木)	3日間	夏休み 熊野大花火大会と高知クルーズ
	8月21日(月)	3日間	夏休み 伊勢志摩クルーズ
	8月27日(日)	4日間	飛んでクルーズ北海道 Aコース
	8月30日(水)	4日間	飛んでクルーズ北海道 Bコース
9月	9月2日(土)	4日間	飛んでクルーズ北海道 Cコース
	9月5日(火)	4日間	飛んでクルーズ北海道 Dコース
	9月8日(金)	4日間	初秋のサハリン・利尻クルーズ
	9月11日(月)	2日間	小樽／秋田クルーズ
	9月16日(土)	3日間	函館／石巻／東京クルーズ
	9月18日(月)	2日間	東京／名古屋クルーズ
	9月19日(火)	4日間	名古屋発着 宮島・宿毛クルーズ ～秋の味覚～

2017年4月～9月のクルーズ一覧

花火と華舞台

北陸の夏を告げる輪島花火と圧倒的な迫力のショーは必見

ぐるりにっぽん 輪島花火とロシアクルーズ
～スペシャルエンターテイメント～

日程	寄港地	入港	出港
5月30日［火］	横浜	—	17:00
5月31日［水］	終日航海	—	—
6月1日［木］	上五島	11:00	17:00
6月2日［金］	隠岐（浦郷）	11:00	18:00
6月3日［土］	輪島	09:00	23:00
6月4日［日］	佐渡島（両津）	09:00	17:00
6月5日［月］	秋田	07:00	23:00
6月6日［火］	終日航海	—	—
6月7日［水］	ウラジオストク	07:00	23:00
6月8日［木］	終日航海	—	—
6月9日［金］	コルサコフ	11:00	18:00
6月10日［土］	終日航海	—	—
6月11日［日］	石巻	09:00	15:00
6月12日［月］	横浜	09:30	—

●食事回数：朝13回、昼12回、夕13回　●最少催行人員：2名
●【隠岐(浦郷)】【佐渡島(両津)】【コルサコフ】では通船により上陸の予定です。天候によっては上陸できない場合もありますので、ご了承ください。通船による乗下船は30分～60分を要します。
●有効期間が2017年12月9日以降の旅券(パスポート)が必要となります。
●ロシアの査証(ビザ)は、ウラジオストクでは不要です。ただし、コルサコフでは、商船三井客船(株)の旅行企画・実施のオプショナルツアー参加者および上陸されないお客様は不要です。それ以外のお客様は必要です(日本国籍の場合)。
●出入国書類は商船三井客船(株)にて無料作成いたします(ただし査証が必要な場合の査証取得費用は除きます)。
●渡航条件は2016年11月1日現在のものです。変更となる場合はご連絡いたします。
●花火大会は天候等により実施されない場合がありますので、ご了承ください。

スペシャルエンターテイメントショー

ジャズシンガー綾戸智恵がお届けする「～DO JAZZ～ in にっぽん丸」。笑いあり、涙ありのステージです！ そして、にっぽん丸初登場の歌手美川憲一。50周年公演も終え、更にパワーアップして多方面で活躍中！ 魅力満載のステージをお送りいたします。

輪島市民大花火大会

1分間の打ち上げ数が日本一の輪島花火。わずか20分で約1万6000発の花火が打ち上がり、最初から最後まで大迫力の光景が楽しめます。

イメージ

●オプショナルツアー（別料金・大人お一人様予定代金）

上五島（6月1日）	にっぽん丸プレミアムツアー ○リゾートホテル マルゲリータのランチと上五島観光	健脚レベル★ 16,500円
輪島（6月3日）	にっぽん丸オリジナルツアー ○にっぽんの職人技をめぐる～輪島編～	健脚レベル★ 9,000円
秋田（6月5日）	にっぽん丸プレミアムツアー ○老舗料亭「濱乃家」でいただく秋田郷土料理とあきた舞妓のおもてなし	健脚レベル★ 20,000円
ウラジオストク（6月7日）	○ウラジオストク半日観光	健脚レベル★ 10,000円
コルサコフ（6月9日）	○ユジノサハリンスク半日観光	健脚レベル★ 14,000円

※隠岐、佐渡島、石巻でもオプショナルツアーを設定しております。
※★印の数で健脚レベルを表します。
※オプショナルツアーは商船三井客船(株)の旅行企画・実施となります。具体的な内容とお申込方法は別途ご案内します。

5月30日出航の詳細パンフレット・その1

2017年5月30日（火）横浜発 ≫6月12日（月）横浜着　14日間

エンターテイナー

横浜〜上五島
綾戸 智恵
（ジャズシンガー）

17歳で単身渡米。帰国後はジャズ・クラブで歌う。1998年デビュー。NHK紅白歌合戦で「テネシー・ワルツ」を熱唱。アルバム『Picture in a Frame』『DO JAZZ』をリリース。

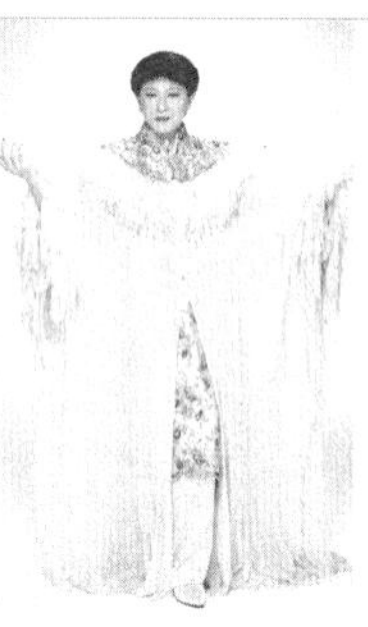

秋田
美川 憲一
（歌手）

「柳ヶ瀬ブルース」「釧路の夜」のミリオンセラー、「さそり座の女」で歌手として不動のものにする。NHK紅白歌合戦出場も26回を数え、様々なジャンルで活躍中。

秋田〜横浜
岸本 力
（歌手）

ロシア音楽・声楽の第一人者として活躍するバスの声楽家。2012年ロシア文化勲章を受章。二期会会員

秋田〜横浜
北川 翔
（バラライカ奏者）

日本人初の国際ロシア民族楽器コンクール優勝、メディアに度々登場しバラライカの普及に努める。

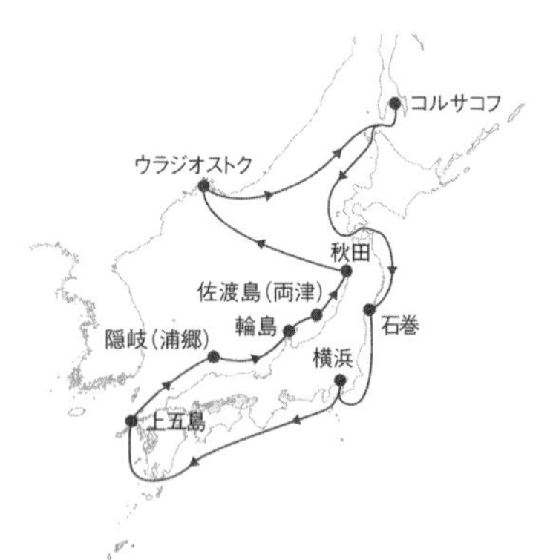

旅行代金（大人お一人様）			熟年割引
ステート	スタンダードステート	507,000円	456,000円
	コンフォートステート	539,000円	485,000円
	スーペリアステート	582,000円	524,000円
デラックス	デラックスツイン	973,000円	876,000円
	デラックスベランダ	1,077,000円	969,000円
	デラックスシングル	1,422,000円	1,280,000円
スイート	ジュニアスイート	1,354,000円	1,219,000円
	ビスタスイート	1,774,000円	1,597,000円
	グランドスイート	2,536,000円	2,282,000円
G3	グループ 3（コンフォートステート）	430,000円	387,000円

※グループ3は、コンフォートステートを3名で利用した場合の代金です。※本クルーズは熟年割引プラン対象クルーズです（満60歳以上）。
※1室1名利用代金、1室3名利用代金、子供代金については裏表紙をご確認ください。

5月30日出航の詳細パンフレット・その2

船上が特等席

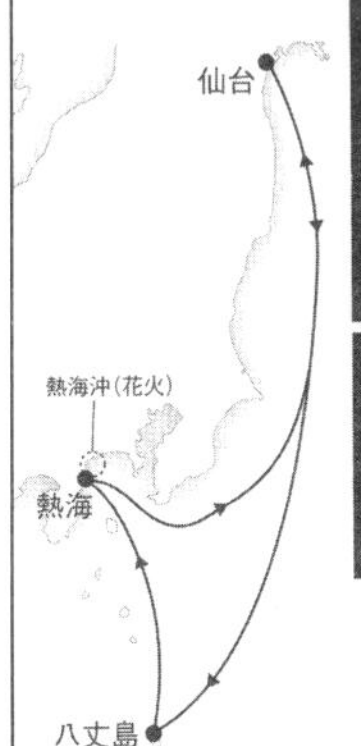

夏休み。島を楽しみ、海・花火
仙台発着、ファミリーでどうぞ

熱海

八丈島ふれあい牧場

熱海海上花火大会（イメージ）

仙台発着 夏休み 八丈島・熱海花火クルーズ

2017年7月21日（金）仙台発 >>> 7月24日（月）仙台着 4日間 通船

7月21日［金］	仙台	—	11:30
7月22日［土］	八丈島	09:00	18:00
7月23日［日］	熱海	08:00	21:30 ※
7月24日［月］	仙台	15:30	—

●食事回数：朝3回、昼4回、夕3回　●最少催行人員：2名
●【八丈島】では通船により上陸の予定です。天候によっては上陸できない場合もありますので、ご了承ください。通船による乗下船は30分～60分を要します。
●花火大会は天候等により実施されない場合がありますので、ご了承ください。
※沖から花火を観覧するために、7月23日は15:00（予定）に離岸いたしますので、その後の乗下船はできません。

●オプショナルツアー（別料金・大人お一人様予定代金）

八丈島（7月22日）	○シュノーケル体験	アクティブ 7,500円
熱海（7月23日）	にっぽん丸プレミアムツアー ○富士屋ホテルのランチと成川美術館	健脚レベル★ 18,000円

※★印の数及びアクティブで健脚レベルを表します。
※オプショナルツアーは商船三井客船（株）の旅行企画・実施となります。具体的な内容とお申込方法は別途ご案内します。

エンターテイナー

仙台～熱海

Dr. レオン（時空のファンタジスタ）

日本人プロとして初めてアメリカのマジック団体F.F.F.FよりDr.の称号（Doctor of Magic Diploma）を授与される。全米NO.1マジシャンであるデビット・カッパーフィールドにマジックを提供したことで海外でも注目を浴び、ラスベガスでの公演、ヨーロッパの世界大会のゲスト出演など、インターナショナルに活躍中。時空を超える不思議な現象を目の前で体験できる。

NIPPON● TOPICS

八丈太鼓

特徴は、太鼓の両面から二人で叩くこと。一人は一定のリズムを刻み、もう一人はそのリズムに合わせてアドリブで打ちます。その自由奔放さが最大の魅力で他の郷土芸能の太鼓とは一線を画しているところです。

熱海の温泉饅頭

古くから温泉地として栄えた熱海には、沢山の温泉まんじゅうのお店があります。にっぽん丸スタッフが実食して推す各店の自慢の逸品を船上にご用意。是非食べ比べてみてください。（航海中1回）

イメージ

旅行代金（大人お一人様）

	スタンダードステート	136,000円
	コンフォートステート	149,000円
	スーペリアステート	161,000円
デラックス	デラックスツイン	235,000円
	デラックスベランダ	260,000円
	デラックスシングル	315,000円
スイート	ジュニアスイート	340,000円
	ビスタスイート	439,000円
	グランドスイート	618,000円
G3	グループ 3（コンフォートステート）	124,000円

※グループ３は、コンフォートステートを３名で利用した場合の代金です。
※1室1名利用代金、1室3名利用代金、子供代金については裏表紙をご確認ください。

7月21日出航の詳細パンフレット

船内新聞「Port & Starboard」

毎晩，キャビンに届けられる船内新聞には，翌日の予定，スケジュール，レストランやバーのオープン時間など，あらゆる情報が記載されている。

〔1日目〕乗船日

NIPPON MARU　2018年1月10日(水)【第1号】

Voy.477　新春初旅 にっぽん丸クルーズ

Port & Starboard

今夜のドレスコード
カジュアル
Dress code for this evening : CASUAL

高級リゾート地を訪れたときのように、決して『ラフになり過ぎない』ことがポイント。カラフルな色使いなど、コーディネートをお楽しみください。

ドレスコードの適用は夕食からドルフィンホールでのショープログラム終了までです（一部公共スペースを除く）。詳しくは、クルーズガイド 10～11ページをご覧ください。

本日の夕食は『ウェルカムディナー（洋食）』です。今航海の夕食は 1回制となります。

航海スケジュール（予定）

気象・海象の条件により、予定航路や運航スケジュールを変更することがあります。

17:00	神戸港 出港 Dep. Kobe
18:30 頃	友ヶ島水道（大阪湾出口） Tomogashima Suido
22:30 頃	室戸岬（高知県） Muroto misaki

日出没（予定）

17:05 頃　日　没　Sunset

避難訓練

本日 16:40 頃から
避難訓練を実施いたします。

避難訓練開始は、非常ベルと放送でご案内いたします。乗船証をお持ちの上、4階プロムナードデッキへお集まりください。4階では乗組員が乗船証を確認し、各集合場所へご案内いたします。
【ご集合場所】
Lifeboat No. が**奇数のお客様**
／4階プロムナードデッキ**右舷側**
Lifeboat No. が**偶数のお客様**
／4階プロムナードデッキ**左舷側**
Lifeboat No. は乗船証に記載されています。

避難訓練終了後、出港のボンボヤージュ・サービスをお楽しみいただけます。
暖かい服装でご参加ください。

Welcome Aboard!

ようこそ「海の上の国にっぽん」へ。
にっぽん丸船上で、お食事、
エンターテインメント、そして
青い空と青い海を満喫してください。
皆様のご乗船を、
乗組員、スタッフ一同、
心より歓迎いたします。

にっぽん丸 船長
久保(くぼ) 滋弘(しげひろ)

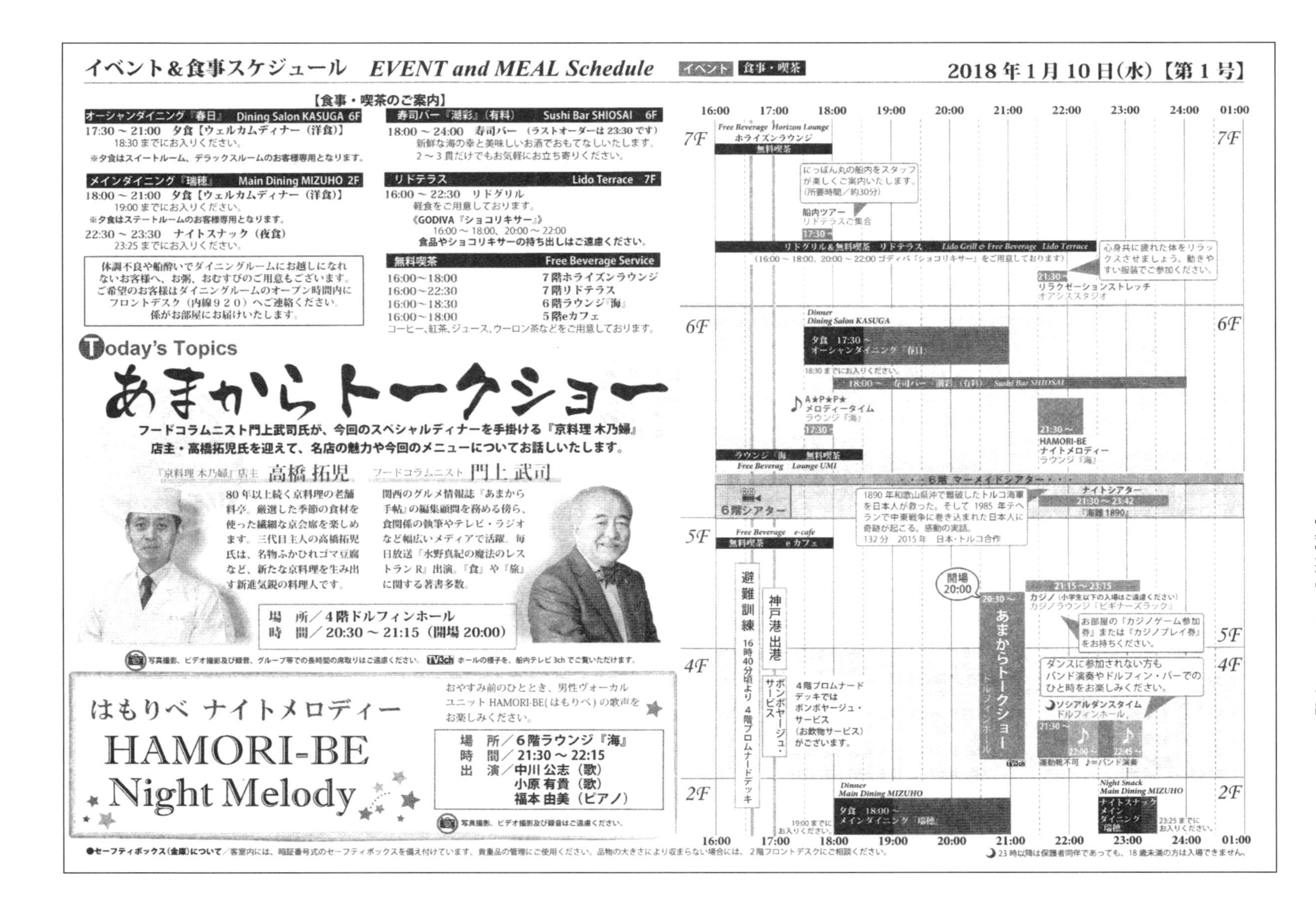

イベント＆食事スケジュール　*EVENT and MEAL Schedule*　イベント　食事・喫茶　2018年1月10日(水)【第1号】

【食事・喫茶のご案内】

オーシャンダイニング『春日』　Dining Salon KASUGA 6F
17:30 ~ 21:00　夕食【ウェルカムディナー（洋食)】
18:30 までにお入りください。
※夕食はスイートルーム、デラックスルームのお客様専用となります。

メインダイニング『瑞穂』　Main Dining MIZUHO 2F
18:00 ~ 21:00　夕食【ウェルカムディナー（洋食)】
19:00 までにお入りください。
※夕食はステートルームのお客様専用となります。
22:30 ~ 23:30　ナイトスナック（夜食）
23:25 までにお入りください。

体調不良や船酔いでダイニングルームにお越しになれないお客様へ、お粥、おむすびのご用意もございます。ご希望のお客様はダイニングルームのオープン時間内にフロントデスク（内線９２０）へご連絡ください。係がお部屋にお届けいたします。

寿司バー『潮彩』（有料）　Sushi Bar SHIOSAI 6F
18:00 ~ 24:00　寿司バー（ラストオーダーは 23:30 です）
新鮮な海の幸と美味しいお酒でおもてなしいたします。
２～３貫だけでもお気軽にお立ち寄りください。

リドテラス　Lido Terrace 7F
16:00 ~ 22:30　リドグリル
軽食をご用意しております。
《GODIVA『ショコリキサー』》
16:00 ~ 18:00、20:00 ~ 22:00
食品やショコリキサーの持ち出しはご遠慮ください。

無料喫茶　Free Beverage Service

16:00~18:00	7階ホライズンラウンジ
16:00~22:30	7階リドテラス
16:00~18:30	6階ラウンジ『海』
16:00~18:00	5階eカフェ

コーヒー、紅茶、ジュース、ウーロン茶などをご用意しております。

Today's Topics

あまからトークショー

フードコラムニスト門上武司氏が、今回のスペシャルディナーを手掛ける『京料理 木乃婦』店主・高橋拓児氏を迎えて、名店の魅力や今回のメニューについてお話しいたします。

『京料理 木乃婦』店主　高橋 拓児

80年以上続く京料理の老舗料亭。厳選した季節の食材を使った繊細な京会席を楽しめます。三代目主人の高橋拓児氏は、名物ふかひれゴマ豆腐など、新たな京料理を生み出す新進気鋭の料理人です。

フードコラムニスト　門上 武司

関西のグルメ情報誌『あまから手帖』の編集顧問を務める傍ら、食関係の執筆やテレビ・ラジオなど幅広いメディアで活躍。毎日放送『水野真紀の魔法のレストラン R』出演。『食』や『旅』に関する著書多数。

場　所／4階ドルフィンホール
時　間／20:30 ~ 21:15（開場 20:00）

写真撮影、ビデオ撮影及び録音、グループ等での長時間の席取りはご遠慮ください。TV3ch ホールの様子を、船内テレビ 3ch でご覧いただけます。

はもりべ　ナイトメロディー
HAMORI-BE Night Melody

おやすみ前のひととき、男性ヴォーカルユニット HAMORI-BE（はもりべ）の歌声をお楽しみください。

場　所／6階ラウンジ『海』
時　間／21:30 ~ 22:15
出　演／中川 公志（歌）
　　　　小原 有貴（歌）
　　　　福本 由美（ピアノ）

写真撮影、ビデオ撮影及び録音はご遠慮ください。

●セーフティボックス（金庫）について／客室内には、暗証番号式のセーフティボックスを備え付けています。貴重品の管理にご使用ください。品物の大きさにより収まらない場合には、2階フロントデスクにご相談ください。

23 時以降は保護者同伴であっても、18 歳未満の方は入場できません。

NIPPON MARU　　2018年1月10日(水)【第1号】

営業時間のご案内

※各バーのラストオーダーは閉店時間の30分前です

フロントデスク　Front Desk　2F（内線920）
00:00～24:00　（24時間オープン）
フロントデスク正面カウンターにクルーズパンフレット、新聞をご用意しております。酔い止め薬をご用意しております。（無料）

クリニック　Clinic　1F（内線899）
16:00～17:30
診療時間外は、2階フロントデスク（内線920）へご連絡ください。※現在服用中のお薬の説明書（お薬手帳）をお持ちください。

ツアーデスク　Tour Desk　3F（内線930）
17:15～18:00、20:30～21:15
オプショナルツアー、寄港地の相談窓口です。

クルーズサロン　Cruise Salon　2F
本日はお休みです
船内生活などクルーズ全般についてのご要望やご質問にお応えします。また、にっぽん丸クルーズのご予約も承ります。

プール　Swimming Pool　7F
16:00～19:00
ライフガードはおりません。ご利用の際は十分にお気をつけください。お子様がご利用の際は、保護者の付き添いをお願いいたします。

プールに行くときは、水着の上にTシャツ、ショートパンツ等を着用してください（プールに更衣室はありません）。

オアシスジム&スタジオ　Oasis Gym & Studio　7F
00:00～24:00　（24時間オープン）
トレーナー不在時は、お怪我のないよう十分にお気をつけください。※フィットネス機器のご利用は、05:00～翌01:00までとさせていただきます。

eカフェ&ライブラリー　e-cafe & Library　5F
00:00～24:00　（24時間オープン）
インターネットをご利用いただけます。（有料）詳しくは、クルーズガイド21ページをご覧ください。

カードルーム　Card Room　5F
00:00～24:00　（24時間オープン）
オセロや麻雀などをご用意しております。（無料）

グランドバス　Grand Bath　3F 船尾部
16:00～翌01:00
タオルはグランドバスにご用意しております。
～にっぽん丸のレジオネラ菌対策について～
官庁指導の塩素濃度基準を遵守した塩素殺菌と、ろ過による管理を行なっています。安心してご利用ください。

オアシスジム、プール、スパ&サロン（7階）、グランドバス（3階）をご利用のお客様へ
酒気を帯びてのご利用は、ご遠慮ください。
また、天候などの事情により船が揺れることがございます。お怪我をなさいませんよう十分にお気をつけください。船の揺れのため、安全上ご利用いただけない場合がございます。

スモーキングラウンジ　Smoking Lounge　2F
00:00～24:00　（24時間オープン）
吸殻やマッチなどは必ず灰皿へ入れてください。火災の予防と喫煙マナーにご協力をお願いいたします。●**客室内は禁煙です。**

スパ&サロン　Spa & Salon（予約制）　7F（内線874）
16:00～24:00　（ご予約受付は22:00まで）
"美と癒し"の総合サロンです。
スパ・トリートメント「Spa de TERRAKE（スパ・ド・テラケ）」、身体をほぐす「ボディケア」、ヘアサロン「ビューティードレッセヤマノ」、手元、足元を美しくする「ネイルコーナー」がございます。

★営業時間外のご予約も承ります。ご希望のお客様は前日までに内線874へどうぞ。

ブティック『アンカー』　Boutique Anchor　5F
16:00～18:00、19:30～22:30
お土産にぴったりなにっぽん丸オリジナルのお菓子類やグッズをはじめ、アクセサリーや衣料品など幅広く取り揃えています。

ショップ『ブイ』　Shop Buoy　5F
16:00～18:00、19:30～22:30
にっぽんマルシェを展開しているほか、日用品、飲料、おつまみなどを販売しております。

フォトショップ『ドルフィン』　Photo Shop Dolphin　3F
専属カメラマンが撮影した写真の展示・販売を行なっております。ご購入は3階販売カウンターにございます注文票にご記入ください。

リド・バー　Lido Bar　7F
16:00～23:00　（リドグリル／22:30まで）
『リドグリル』メニュー（軽食）は無料でお召し上がりいただけます。

ミッドシップ・バー　Midship Bar　6F
16:00～24:00
寿司バー営業時間中、寿司バー『潮彩』メニューの出前も承ります。※夕食以降は、小学生以下のお子様のご入場はできませんのでご了承ください。

ネプチューン・バー　Neptune Bar　5F
17:00～18:30、20:30～翌01:00
20:30以降は、小学生以下のお子様のご入場はできませんのでご了承ください。

ドルフィン・バー　Dolphin Bar　4F
21:30～23:15
ダンスタイム・カジノの合間など、お気軽にご利用ください。

カジノゲームの遊び方
カジノオープン時間に
お部屋の『カジノゲーム参加券』をお持ちください。
チップ20枚と交換をいたします。

テレビシアターのご案内

6ch『にっぽん丸からのお願いとご案内』
救命胴衣の着用方法や緊急避難時の対応について、また安全に関するご案内や船内設備・サービスについてのご案内を終日放映しております。

7ch『摩天楼はバラ色に』
主演／マイケル・J・フォックス
110分　1987年　字幕
《あらすじ》
カンザス州から出てきた田舎者がニューヨーク州マンハッタンの大企業で経営陣に上り詰めるまでを描いたコメディ。
《放映時間》21:30～23:20

簡単な登録手続きで、フロントでの精算手続きが必要なくなります。
クレジットカードでお支払い予定のお客様は、あらかじめ2階フロントデスクでクレジットカードを登録されますと、最終下船日に領収書を確認いただくだけで、精算が終了となります。※登録手続きは13日午前1時まで承ります。
● ご利用いただけるクレジットカード
アメリカンエキスプレス、ダイナース、JCB、VISA、マスターカード

今航海の有料ランドリーサービスは11日午前9時の受付で終了いたします。※3階のセルフランドリールーム（無料）は、皆様ご利用いただけます。

〔2日目〕

NIPPON MARU　**2018年1月11日(木)【第2号】**

Voy.477　新春初旅 にっぽん丸クルーズ

Port & Starboard

今夜のドレスコード

セミフォーマル

Dress code for this evening : SEMI FORMAL

目安は、おしゃれなレストランやクラシックコンサートへ出かけるようなスタイル。男性はいつものネクタイに、カラーシャツを合わせてみても。女性は、襟元の華やかなブラウスを選んだり、小物をアレンジして幅のある装いを。

今夕のドレスコードの適用は夕食からドルフィンホールショープログラム終了までです（一部公共スペースを除く）。詳しくは、クルーズガイド10～11ページをご覧ください。

本日の夕食は『京料理 木乃婦 スペシャルディナー（和食）』です。

航海スケジュール（予定）

気象・海象の条件により、予定航路や運航スケジュールを変更することがあります。

11:00頃	油津港 入港 Arr. Aburatsu
17:00	油津港 出港 Dep. Aburatsu

日出没（予定）

07:10頃	日　出	Sunrise
17:25頃	日　没	Sunset

《1月11日》
油津の天気予報
晴れ

降水確率／ 10%
最高気温／　6℃
最低気温／　3℃

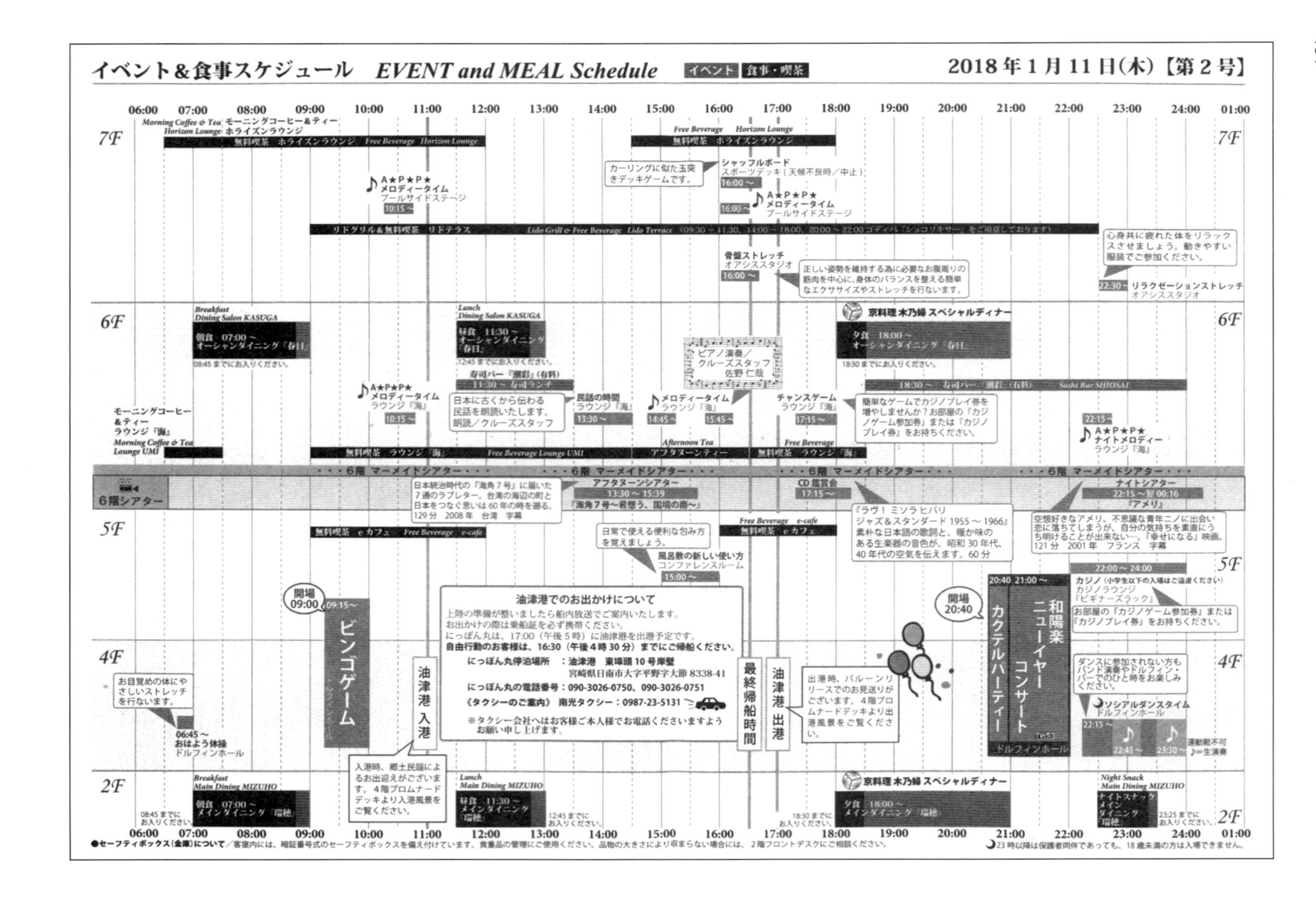

イベント＆食事スケジュール　EVENT and MEAL Schedule
イベント　食事・喫茶
2018年1月11日(木)【第2号】
06:00　07:00　08:00　09:00　10:00　11:00　12:00　13:00　14:00　15:00　16:00　17:00　18:00　19:00　20:00　21:00　22:00　23:00　24:00　01:00
7F
Morning Coffee & Tea Horizon Lounge モーニングコーヒー＆ティー ホライズンラウンジ
無料喫茶　ホライズンラウンジ　Free Beverage Horizon Lounge
Free Beverage Horizon Lounge 無料喫茶　ホライズンラウンジ
カーリングに似た玉突きデッキゲームです。
シャッフルボード
スポーツデッキ（天候不良時／中止）
16:00～
A★P★P★ メロディータイム
プールサイドステージ
10:15～
16:00～
リドグリル＆無料喫茶　リドテラス　Lido Grill & Free Beverage Lido Terrace
心身共に疲れた体をリラックスさせましょう。動きやすい服装でご参加ください。
骨盤ストレッチ
オアシススタジオ
正しい姿勢を維持する為に必要なお腹周りの筋肉を中心に、身体のバランスを整える簡単なエクササイズやストレッチを行ないます。
22:30～ リラクゼーションストレッチ
オアシススタジオ
6F
Breakfast Dining Salon KASUGA
朝食　07:00～
オーシャンダイニング『春日』
08:45までにお入りください。
Lunch Dining Salon KASUGA
昼食　11:30～
オーシャンダイニング『春日』
12:45までにお入りください。
京料理 木乃婦 スペシャルディナー
夕食　18:00～
オーシャンダイニング『春日』
18:30までにお入りください。
ピアノ演奏／クルーズスタッフ 佐野仁哉
寿司バー『潮彩』(有料)
11:30～ 寿司ランチ
18:30～ 寿司バー『潮彩』(有料) Sushi Bar SHIOSAI
A★P★P★ メロディータイム
ラウンジ『海』
10:15～
日本に古くから伝わる民話を朗読いたします。
朗読／クルーズスタッフ
民話の時間
ラウンジ『海』
13:30～
メロディータイム
ラウンジ『海』
14:45～
15:45～
チャンスゲーム
ラウンジ『海』
17:15～
簡単なゲームでカジノプレイ券を増やしませんか？お部屋の『カジノゲーム参加券』または『カジノプレイ券』をお持ちください。
22:15～
A★P★P★ ナイトメロディー
ラウンジ『海』
モーニングコーヒー＆ティー
ラウンジ『海』
Morning Coffee & Tea Lounge UMI
無料喫茶　ラウンジ『海』　Free Beverage Lounge UMI
Afternoon Tea アフタヌーンティー
Free Beverage 無料喫茶　ラウンジ『海』
6階シアター
・・・6階 マーメイドシアター・・・
日本統治時代の『海角7号』に届いた7通のラブレター。台湾の海辺の町と日本をつなぐ思いは60年の時を遡る。
129分　2008年　台湾　字幕
アフタヌーンシアター
13:30～15:39
『海角7号～君想う、国境の南～』
CD鑑賞会
17:15～
『ラヴ！ミソラヒバリ ジャズ＆スタンダード1955～1966』
素朴な日本語の歌詞と、暖か味のある生楽器の音色が、昭和30年代、40年代の空気を伝えます。60分
ナイトシアター
22:15～翌00:16
『アメリ』
空想好きなアメリ。不思議な青年ニノに出会い恋に落ちてしまうが、自分の気持ちを素直にうち明けることが出来ない…。『幸せになる』映画。
121分　2001年　フランス　字幕
5F
無料喫茶　eカフェ　Free Beverage e-cafe
日常で使える便利な包み方を覚えましょう。
風呂敷の新しい使い方
コンファレンスルーム
15:00～
Free Beverage e-cafe
無料喫茶　eカフェ
22:00～24:00
カジノ（小学生以下の入場はご遠慮ください）
カジノラウンジ
『ビギナーズラック』
お部屋の『カジノゲーム参加券』または『カジノプレイ券』をお持ちください。
開場 09:00
09:15～
ビンゴゲーム
ドルフィンホール
油津港でのお出かけについて
上陸の準備が整いましたら船内放送でご案内いたします。
お出かけの際は乗船証を必ず携帯ください。
にっぽん丸は、17:00（午後5時）に油津港を出港予定です。
自由行動のお客様は、16:30（午後4時30分）までにご帰船ください。
にっぽん丸停泊場所　：油津港　東埠頭10号岸壁
宮崎県日南市大字平野字大節8338-41
にっぽん丸の電話番号：090-3026-0750、090-3026-0751
《タクシーのご案内》　南光タクシー：0987-23-5131
※タクシー会社へはお客様ご本人様でお電話くださいますようお願い申し上げます。
油津港 入港
最終帰船時間
油津港 出港
出港時、バルーンリリースでのお見送りがございます。4階プロムナードデッキより出港風景をご覧ください。
開場 20:40
20:40　21:00～
カクテルパーティー
和陽楽 ニューイヤーコンサート
ドルフィンホール
4F
お目覚めの体にやさしいストレッチを行ないます。
06:45～
おはよう体操
ドルフィンホール
入港時、郷土民謡によるお出迎えがございます。4階プロムナードデッキより入港風景をご覧ください。
ダンスに参加されない方もバンド演奏やドルフィン・バーでのひと時をお楽しみください。
ソシアルダンスタイム
ドルフィンホール
22:15～
22:45～
23:30～
運動靴不可
♪＝生演奏
2F
Breakfast Main Dining MIZUHO
朝食　07:00～
メインダイニング『瑞穂』
08:45までにお入りください。
Lunch Main Dining MIZUHO
昼食　11:30～
メインダイニング『瑞穂』
12:45までにお入りください。
京料理 木乃婦 スペシャルディナー
夕食　18:00～
メインダイニング『瑞穂』
18:30までにお入りください。
Night Snack Main Dining MIZUHO
ナイトスナック
メインダイニング『瑞穂』
23:25までにお入りください。
●セーフティボックス(金庫)について／客室内には、暗証番号式のセーフティボックスを備え付けています。貴重品の管理にご使用ください。品物の大きさにより収まらない場合には、2階フロントデスクにご相談ください。
23時以降は保護者同伴であっても、18歳未満の方は入場できません。

NIPPON MARU　2018年1月11日(木)【第2号】

営業時間のご案内

※各バーのラストオーダーは閉店時間の30分前です

フロントデスク　Front Desk　2F（内線920）
00:00～24:00　(24時間オープン)
フロントデスク正面カウンターにクルーズパンフレット、新聞をご用意しております。酔い止め薬をご用意しております。(無料)

クリニック　Clinic　1F（内線899）
09:00～11:00
診療時間外は、2階フロントデスク（内線920）へご連絡ください。※現在服用中のお薬の説明書（お薬手帳）をお持ちください。

ツアーデスク　Tour Desk　3F（内線930）
09:00～10:00、17:15～18:00
20:00～21:30
オプショナルツアー、寄港地の相談窓口です。

クルーズサロン　Cruise Salon　2F
09:00～12:00、15:30～17:00
船内生活などクルーズ全般についてのご要望やご質問にお応えします。また、にっぽん丸クルーズのご予約も承ります。

プール　Swimming Pool　7F
07:00～19:00
ライフガードはおりません。ご利用の際は十分にお気をつけください。お子様がご利用の際は、保護者の付き添いをお願いいたします。

プールに行くときは、水着の上にTシャツ、ショートパンツ等を着用してください（プールに更衣室はありません）。

オアシスジム＆スタジオ　Oasis Gym & Studio　7F
00:00～24:00　(24時間オープン)
トレーナー不在時は、お怪我のないよう十分にお気をつけください。※フィットネス機器のご利用は、05:00～翌01:00までとさせていただきます。

eカフェ＆ライブラリー　e-cafe & Library　5F
00:00～24:00　(24時間オープン)
インターネットをご利用いただけます（有料）。詳しくは、クルーズガイド21ページをご覧ください。

カードルーム　Card Room　5F
00:00～24:00　(24時間オープン)

グランドバス　Grand Bath　3F 船尾部
06:00～12:00、15:00～翌01:00
タオルはグランドバスにご用意しております。
～にっぽん丸のレジオネラ菌対策について～
官庁指導の塩素濃度基準を遵守した塩素殺菌と、ろ過による管理を行なっています。安心してご利用ください。

オアシスジム、プール、スパ&サロン（7階）、グランドバス（3階）をご利用のお客様へ
酒気を帯びてのご利用は、ご遠慮ください。
また、天候などの事情により船が揺れることがございます。お怪我をなさいませんよう十分にお気をつけください。船の揺れのため、安全上ご利用いただけない場合がございます。

スモーキングラウンジ　Smoking Lounge　2F
00:00～24:00　(24時間オープン)
吸殻やマッチなどは必ず灰皿へ入れてください。火災の予防と喫煙マナーにご協力をお願いいたします。●客室内は禁煙です。

スパ&サロン　Spa & Salon（予約制）　7F（内線874）
16:00～24:00　(ご予約受付は22:00まで)
"美と癒し"の総合サロンです。
スパ・トリートメント「Spa de TERRAKE（スパ・ド・テラケ）」、身体をほぐす「ボディケア」、ヘアサロン「ビューティードレッセヤマノ」、手元、足元を美しくする「ネイルコーナー」がございます。

★営業時間外のご予約も承ります。ご希望のお客様は前日までに内線874へどうぞ。

ブティック『アンカー』　Boutique Anchor　5F
08:00～11:00、15:00～18:00
19:30～22:30
お土産にぴったりなにっぽん丸オリジナルのお菓子類やグッズをはじめ、アクセサリーや衣料品など幅広く取り揃えています。

ショップ『ブイ』　Shop Buoy　5F
08:00～11:00、15:00～18:00
19:30～22:30
にっぽんマルシェを展開しているほか、日用品、飲料、おつまみなどを販売しております。

フォトショップ『ドルフィン』　Photo Shop Dolphin　3F
専属カメラマンが撮影した写真の展示・販売を行なっております。ご購入は3階販売カウンターにございます注文票にご記入ください。

リド・バー　Lido Bar　7F
09:00～23:00（リドグリル／22:30まで）
『リドグリル』メニュー（軽食）は無料でお召し上がりいただけます。

ミッドシップ・バー　Midship Bar　6F
09:00～24:00
寿司バー営業時間中、寿司バー『潮彩』メニューの出前も承ります。※夕食以降は、小学生以下のお子様のご入場はできませんのでご了承ください。

ネプチューン・バー　Neptune Bar　5F
17:00～18:30、20:30～翌01:00
20:30以降は、小学生以下のお子様のご入場はできませんのでご了承ください。

ドルフィン・バー　Dolphin Bar　4F
22:15～24:00
ダンスタイム・カジノの合間など、お気軽にご利用ください。

カジノゲームの遊び方
カジノオープン時間にお部屋の『カジノゲーム参加券』をお持ちください。
チップ20枚と交換をいたします。

テレビシアターのご案内

6ch『カサブランカ』
出演／ハンフリー・ボガート
イングリッド・バーグマン
102分　1942年　字幕
《あらすじ》「君の瞳に乾杯」という名ゼリフを生んだ、不朽の名作。第二次大戦中、仏領モロッコのカサブランカにはアメリカへの亡命を望む人々が集まっていた。かつて恋人だったイルザと再会したリックは、彼女とその夫の国外脱出に手を貸す決意をする。
①09:00～10:42 ②13:30～15:12
③16:00～17:42 ④22:15～23:57

7ch『夜霧よ今夜も有難う』
出演／石原 裕次郎　浅丘 ルリ子
93分　1967年
《あらすじ》
名画『カサブランカ』の舞台を1960年代の横浜に置換えた傑作ロマンチックアクション。
①09:00～10:33 ②13:30～15:03
③16:00～17:33 ④22:15～23:48

岸壁イベントのご案内

・観光案内デスク
入港後準備が整い次第～14:00頃

・特産品販売
入港後準備が整い次第～16:30頃
焼酎、海産加工品、郷土銘菓、他の販売

・魚うどんの振舞い
14:00頃～16:30頃

1階～4階のお部屋のお客様へ：靴べら、ランドリーバッグ、洋服ブラシは引き出しの最下段にございます（赤い●シールが目印です）

NIPPON MARU　2018年1月11日(木)【第2号】

【食事・喫茶のご案内】

オーシャンダイニング『春日』Dining Salon KASUGA　6F

07:00 ～ 09:00　朝食【洋食】
08:45 までにお入りください。

11:30 ～ 13:00　昼食【洋食】
12:45 までにお入りください。

18:00 ～ 21:00　夕食【京料理 木乃婦 スペシャルディナー】
18:30 までにお入りください。
※夕食はスイートルーム、デラックスルームのお客様専用となります。

メインダイニング『瑞穂』Main Dining MIZUHO　2F

07:00 ～ 09:00　朝食【和食膳と洋食ビュッフェ】
08:45 までにお入りください。

11:30 ～ 13:00　昼食【和食】
12:45 までにお入りください。

18:00 ～ 21:00　夕食【京料理 木乃婦 スペシャルディナー】
18:30 までにお入りください。
※夕食はステートルームのお客様専用となります。

22:30 ～ 23:30　ナイトスナック（夜食）
23:25 までにお入りください。

寿司バー『潮彩』（有料）　Sushi Bar SHIOSAI　6F

11:30 ～ 13:30　寿司ランチ
お手頃な『ランチにぎり』、『海鮮丼ランチ』（各 2,160 円）をお試しください。

18:30 ～ 24:00
寿司バー　（ラストオーダーは 23:30 です）
新鮮な海の幸と美味しいお酒でおもてなしいたします。
2 ～ 3 貫だけでもお気軽にお立ち寄りください。

体調不良や船酔いでダイニングルームにお越しになれないお客様へ、お粥、おむすびのご用意もございます。ご希望のお客様はダイニングルームのオープン時間内にフロントデスク（内線９２０）へご連絡ください。係がお部屋にお届けいたします。

リドテラス　Lido Terrace　7F

09:00 ～ 22:30　リドグリル
軽食をご用意しております。

《GODIVA『ショコリキサー』》
09:30 ～ 11:30、14:00 ～ 18:00、20:00 ～ 22:00
《アイスクリーム・シャーベット》
13:30 ～ 16:30

09:30 頃、焼きたてのパンがリドテラスに届きますので、どうぞお召し上がりください。
※数に限りがありますので、品切れの際はご容赦願います。**食品の持ち出しはご遠慮ください。**

モーニングコーヒー & ティー　Morning Coffee & Tea

06:30 ～ 07:30
7 階ホライズンラウンジ、6 階ラウンジ『海』
コーヒー、紅茶（にっぽん丸オリジナルティー）、
ジュース、ウーロン茶、
デニッシュ・ペイストリー等をご用意しております。

アフタヌーンティー　Afternoon Tea

14:30 ～ 16:30
7 階ホライズンラウンジ、6 階ラウンジ『海』
コーヒー、紅茶 (日替わり)、お菓子をご用意しております。

無料喫茶　Free Beverage Service

06:30～12:00、14:30～18:00	7 階ホライズンラウンジ
09:00～22:30	7 階リドテラス
06:30～07:30、09:00～18:30	6 階ラウンジ『海』
09:00～12:00、16:00～18:00	5 階eカフェ

コーヒー、紅茶、ジュース、ウーロン茶などをご用意しております。

お客様へ、スペシャルディナーをより美味しく お召し上がりいただく為のお願い

- 料理が冷めないうちにお召し上がりいただく為、ダイニングルームにお入りいただいた順にクルーがご案内するテーブルにご着席願います。窓側テーブルなどの希望はお受けすることができません。また相席をお願いする場合がございますので、何卒ご了承願います。
- 一品ずつ料理をお出ししますので、すべての料理をお出しするまで時間がかかります。
あらかじめご理解をいただきますようお願い申し上げます。

今夜のドレスコードは『セミフォーマル』です。

目安は、おしゃれなレストランやクラシックコンサートへ出かけるようなスタイルです。
※ドレスコードの適用時間帯は、夕食からドルフィンホールでのショープログラム終了時までとさせていただきます。
詳しくは、クルーズガイド 10 ～ 11 ページをご覧ください。

ベーシックなスーツ、ブレザー、サマースーツ、ネクタイ、チーフ、アスコットタイ、ループタイ、革靴（上着、タイはご着用ください）。

スーツ、ワンピース、パンツスーツ、またはブラウスにスカートの組み合わせにパンプスなど

和陽楽
ニューイヤーコンサート

山田 路子
（篠笛・能管）

能楽師一噌流笛方・一噌幸弘氏に師事。『AUN J クラシックオーケストラ』として、世界遺産公演やメディア出演多数。年数回の自主公演や『竹弦囃子』『打花打火』としても活動。

喜羽 美帆
（二十五絃箏）

7歳より箏・二十五絃箏・三絃を師事。生田流地唄箏曲松の実會師範。NHK 邦楽技能者育成会第 42 期卒。受賞歴も多く、和楽器・洋楽器演劇・ダンス等他ジャンルとの表現・作曲活動を展開。

岸 淑香
（ピアノ）

4歳よりピアノ・エレクトーンを始める。ヤマハ音楽院エレクトーン科在学中 Jazz に目覚め、Jazz organ を佐々木昭雄氏、Jazz piano を嶋津健一氏に師事。

高木 てん
（和太鼓）

24 歳の時初めて和太鼓と出会い、その後世界的和太鼓集団『ＴＡＯ』に入団。退団後は 2008 年よりソロ活動やユニット活動を開始し、現在『ZI-PANG』にも所属。

篠笛・二十五絃箏・和太鼓・ピアノの女性４名が織り成す、新春のパワフルステージ。お正月に聞きたいあの名曲や、懐かしい曲、オリジナル曲などを演奏します。情感豊かで躍動感あるステージをどうぞお楽しみください。

場　所／４階ドルフィンホール
時　間／21:00～22:00（開場20:40）
※コンサート終了後に CD・DVD の販売を行ないます。
（現金のみのお取り扱いとなります。）

写真撮影、ビデオ撮影及び録音、グループ等での長時間の席取りはご遠慮ください。　TV3ch ホールの様子を、船内テレビ 3ch でご覧いただけます。

カクテルパーティー　時間／ 20:40 ～ 21:00
コンサート前ににっぽん丸バーテンダーによるオリジナルカクテルをお楽しみください。
※カクテルのご提供はコンサート開始までです。

ビンゴゲームで盛り上がりましょう！！
ビンゴゲーム
BINGO!
場　所／４階ドルフィンホール
時　間／ 09:15 ～ 10:00（開場 09:00）

〔3日目〕

NIPPON MARU　**2018年1月12日(金)【第3号】**

Voy.477　新春初旅 にっぽん丸クルーズ

Port & Starboard

今夜のドレスコード

カジュアル

Dress code for this evening : CASUAL

高級リゾート地を訪れたときのように、決して『ラフになり過ぎない』ことがポイント。カラフルな色使いなど、コーディネートをお楽しみください。詳しくは、クルーズガイド 10～11 ページをご覧ください。

ドレスコードの適用は夕食からドルフィンホールでのショープログラム終了までです。

本日の夕食は『フェアウェルディナー（洋食)』です。

航海スケジュール（予定）

気象・海象の条件により、予定航路や運航スケジュールを変更することがあります。

16:00 頃	岸壁発最終通船 The Last Tender Boat Service from the Wharf
17:00	柳井港沖 出港 Dep. Yanai off.
21:30 頃	来島海峡大橋 Kurushima Bridge
翌 01:30 頃	瀬戸大橋 Seto Bridge

日出没（予定）

07:15 頃　日　出　Sunrise
17:20 頃　日　没　Sunset

《1月12日》
柳井の天気予報
晴れ
降水確率／ 20%
最高気温／　3℃
最低気温／ -4℃

錦帯橋

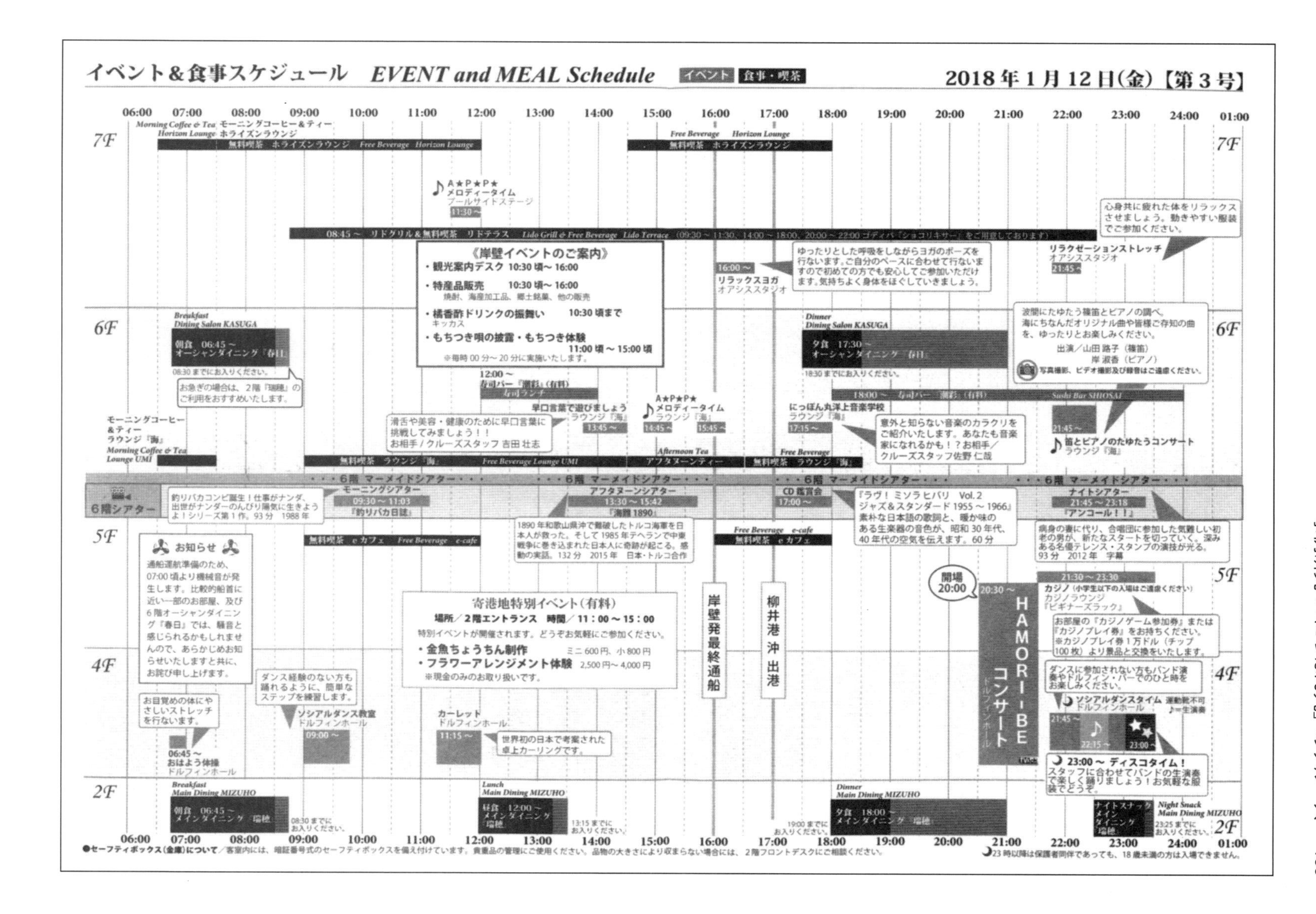
イベント＆食事スケジュール　EVENT and MEAL Schedule
イベント　食事・喫茶
2018年1月12日(金)【第3号】
06:00 07:00 08:00 09:00 10:00 11:00 12:00 13:00 14:00 15:00 16:00 17:00 18:00 19:00 20:00 21:00 22:00 23:00 24:00 01:00
7F
Morning Coffee & Tea モーニングコーヒー＆ティー
Horizon Lounge ホライズンラウンジ
無料喫茶　ホライズンラウンジ　Free Beverage　Horizon Lounge
Free Beverage　Horizon Lounge
無料喫茶　ホライズンラウンジ
A★P★P★
メロディータイム
プールサイドステージ
11:30～
08:45～　リドグリル＆無料喫茶　リドテラス　Lido Grill & Free Beverage　Lido Terrace　(09:30～11:30、14:00～18:00、20:00～22:00 ゴディバ「ショコリキサー」をご用意しております)
《岸壁イベントのご案内》
・観光案内デスク 10:30頃～16:00
・特産品販売　10:30頃～16:00
焼酎、海産加工品、郷土銘菓、他の販売
・橘香酢ドリンクの振舞い　10:30頃まで
キッカス
・もちつき唄の披露・もちつき体験
11:00頃～15:00頃
※毎時00分～20分に実施いたします。
16:00～
リラックスヨガ
オアシススタジオ
ゆったりとした呼吸をしながらヨガのポーズを行ないます。ご自分のペースに合わせて行ないますので初めての方でも安心してご参加いただけます。気持ちよく身体をほぐしていきましょう。
リラクゼーションストレッチ
オアシススタジオ
21:45～
心身共に疲れた体をリラックスさせましょう。動きやすい服装でご参加ください。
6F
Breakfast
Dining Salon KASUGA
朝食　06:45～
オーシャンダイニング『春日』
08:30までにお入りください。
お急ぎの場合は、2階『瑞穂』のご利用をおすすめいたします。
12:00～
寿司バー『潮彩』(有料)
寿司ランチ
早口言葉で遊びましょう
ラウンジ『海』
13:45～
滑舌や美容・健康のために早口言葉に挑戦してみましょう！！
お相手／クルーズスタッフ 吉田 壮志
A★P★P★
メロディータイム
ラウンジ『海』
14:45～　15:45～
Dinner
Dining Salon KASUGA
夕食　17:30～
オーシャンダイニング『春日』
18:30までにお入りください。
18:00～　寿司バー『潮彩』(有料)　Sushi Bar SHIOSAI
にっぽん丸洋上音楽学校
ラウンジ『海』
17:15～
意外と知らない音楽のカラクリをご紹介いたします。あなたも音楽家になれるかも！？お相手／クルーズスタッフ佐野 仁哉
波間にたゆたう篠笛とピアノの調べ。
海にちなんだオリジナル曲や皆様ご存知の曲を、ゆったりとお楽しみください。
出演／山田 路子（篠笛）
岸 淑香（ピアノ）
写真撮影、ビデオ撮影及び録音はご遠慮ください。
21:45～
笛とピアノのたゆたうコンサート
ラウンジ『海』
モーニングコーヒー＆ティー
ラウンジ『海』
Morning Coffee & Tea
Lounge UMI
無料喫茶　ラウンジ『海』　Free Beverage Lounge UMI
Afternoon Tea
アフタヌーンティー
Free Beverage
無料喫茶　ラウンジ『海』
・・・6階 マーメイドシアター・・・
6階シアター
釣りバカコンビ誕生！仕事がナンダ、出世がナンダーのんびり陽気に生きようよ！シリーズ第1作。93分　1988年
モーニングシアター
09:30～11:03
『釣りバカ日誌』
アフタヌーンシアター
13:30～15:42
『海難1890』
1890年和歌山県沖で難破したトルコ海軍を日本人が救った。そして1985年テヘランで中東戦争に巻き込まれた日本人に奇跡が起こる。感動の実話。132分　2015年　日本・トルコ合作
CD鑑賞会
17:00～
『ラヴ！ミソラヒバリ　Vol.2
ジャズ＆スタンダード 1955～1966』
素朴な日本語の歌詞と、暖か味のある生楽器の音色が、昭和30年代、40年代の空気を伝えます。60分
ナイトシアター
21:45～23:18
『アンコール！！』
病身の妻に代り、合唱団に参加した気難しい初老の男が、新たなスタートを切っていく。深みある名優テレンス・スタンプの演技が光る。93分　2012年　字幕
5F
お知らせ
通船運航準備のため、07:00頃より機械音が発生します。比較的船首に近い一部のお部屋、及び6階オーシャンダイニング『春日』では、騒音と感じられるかもしれませんので、あらかじめお知らせいたしますと共に、お詫び申し上げます。
無料喫茶　eカフェ　Free Beverage　e-cafe
Free Beverage　e-cafe
無料喫茶　eカフェ
岸壁発最終通船
柳井港沖出港
開場
20:00
20:30～
HAMORI-BE
コンサート
ドルフィンホール
21:30～23:30
カジノ（小学生以下の入場はご遠慮ください）
カジノラウンジ
『ビギナーズラック』
お部屋の『カジノゲーム参加券』または『カジノプレイ券』をお持ちください。
※カジノプレイ券1万ドル（チップ100枚）より景品と交換をいたします。
4F
寄港地特別イベント（有料）
場所／2階エントランス　時間／11：00～15：00
特別イベントが開催されます。どうぞお気軽にご参加ください。
・金魚ちょうちん制作　ミニ600円、小800円
・フラワーアレンジメント体験　2,500円～4,000円
※現金のみのお取り扱いです。
お目覚めの体にやさしいストレッチを行ないます。
06:45～
おはよう体操
ドルフィンホール
ダンス経験のない方も踊れるように、簡単なステップを練習します。
ソシアルダンス教室
ドルフィンホール
09:00～
カーレット
ドルフィンホール
11:15～
世界初の日本で考案された卓上カーリングです。
ダンスに参加されない方もバンド演奏やドルフィン・バーでのひと時をお楽しみください。
ソシアルダンスタイム　運動靴不可
ドルフィンホール　♪＝生演奏
21:45～　22:15～　23:00～
23:00～ディスコタイム！
スタッフに合わせてバンドの生演奏で楽しく踊りましょう！お気軽な服装でどうぞ。
2F
Breakfast
Main Dining MIZUHO
朝食　06:45～
メインダイニング『瑞穂』
08:30までにお入りください。
Lunch
Main Dining MIZUHO
昼食　12:00～
メインダイニング『瑞穂』
13:15までにお入りください。
Dinner
Main Dining MIZUHO
夕食　18:00～
メインダイニング『瑞穂』
19:00までにお入りください。
ナイトスナック
メインダイニング『瑞穂』
Night Snack
Main Dining MIZUHO
23:25までにお入りください。
●セーフティボックス（金庫）について／客室内には、暗証番号式のセーフティボックスを備え付けています。貴重品の管理にご使用ください。品物の大きさにより収まらない場合には、2階フロントデスクにご相談ください。
23時以降は保護者同伴であっても、18歳未満の方は入場できません。

NIPPON MARU　2018年1月12日(金)【第3号】

営業時間のご案内

※各バーのラストオーダーは閉店時間の30分前です

フロントデスク　Front Desk　2F（内線920）
00:00～24:00　(24時間オープン)
フロントデスク正面カウンターにクルーズパンフレット、新聞をご用意しております。酔い止め薬をご用意しております。(無料)

クリニック　Clinic　1F（内線899）
16:00～17:30
診療時間外は、2階フロントデスク（内線920）へご連絡ください。※**現在服用中のお薬の説明書（お薬手帳）をお持ちください。**

ツアーデスク　Tour Desk　3F（内線930）
19:30～20:30
オプショナルツアー、寄港地の相談窓口です。

クルーズサロン　Cruise Salon　2F
08:00～11:00、15:30～17:00
船内生活などクルーズ全般についてのご要望やご質問にお応えします。また、にっぽん丸クルーズのご予約も承ります。

プール　Swimming Pool　7F
07:00～19:00
ライフガードはおりません。ご利用の際は十分にお気をつけください。お子様がご利用の際は、保護者の付き添いをお願いいたします。

プールに行くときは、水着の上にTシャツ、ショートパンツ等を着用してください（プールに更衣室はありません）。

オアシスジム＆スタジオ　Oasis Gym & Studio　7F
00:00～24:00　(24時間オープン)
トレーナー不在時は、お怪我のないよう十分にお気をつけください。※フィットネス機器のご利用は、05:00～翌01:00までとさせていただきます。

eカフェ＆ライブラリー　e-cafe & Library　5F
00:00～24:00　(24時間オープン)
インターネットをご利用いただけます（有料）。詳しくは、クルーズガイド21ページをご覧ください。

カードルーム　Card Room　5F
00:00～24:00　(24時間オープン)
オセロや麻雀などをご用意しております。(無料)

グランドバス　Grand Bath　3F 船尾部
06:00～12:00、15:00～翌01:00
タオルはグランドバスにご用意しております。
～にっぽん丸のレジオネラ菌対策について～
官庁指導の塩素濃度基準を遵守した塩素殺菌と、ろ過による管理を行なっています。安心してご利用ください。

オアシスジム、プール、スパ＆サロン（7階）、グランドバス（3階）をご利用のお客様へ
酒気を帯びてのご利用は、ご遠慮ください。
また、天候などの事情により船が揺れることがございます。お怪我をなさいませんよう十分にお気をつけください。船の揺れのため、安全上ご利用いただけない場合がございます。

スモーキングラウンジ　Smoking Lounge　2F
00:00～24:00　(24時間オープン)
吸殻やマッチなどは必ず灰皿へ入れてください。火災の予防と喫煙マナーにご協力をお願いいたします。**●客室内は禁煙です。**

スパ＆サロン　Spa & Salon（予約制）　7F（内線874）
15:00～24:00　(ご予約受付は22:00まで)
"美と癒し"の総合サロンです。
スパ・トリートメント「Spa de TERRAKE（スパ・ド・テラケ）」、身体をほぐす「ボディケア」、ヘアサロン「ビューティードレッセヤマノ」、手元、足元を美しくする「ネイルコーナー」がございます。

★営業時間外のご予約も承ります。ご希望のお客様は前日までに内線874へどうぞ。

ブティック『アンカー』　Boutique Anchor　5F
08:00～10:00、15:00～18:00
19:30～22:30
お土産にぴったりなにっぽん丸オリジナルのお菓子類やグッズをはじめ、アクセサリーや衣料品など幅広く取り揃えています。

ショップ『ブイ』　Shop Buoy　5F
08:00～10:00、15:00～18:00
19:30～22:30
にっぽんマルシェを展開しているほか、日用品、飲料、おつまみなどを販売しております。

フォトショップ『ドルフィン』　Photo Shop Dolphin　3F
専属カメラマンが撮影した写真の展示・販売を行なっております。ご購入は3階販売カウンターにございます注文票にご記入ください。

リド・バー　Lido Bar　7F
08:45～23:00（リドグリル／22:30まで）
『リドグリル』メニュー（軽食）は無料でお召し上がりいただけます。

ミッドシップ・バー　Midship Bar　6F
09:00～24:00
寿司バー営業時間中、寿司バー『潮彩』メニューの出前も承ります。

ネプチューン・バー　Neptune Bar　5F
17:00～18:30、20:30～翌01:00
20:30以降は、小学生以下のお子様のご入場はできませんのでご了承ください。

ドルフィン・バー　Dolphin Bar　4F
21:45～23:30
ダンスタイム・カジノの合間など、お気軽にご利用ください。

下船時のお荷物について

詳しくは13日付け船内新聞をご覧ください

宅急便ご利用の方へ（着払い・往復便）

ネームタグ／着払伝票／ガムテープ・荷造り紐の貸出 …………2階フロントデスク
段ボール・スーツケース・ボストンバッグのカバーの販売 …………5階ショップ『ブイ』

13日**07:00～08:00**の間にお部屋の前の通路（部屋扉横）に伝票を添えてお出しください。**08:00**より係員が集荷しますので、直接お渡しいただく必要はありません。
伝票には必ず『**品名**』を詳しくご記入ください。往復便は復路伝票をつけたまま、荷物をお出しください。
※『元払発送』『クール宅急便』のお取扱いはありません。
※配送日時指定をご希望の方は、配送先によりご希望に添えない場合がございます。

テレビシアターのご案内

6ch『雨に唄えば』	7ch『男はつらいよ』
出演／ジーン・ケリー 99分　1952年　字幕	監督／山田洋次 主演／渥美 清　マドンナ／光本幸子 91分　1969年
トーキー映画到来期のハリウッドを舞台に綴られる映画スターの恋…。ジーン・ケリーがどしゃ振りの雨の中で踊るシーンが有名なMGMミュージカルの傑作！	日本中をさすらい続ける永遠のヒーロー、フーテンの寅さんが銀幕初登場！
①09:00～10:39 ②13:30～15:09 ③16:00～17:39 ④21:45～23:24	①09:00～10:31 ②13:30～15:01 ③16:00～17:31 ④21:45～23:16

1階～4階のお部屋のお客様へ:靴べら、ランドリーバッグ、洋服ブラシは引き出しの最下段にございます(赤い●シールが目印です)

【食事・喫茶のご案内】

オーシャンダイニング『春日』Dining Salon KASUGA 6F

06:45 ～ 08:45　朝食【洋食】
08:30 までにお入りください。
ご用意にお時間が かかる場合がございます。
お急ぎのお客様は 2 階メインダイニング『瑞穂』の
ご利用をおすすめいたします。

17:30 ～ 21:00　夕食【フェアウェルディナー（洋食）】
18:30 までにお入りください。
※夕食はスイートルーム、デラックスルームのお客様専用となります。

メインダイニング『瑞穂』Main Dining MIZUHO 2F

06:45 ～ 08:45　朝食【和食膳と洋食ビュッフェ】
08:30 までにお入りください。

12:00 ～ 13:30　昼食【ビュッフェ】
13:15 までにお入りください。

18:00 ～ 21:00　夕食【フェアウェルディナー（洋食）】
19:00 までにお入りください。
※夕食はステートルームのお客様専用となります。

22:30 ～ 23:30　ナイトスナック（夜食）
23:25 までにお入りください。

寿司バー『潮彩』（有料）Sushi Bar SHIOSAI 6F

12:00 ～ 14:00　寿司ランチ
お手頃な『ランチにぎり』、『海鮮丼ランチ』（各 2,160 円）
をお試しください。

18:00 ～ 24:00　寿司バー（ラストオーダーは 23:30 です）
新鮮な海の幸と美味しいお酒でおもてなしいたします。
2 ～ 3 貫だけでもお気軽にお立ち寄りください。

体調不良や船酔いでダイニングルームにお越しになれないお客様へ、お粥、おむすびのご用意もございます。ご希望のお客様はダイニングルームのオープン時間内にフロントデスク（内線９２０）へご連絡ください。係がお部屋にお届けいたします。

リドテラス Lido Terrace 7F

08:45 ～ 22:30　リドグリル
軽食をご用意しております。
《GODIVA『ショコリキサー』》
09:30 ～ 11:30、14:00 ～ 18:00、20:00 ～ 22:00
《アイスクリーム・シャーベット》
13:30 ～ 16:30
09:30 頃、焼きたてのパンがリドテラスに届きますので、
どうぞお召し上がりください。
※数に限りがありますので、品切れの際はご容赦願います。**食品の持ち出しはご遠慮ください。**

モーニングコーヒー & ティー Morning Coffee & Tea

06:30 ～ 07:30
7 階ホライズンラウンジ、6 階ラウンジ『海』
コーヒー、紅茶（にっぽん丸オリジナルティー）、
ジュース、ウーロン茶、
デニッシュ・ペイストリー等をご用意しております。

アフタヌーンティー Afternoon Tea

14:30 ～ 16:30　6 階ラウンジ『海』
コーヒー、紅茶（日替わり）、お菓子をご用意しております。

無料喫茶 Free Beverage Service

06:30～12:00、14:30～18:00　7 階ホライズンラウンジ
08:45～22:30　7 階リドテラス
06:30～07:30、09:00～18:30　6 階ラウンジ『海』
09:00～12:00、16:00～18:00　5 階eカフェ
コーヒー、紅茶、ジュース、ウーロン茶などをご用意しております。

Today's Topic

HAMORI-BE（はもりべ）コンサート

2008 年、由紀さおり・安田祥子姉妹のようなデュエットを目指し活動を開始。その後、安田姉妹に見出され、2011 年より由紀さおり・安田祥子姉妹の童謡コンサートにて共演を重ね、全国に歌声を届けてきた。ユニット名の HAMORI-BE（はもりべ）は、作詞家の故山川啓介氏による命名。
『日本の美しい歌を後世に歌い継ぎたい』という思いのもと、唱歌・童謡や愛唱歌など、その日本語の美しさを大切に歌っている。
『懐かしい、唱歌・童謡の数々を聴きながら、ゆったりと穏やかな時間をお過ごしいただければと思います。』

場　所／4 階ドルフィンホール
時　間／20:30 ～ 21:30（開場 20:00）
出　演／中川 公志（歌）
小原 有貴（歌）
福本 由美（ピアノ）

 写真撮影、ビデオ撮影及び録音、グループ等での長時間の席取りはご遠慮ください。　TV3ch ホールの様子を、船内テレビ 3ch でご覧いただけます。

NIPPON MARU　　2018年1月12日(金)【第3号】

12日は『柳井港（やない）』沖に到着いたします。

《お出かけのお客様へ》

にっぽん丸と柳井港岸壁間は**通船**（小型ボート）で送迎します。準備の都合上、直ぐにご上陸できません。気象･海象状況により、航海スケジュール、通船運航スケジュールが変更される場合がございます。また、安全な運航ができない場合は、船長の判断により通船の運航を中止させていただくこともございます。あらかじめご了承ください。お出かけの際は**乗船証**を必ずご携帯ください。通船乗り場（岸壁）／にっぽん丸船内 で乗船証を係員へご提示ください。

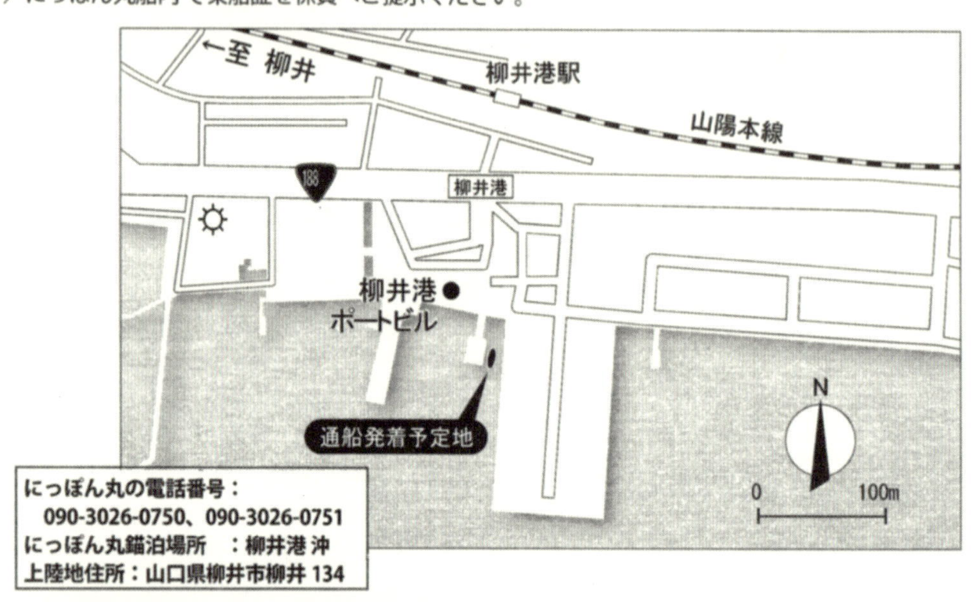

【通船運航時刻表】（所要時間／約 10 分） ＊スケジュールは変更される場合がございます。

行き にっぽん丸 ⇒油津 通船発着所	
9 時	30
10 時	00　30
11 時	00　30
12 時	30
13 時	00　30
14 時	30
15 時	15

帰り 油津 通船発着所⇒ にっぽん丸	
10 時	5　30
11 時	00　30
12 時	00　15　45
13 時	00 ※　30
14 時	00
15 時	00　30
16 時	00（岸壁発最終通船）

※ 船内でご昼食をお召し上がりになるお客様は、13 時 00 分までに岸壁発の通船にお乗りください。

- 通船は定員制です。ご希望の時間にご乗艇いただけない場合がございます。
- 最終通船時間の間近は大変混雑します。余裕を持って早目の帰船便をご利用くださいますようお願いいたします。
- 乗艇時の安全のためにハイヒール、サンダルなど滑りやすい履物はご遠慮ください。
- **車いす では通船に乗艇できません。また、通船の乗艇経路には階段があります。**
  階段の昇降ができない方は、安全上の理由により乗艇をお断りする場合があります。

〔4日目〕下船日

NIPPON MARU　**2018年1月13日(土)【第4号・最終号】**

Voy.477　新春初旅 にっぽん丸クルーズ

Port & Starboard

航海スケジュール（予定）	
気象・海象の条件により、予定航路や運航スケジュールを変更することがあります。	
02:00 頃	瀬戸大橋 Seto Bridge
08:00 頃	明石海峡大橋 Akashi Bridge
10:00 頃	神戸港 入港 Arr. Kobe
日出没（予定）	
07:05 頃	日　出　Sunrise

《1月13日》
神戸の天気予報
曇り時々晴れ

降水確率／ 40%
最高気温／ 4℃
最低気温／ -2℃

We appreciate it!

この度は、にっぽん丸
「新春初旅 にっぽん丸クルーズ」にご乗船
いただき、誠にありがとうございました。
乗組員・スタッフ一同、
またのご乗船を心よりお待ちしております。

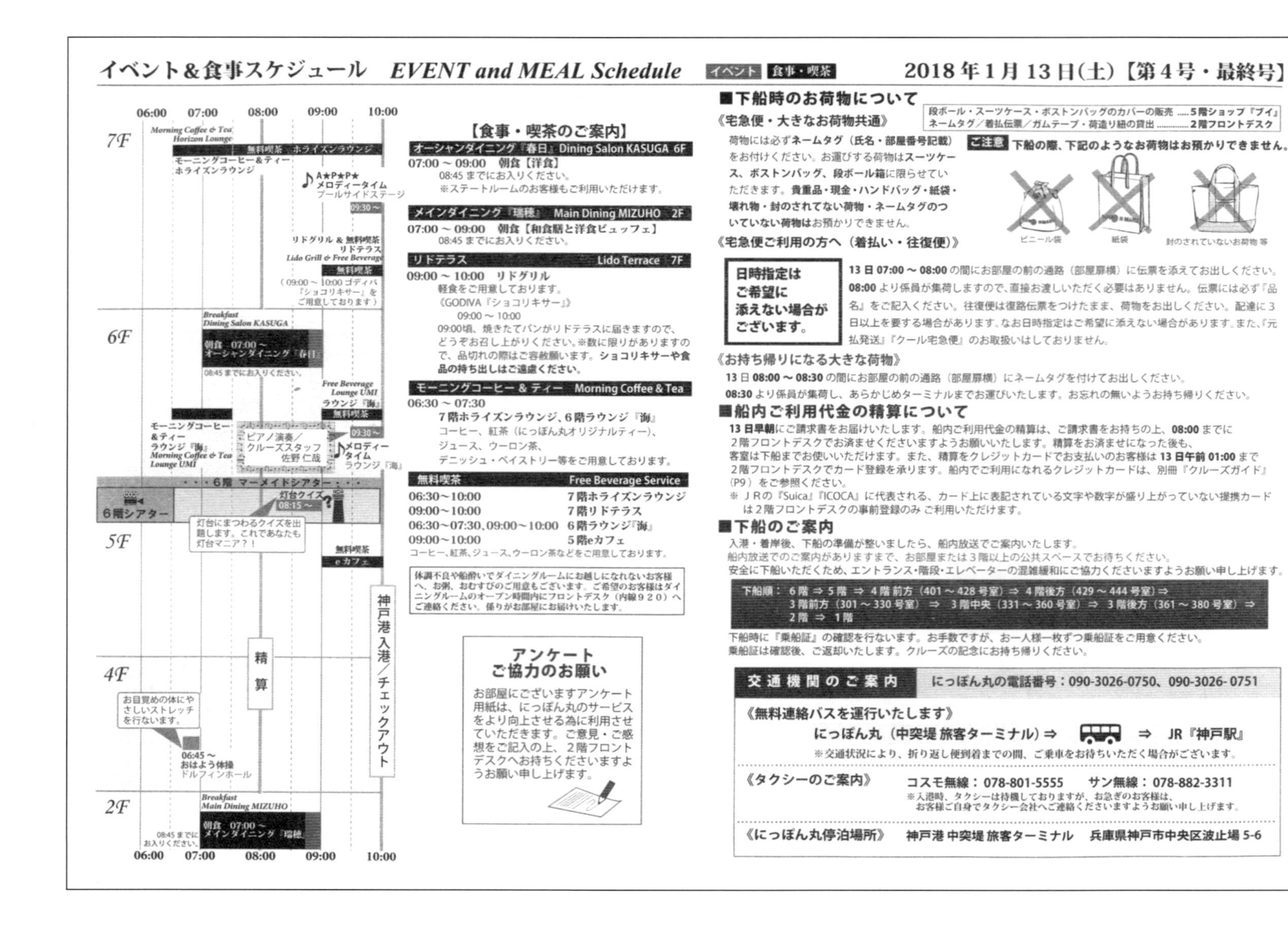

イベント＆食事スケジュール　EVENT and MEAL Schedule

イベント　食事・喫茶

2018年1月13日(土)【第4号・最終号】

06:00　07:00　08:00　09:00　10:00

7F
Morning Coffee & Tea Horizon Lounge
無料喫茶　ホライズンラウンジ
モーニングコーヒー＆ティー　ホライズンラウンジ
♪A★P★P★ メロディータイム　プールサイドステージ　09:30～
リドグリル ＆ 無料喫茶　リドテラス　Lido Grill & Free Beverage
無料喫茶
（09:00～10:00 ゴディバ『ショコリキサー』をご用意しております）

6F
Breakfast Dining Salon KASUGA
朝食　07:00～ オーシャンダイニング『春日』
08:45までにお入りください。
Free Beverage Lounge UMI ラウンジ『海』
無料喫茶
モーニングコーヒー＆ティー　ラウンジ『海』　Morning Coffee & Tea Lounge UMI
ピアノ演奏／クルーズスタッフ　佐野 仁哉
09:30～ ♪メロディータイム　ラウンジ『海』
・・・6階 マーメイドシアター・・・
6階シアター
灯台クイズ 08:15～
灯台にまつわるクイズを出題します。これであなたも灯台マニア？！

5F
無料喫茶　eカフェ
神戸港入港／チェックアウト
精算

4F
お目覚めの体にやさしいストレッチを行ないます。
06:45～ おはよう体操　ドルフィンホール

2F
Breakfast Main Dining MIZUHO
朝食　07:00～ メインダイニング『瑞穂』
08:45までにお入りください。

06:00　07:00　08:00　09:00　10:00

【食事・喫茶のご案内】

オーシャンダイニング『春日』 Dining Salon KASUGA 6F

07:00～09:00　朝食【洋食】
08:45までにお入りください。
※ステートルームのお客様もご利用いただけます。

メインダイニング『瑞穂』 Main Dining MIZUHO 2F

07:00～09:00　朝食【和食膳と洋食ビュッフェ】
08:45までにお入りください。

リドテラス　Lido Terrace 7F

09:00～10:00　リドグリル
軽食をご用意しております。
《GODIVA『ショコリキサー』》
09:00～10:00
09:00頃、焼きたてパンがリドテラスに届きますので、どうぞお召し上がりください。※数に限りがありますので、品切れの際はご容赦願います。**ショコリキサーや食品の持ち出しはご遠慮ください。**

モーニングコーヒー＆ティー　Morning Coffee & Tea

06:30～07:30
7階ホライズンラウンジ、6階ラウンジ『海』
コーヒー、紅茶（にっぽん丸オリジナルティー）、ジュース、ウーロン茶、デニッシュ・ペイストリー等をご用意しております。

無料喫茶　Free Beverage Service

06:30～10:00	7階ホライズンラウンジ
09:00～10:00	7階リドテラス
06:30～07:30、09:00～10:00	6階ラウンジ『海』
09:00～10:00	5階eカフェ

コーヒー、紅茶、ジュース、ウーロン茶などをご用意しております。

体調不良や船酔いでダイニングルームにお越しになれないお客様へ、お粥、おむすびのご用意もございます。ご希望のお客様はダイニングルームのオープン時間内にフロントデスク（内線９２０）へご連絡ください。係りがお部屋にお届けいたします。

アンケート ご協力のお願い

お部屋にございますアンケート用紙は、にっぽん丸のサービスをより向上させる為に利用させていただきます。ご意見・ご感想をご記入の上、２階フロントデスクへお持ちくださいますようお願い申し上げます。

■下船時のお荷物について

段ボール・スーツケース・ボストンバッグのカバーの販売 ..... 5階ショップ『ブイ』
ネームタグ／着払伝票／ガムテープ・荷造り紐の貸出 2階フロントデスク

《宅急便・大きなお荷物共通》

荷物には必ずネームタグ（氏名・部屋番号記載）をお付けください。お運びする荷物は**スーツケース、ボストンバッグ、段ボール箱**に限らせていただきます。**貴重品・現金・ハンドバッグ・紙袋・壊れ物・封のされてない荷物・ネームタグのついていない荷物**はお預かりできません。

ご注意　下船の際、下記のようなお荷物はお預かりできません。

ビニール袋　紙袋　封のされていないお荷物 等

《宅急便ご利用の方へ（着払い・往復便）》

日時指定はご希望に添えない場合がございます。

13日 07:00～08:00 の間にお部屋の前の通路（部屋扉横）に伝票を添えてお出しください。**08:00** より係員が集荷しますので、直接お渡しいただく必要はありません。伝票には必ず『品名』をご記入ください。往復便は復路伝票をつけたまま、荷物をお出しください。配達に3日以上を要する場合があります。なお日時指定はご希望に添えない場合があります。また、『元払発送』『クール宅急便』のお取扱いはしておりません。

《お持ち帰りになる大きな荷物》

13日 **08:00～08:30** の間にお部屋の前の通路（部屋扉横）にネームタグを付けてお出しください。
08:30 より係員が集荷し、あらかじめターミナルまでお運びいたします。お忘れの無いようお持ち帰りください。

■船内ご利用代金の精算について

13日早朝にご請求書をお届けいたします。船内ご利用代金の精算は、ご請求書をお持ちの上、**08:00** までに２階フロントデスクでお済ませくださいますようお願いいたします。精算をお済ませになった後も、客室は下船までお使いいただけます。また、精算をクレジットカードでお支払いのお客様は **13日午前 01:00** まで２階フロントデスクでカード登録を承ります。船内でご利用になれるクレジットカードは、別冊『クルーズガイド』（P9）をご参照ください。
※ ＪＲの『Suica』『ICOCA』に代表される、カード上に表記されている文字や数字が盛り上がっていない提携カードは２階フロントデスクの事前登録のみ ご利用いただけます。

■下船のご案内

入港・着岸後、下船の準備が整いましたら、船内放送でご案内いたします。
船内放送でのご案内がありますまで、お部屋または３階以上の公共スペースでお待ちください。
安全に下船いただくため、エントランス・階段・エレベーターの混雑緩和にご協力くださいますようお願い申し上げます。

下船順：６階 ⇒ ５階 ⇒ ４階前方（401～428号室）⇒ ４階後方（429～444号室）⇒ ３階前方（301～330号室）⇒ ３階中央（331～360号室）⇒ ３階後方（361～380号室）⇒ ２階 ⇒ １階

下船時に『乗船証』の確認を行ないます。お手数ですが、お一人様一枚ずつ乗船証をご用意ください。
乗船証は確認後、ご返却いたします。クルーズの記念にお持ち帰りください。

交通機関のご案内　にっぽん丸の電話番号：090-3026-0750、090-3026-0751

《無料連絡バスを運行いたします》
にっぽん丸（中突堤 旅客ターミナル）⇒ ⇒ JR『神戸駅』
※交通状況により、折り返し便到着までの間、ご乗車をお待ちいただく場合がございます。

《タクシーのご案内》　コスモ無線：078-801-5555　サン無線：078-882-3311
※入港時、タクシーは待機しておりますが、お急ぎのお客様は、お客様ご自身でタクシー会社へご連絡くださいますようお願い申し上げます。

《にっぽん丸停泊場所》　神戸港 中突堤 旅客ターミナル　兵庫県神戸市中央区波止場 5-6

2018年1月13日(土)【第4号・最終号】

営業時間のご案内

※各バーのラストオーダーは閉店時間の30分前です

フロントデスク　Front Desk　2F（内線920）
00:00～24:00　（24時間オープン）
フロントデスク正面カウンターにクルーズパンフレット、新聞をご用意しております。酔い止め薬をご用意しております。（無料）

クリニック　Clinic　1F（内線899）
09:00～10:00
診療時間外は、2階フロントデスク（内線920）へご連絡ください。**※現在服用中のお薬の説明書（お薬手帳）をお持ちください。**

ツアーデスク　Tour Desk　3F（内線930）
08:00～09:00

クルーズサロン　Cruise Salon　2F
08:30～09:30
船内生活などクルーズ全般についてのご要望やご質問にお応えします。また、にっぽん丸クルーズのご予約も承ります。

プール　Swimming Pool　7F
07:00～10:00
ライフガードはおりません。ご利用の際は十分にお気をつけください。お子様がご利用の際は、保護者の付き添いをお願いいたします。

プールに行くときは、水着の上にTシャツ、ショートパンツ等を着用してください（プールに更衣室はありません）。

オアシスジム&スタジオ　Oasis Gym & Studio　7F
00:00～24:00　（24時間オープン）
トレーナー不在時は、お怪我のないよう十分にお気をつけください。※フィットネス機器のご利用は、05:00～翌01:00までとさせていただきます。

カードルーム　Card Room　5F
00:00～24:00　（24時間オープン）
オセロや麻雀などをご用意しております。（無料）

eカフェ&ライブラリー　e-cafe & Library　5F
00:00～24:00　（24時間オープン）
インターネットをご利用いただけます（有料）。詳しくは、クルーズガイド21ページをご覧ください。

グランドバス　Grand Bath　3F船尾部
06:00～10:00
タオルはグランドバスにご用意しております。
～にっぽん丸のレジオネラ菌対策について～
官庁指導の塩素濃度基準を遵守した塩素殺菌と、ろ過による管理を行なっています。安心してご利用ください。

オアシスジム、プール、スパ&サロン（7階）、グランドバス（3階）をご利用のお客様へ
酒気を帯びてのご利用は、ご遠慮ください。
また、天候などの事情により船が揺れることがございます。お怪我をなさいませんよう十分にお気をつけください。船の揺れのため、安全上ご利用いただけない場合がございます。

スモーキングラウンジ　Smoking Lounge　2F
00:00～24:00　（24時間オープン）
吸殻やマッチなどは必ず灰皿へ入れてください。火災の予防と喫煙マナーにご協力をお願いいたします。**●客室内は禁煙です。**

スパ&サロン　Spa & Salon（予約制）　7F（内線874）
本日の営業はございません。

ブティック『アンカー』　Boutique Anchor　5F
08:00～10:00
お土産にぴったりなにっぽん丸オリジナルのお菓子類やグッズをはじめ、アクセサリーや衣料品など幅広く取り揃えています。

ショップ『ブイ』　Shop Buoy　5F
08:00～10:00
にっぽんマルシェを展開しているほか、日用品、飲料、おつまみなどを販売しております。

フォトショップ『ドルフィン』　Photo Shop Dolphin　3F
09:00～10:00
専属カメラマンが撮影した写真の展示・販売を行なっております。

リド・バー　Lido Bar　7F
09:00～10:00
『リドグリル』メニュー（軽食）は無料でお召し上がりいただけます。

ミッドシップ・バー　Midship Bar　6F
09:00～10:00

ネプチューン・バー　Neptune Bar　5F
本日の営業はございません。

ドルフィン・バー　Dolphin Bar　4F
本日の営業はございません。

この度はご利用いただき、誠にありがとうございました。

柳原良平画伯
『宝船』展示

にっぽん丸が皆様にとりましての『宝船』となれることを願い、アンクルトリスでお馴染みの柳原良平画伯の縁起の良い作品を、5階エレベータースペースに展示しております。どうぞご覧ください。

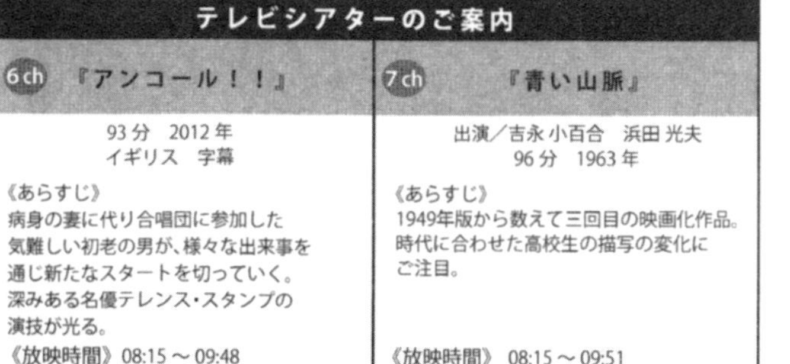

テレビシアターのご案内

6ch　『アンコール！！』	7ch　『青い山脈』
93分　2012年 イギリス　字幕	出演／吉永小百合　浜田光夫 96分　1963年
《あらすじ》 病身の妻に代り合唱団に参加した気難しい初老の男が、様々な出来事を通じ新たなスタートを切っていく。深みある名優テレンス・スタンプの演技が光る。	《あらすじ》 1949年版から数えて三回目の映画化作品。時代に合わせた高校生の描写の変化にご注目。
《放映時間》08:15～09:48	《放映時間》08:15～09:51

5階ショップ『ブイ』では小さな洋上市場"にっぽんマルシェ"を展開中。ぜひ一度お立ち寄りください。

船体の動揺等により、お部屋の出入り口ドアが完全に閉まらない場合がございますので、カチッと音がするまで、しっかりお閉めください。

食事メニュー

2F メインダイニング『瑞穂』　Main Dining “MIZUHO”

朝 食 Breakfast

2017年12月27日（水）　Voy.474　ニューイヤーグアム・サイパンクルーズ

～ 和 食 Washoku ～

じゃが芋と玉葱の味噌汁

金華鯖の火取り

温度玉子

焼き豆腐の土佐煮と茄子の煮物

つるなの胡麻和え

豚肉と根菜の金平

納豆、味付け海苔、香の物

① お粥 ②本日のお粥（十六穀米粥）③ 御飯

御飯またはお粥をお選びください。

日本茶

※食卓の醤油は減塩醤油を使用しております。
※食品衛生上、生卵はお出しする事ができませんのでご了承ください。
※ダイニングルームでお出ししているお米は富山県産こしひかりを使用しております。

～ 洋風ビュッフェ Buffet ～

ベジタブルクリームスープ

カットフルーツ　ドライフルーツシロップ漬け

スパニッシュオムレツ
ボイルドエッグ

鶏肉と野菜の炒め物

野菜のトマトグラタン

グリルドベーコン　グリルドポークソーセージ
グリルドボローニャソーセージ

フレッシュサラダ　ポテトサラダ

チーズ各種　乳製品各種

パン各種　バター、ジャム、マーマレード

シリアル各種、ミルク、ジュース各種
オートミール（お気軽にお申し付けください）

伝統のにっぽん丸ビーフカレー

～ エッグ・ステーション Egg station ～
トッピング
（トマト、チーズ、ミックスハム、オニオン、ツナ）

コーヒー、紅茶、エスプレッソ

朝食のメニュー（和食か洋食が選択できる）

和朝食

昼　食

二階　メインダイニング『瑞穂』

あんかけ焼きそば

花イカ焼売

クラゲと胡瓜の胡麻酢和え

青ザーサイの混ぜ御飯
※白御飯もございます。係にお申し付けください。

ハムと白菜の中華スープ

香の物　桜大根漬け

杏仁豆腐

日本茶
※コーヒー、エスプレッソ、紅茶を
ご希望のお客様は、係にお申し付けください。

※メニューに記載のないお飲み物は係にお申し付けください（有料）。
※ダイニングルームでお出ししているお米は、富山県産コシヒカリを使用しております。

二〇一七年　十二月二十七日(水)
第474次航　ニューイヤーグアム・サイパンクルーズ

6F オーシャンダイニング『春日』　　Dining Salon KASUGA

昼　食　～ **Lunch** ～

2017年12月27日（水）　Voy.474　ニューイヤーグアム・サイパンクルーズ

クラムチャウダー
Soup

味噌カツサンド
Main dish

サラダバー
Salad Bar
レタス　トマト　胡瓜
人参　茸　わわ菜

カットフルーツ
Cut Fruits

～お飲み物～
Beverage

コーヒー（ブレンド、アメリカン、カプチーノ、エスプレッソ）
紅　茶（ディクサム）
Coffee (Blended, American, Cappuccino, Espresso),Tea

メニューに記載のないお飲み物は係にお申し付けください（有料）。

昼食のメニュー（上は和食，下は洋食風の軽食）

ウェルカムディナー　*Welcome Dinner*　　2017年12月27日（水）　Voy.474　ニューイヤーグアム・サイパンクルーズ

2F メインダイニング『瑞穂』　Main Dining MIZUHO

〜 オードブル 〜
燻製の香りをつけた炙り真鯛とホワイトアスパラガス
カリフラワーのソースで
三色大根

〜 スープ 〜
オニオングラタンスープ

〜 魚料理 〜
平目と海老のヴァプール
シャンピニヨンソース
チャービル　たもぎ茸　絹さや

〜 お口直しの氷菓 〜
オレンジのグラニテ

〜 肉料理 〜
牛フィレ肉のパイ包み焼き
エシャロットのコンポート入りマデイラソース
プチベール　ダッチポテト　ミニトマト　イエロー人参　おかひじき

〜 サラダ 〜
クリスタルリーフと渦巻きビーツのサラダ
黒酢玉葱ドレッシング
ベビーリーフ

〜 パン 〜
にっぽん丸特製パン　バター
（オリーブオイルもございます）

〜 デザート 〜
アールスメロンとマスカルポーネのムース
ナッツトゥーユーアイスクリーム

〜 お飲み物 〜
コーヒー、エスプレッソ
今夕の紅茶（ニルギリ）
※メニューに記載のないお飲み物は係にお申し付けください。（有料）

ウェルカムディナー　*Welcome Dinner*　　2017年12月27日（水）　Voy.474　ニューイヤーグアム・サイパンクルーズ

2F メインダイニング『瑞穂』　Main Dining MIZUHO

ヘルシーコース

〜 オードブル 〜
カプレーゼ
小豆島産オリーブオイルで

〜 スープ 〜
オニオンスープ

〜 魚料理 〜
金目鯛のファルシ
牛蒡のソース
チャービル

〜 お口直しの氷菓 〜
オレンジのグラニテ

〜 肉料理 〜
豚フィレ肉のポシェと野菜のスープ煮
ポテト　人参　絹さや　ブロッコリー　蕪

〜 サラダ 〜
クリスタルリーフと渦巻きビーツのサラダ
黒酢玉葱ドレッシング
ベビーリーフ

〜 パン 〜
にっぽん丸特製パン　バター
（オリーブオイルもございます）

〜 デザート 〜
フルーツ盛り合わせ

〜 お飲み物 〜
コーヒー、エスプレッソ
今夕の紅茶（ニルギリ）
※メニューに記載のないお飲み物は係にお申し付けください。（有料）

夕食のメニュー（ウェルカムディナー当日のメニューで，下はヘルシーディナーのもの）

ナイトスナック **Night Snack** *2F メインダイニング『瑞穂』* **Main Dining MIZUHO**

水団汁

きんつば

フルーツコーナー

香の物

日本茶

ウーロン茶、ジュース

コーヒー、紅茶

今夜のおすすめ

ソース焼きそば

２０１７年１２月２７日（水）

２２：３０～２３：３０

※このメニューに記載のないお飲み物は
係にお申し付けください。（有料）

※コーヒー、紅茶のご注文はレストランスタッフにお申し付けください。

夜食のメニュー

メニューの表紙は毎日変わり，乗客の関心を集める。この表紙は，商船三井の前身である大阪商船の南米航路客船「ぶらじる丸」と「あるぜんちな丸」の姿をあしらっている。

事例研究（外国籍船）

「フリーダム・オブ・ザ・シーズ」の1週間西カリブ海クルーズ

この2つ目の事例研究は，海外のクルーズであり，ほぼすべて英文のままの資料を掲載した。辞書を使って十分に内容を理解して，クルーズを楽しんでもらいたい。

マイアミ起点の1週間西カリブ海クルーズである。船は16万総トンの「フリーダム・オブ・ザ・シーズ」（FREEDOM OF THE SEAS）。現代クルーズにおける超大型船の魅力は，なんといっても，船内での「選択の自由度」が極めて大きいことだ。この船の船名の「フリーダム」も，この「自由度」を意味している。

マイアミ港で乗船

マイアミ港のクルーズ客船ターミナルに停泊する「フリーダム・オブ・ザ・シーズ」

セントラルアーケード

アイススケートリンク

カジノ

メイン・ダイニングルームでの誕生祝

船長挨拶

船長主催カクテルパーティにて

寄港地①　コスメル（メキシコ）

コスメル市内で自由時間を楽しむ

コスメルで世界一周途上の「にっぽん丸」と会う

寄港地②　ジョージタウン（イギリス領グランドケイマン島）

ケイマン島沖に停泊するクルーズ客船

海亀の養殖場へのオプショナルツアー

船上の楽しみ

サーフィンプール

幼児用プール

船上のロッククライミングも人気

寄港地③　モンテゴベイ（ジャマイカ）

プランテーションを訪れるオプショナルツアー

レゲエの神様の博物館も訪問

ガラディナー

メイン・ダイニングルームでのガラディナー（特別なディナーのこと）。サービス要員が総出でご挨拶。

寄港地④　ラバディ（ハイチ）

クルーズ客を運ぶテンダーボート

ラバディの沖合に停泊する「フリーダム・オブ・ザ・シーズ」

フェアウェルショー

ミュージカルの上演

日本の代理店が作成した日本語版のパンフレット

2006年6月
いよいよ就航

世界最大客船 158,000トン
ウルトラ・ボイジャーシリーズ

完成イメージ

フリーダム・オブ・ザ・シーズ
（ウルトラ・ボイジャーシリーズ）

2006年6月4日デビュー
総トン数　158,000トン（世界最大）
乗客定員　3,634人
乗組員定員　1,360人
客室数　1,817室
全長　339M　全幅　38,6M
巡航速度　21.6ノット

洋上最大のウォーターパーク登場

海に浮かぶ大型ジャグジー
ソラリウムの両端に位置する海に突き出た展望ジャグジーでは刻々と色を変える景色や海をゆっくり眺めていただけます。

フローライダー
船上初のサーフィン・パーク、フローライダー。幅9.6m、長さ12mの波に乗り、サーフィン、ボギーボードに挑戦してみてはいかがでしょう。

夜のダンスフロアー
スポーツプールの夜はライブミュージックやDJ付のナイトクラブになります。星空を眺めながら、広いダンスフロアーでロマンチックなダンスのひとときを。

ファミリー向け H20 ゾーン
ファミリーのためのファンタジーランド、巨大なオブジェから噴水や水噴射が飛び出る、浅い子供用プールがあります。上階のパノラマブリッジからは大きな滝が流れています。

大人のためのソラリウム
ソラリウムのプール内ではくつろぎの音楽が流れ海底をイメージしたサンゴ礁の映像が映ります。熱帯雨林をイメージした雰囲気の中、ハンモッグでのんびりお過ごしください。夜はダンスフロアーになります。

スポーツプール
メインプールはスポーツプールとしてさまざまなウォータースポーツが楽しめます。シンクロナイズド・スイミング、ウォーターバレー、フローティングゴルフ、水中バスケットなど眺めるだけでも楽しめます。

洋上のアミューズメントパーク

※ ボイジャーシリーズ、ウルトラ・ボイジャーシリーズ共通施設

メインダイニング
3食フルコースメニューのメインダイニングルームは豪華な3階建て。他にビュッフェレストラン、イタリアン、ステーキハウス、ハンバーガーなど種類も船上最大規模です。

プロムナード
船内には130m以上のアーケードがあり、両側には多数のショップ、カフェが並んでいます。ここでは大道芸人が通行人を笑わせ、夜のパレードショーも開催されます。

スケート
マリンスポーツでほてった体を冷やすためアイススケートはいかが。夜にはスケートショーも開催され見ごたえ十分です。プールデッキではロッククライミングもできます。

※写真は全てイメージです。

ロイヤル・カリビアン・インターナショナル日本総代理店　（株）ミキツーリスト　クルーズセンター　TEL 03-5404-8813　www.mikicruise.jp

コース日程表

西カリブ海クルーズ　7泊8日

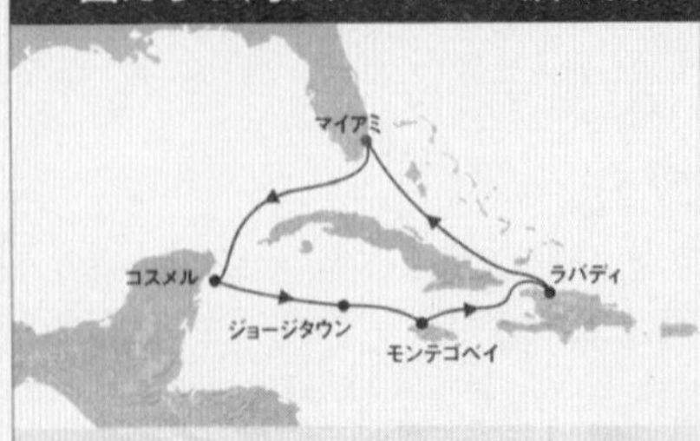

日次	曜日	寄港地		入港	出港
1	日	マイアミ	フロリダ州		17:00
2	月	終日クルージング		—	—
3	火	コスメル	メキシコ	07:00	16:00
4	水	ジョージタウン	グランドケイマン	09:00	17:30
5	木	モンテゴベイ	ジャマイカ	07:00	16:00
6	金	ラバディ	ハイチ	09:00	17:00
7	土	終日クルージング		—	—
8	日	マイアミ	フロリダ州	07:00	

出航日：毎週日曜日（2006年6月4日～）

※デビュークルーズのみ日本人コーディネーターが乗船します。
※各出港日、マイマミ出港時に日本人スタッフによる船内説明会を行います。

西カリブ海クルーズ

世界最大客船で巡るのは数あるコースの中でも一番人気のカリブ海クルーズ
エメラルドグリーンの海と白い砂浜、常夏の楽園のカリブ海へ
船と寄港地とも充実のコースです。

マイアミ（フロリダ州）

全米一の避寒地で、カリブ海や中南米地域と経済・文化的にも深い繋がりを持ち、またカリブ海クルーズのゲートウェイとなっています。

ジョージタウン（グランドケイマン）

ダイビングで有名な美しい諸島グランドケイマン。島のシンボル、アカエイと一緒に泳ぐツアーが人気です。

モンテゴベイ（ジャマイカ）

ジャマイカで一番のリゾート地。植民地時代の豪邸を訪ねたり、近郊には熱帯ジャングルを流れるいかだ下りツアーなどがあります。

コスメル（メキシコ）

世界屈指の透明度の高い海はマリンスポーツの天国。神秘のマヤ文明を偲ぶトゥルムの遺跡も必見です。

ラバディ（ハイチ）

ロイヤル・カリビアンのお客様専用のプライベート半島。民族舞踊、バーベキューランチのサービスがありますので木陰でのんびりお過ごしください。

選べる客室カテゴリー

	カテゴリー	面積（キャビン＋バルコニー）
RS	ロイヤルスイート	110.3㎡＋15.7㎡
PS	プレジデンシャルファミリースイート	111.4㎡＋71.5㎡
OS	オーナーズスイート	47.0㎡＋5.9㎡
FS	ロイヤルファミリースイート	56.6㎡＋21.7㎡
GS	グランドスイート	35.3㎡＋8.8㎡
JS	ジュニアスイート	25.7㎡＋6.4㎡
D1/D2	スーペリアバルコニー	17.4㎡＋4.6㎡
E1/E2	デラックスバルコニー	16.0㎡＋4.3㎡
FO	ファミリー海側	24.6㎡
F	ラージャー海側	19.6㎡
H/I	スタンダード海側	16.7㎡
FI	プロムナードファミリー	27.7㎡
PR	プロムナード	15.5㎡
L/M/N/Q	スタンダード内側	14.8㎡

〈客室内標準設備〉
フラットテレビ、電話、ラジオ、ミニバー、化粧台、ドライヤー、クローゼット
室内温度調節器、アメニティ（タオル、石鹸、リンスインシャンプー）、金庫
110V（日本同様A型）及び220Vの電源

客室イメージ写真

JS ジュニアスイート

E1 E2 デラックスバルコニー

H I スタンダード海側

PR プロムナード

JS ジュニアスイート

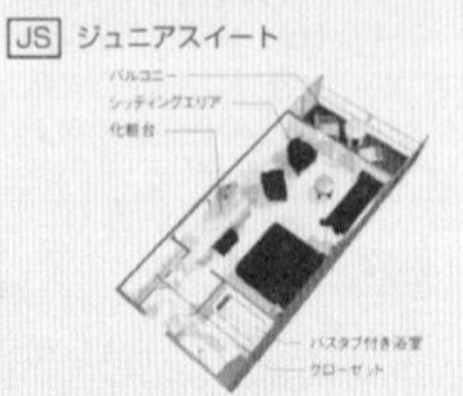

H I スタンダード海側

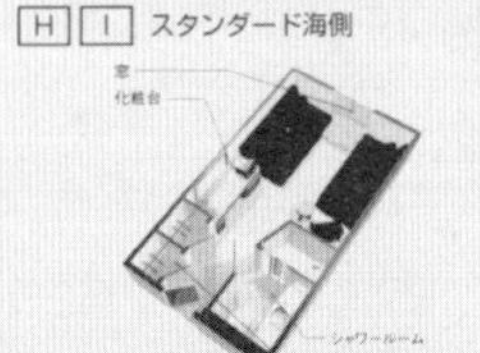

※写真は全てイメージです。

船内で配布されるもの

西カリブ海１週間クルーズで，船内で配布されるものをほぼすべて原文のまま掲載する。海外でのクルーズを楽しむためには，これらの資料に目を通し，理解することが必要となる。ただし，最近は日本語版の資料を用意している船も多くなっている。

総合案内

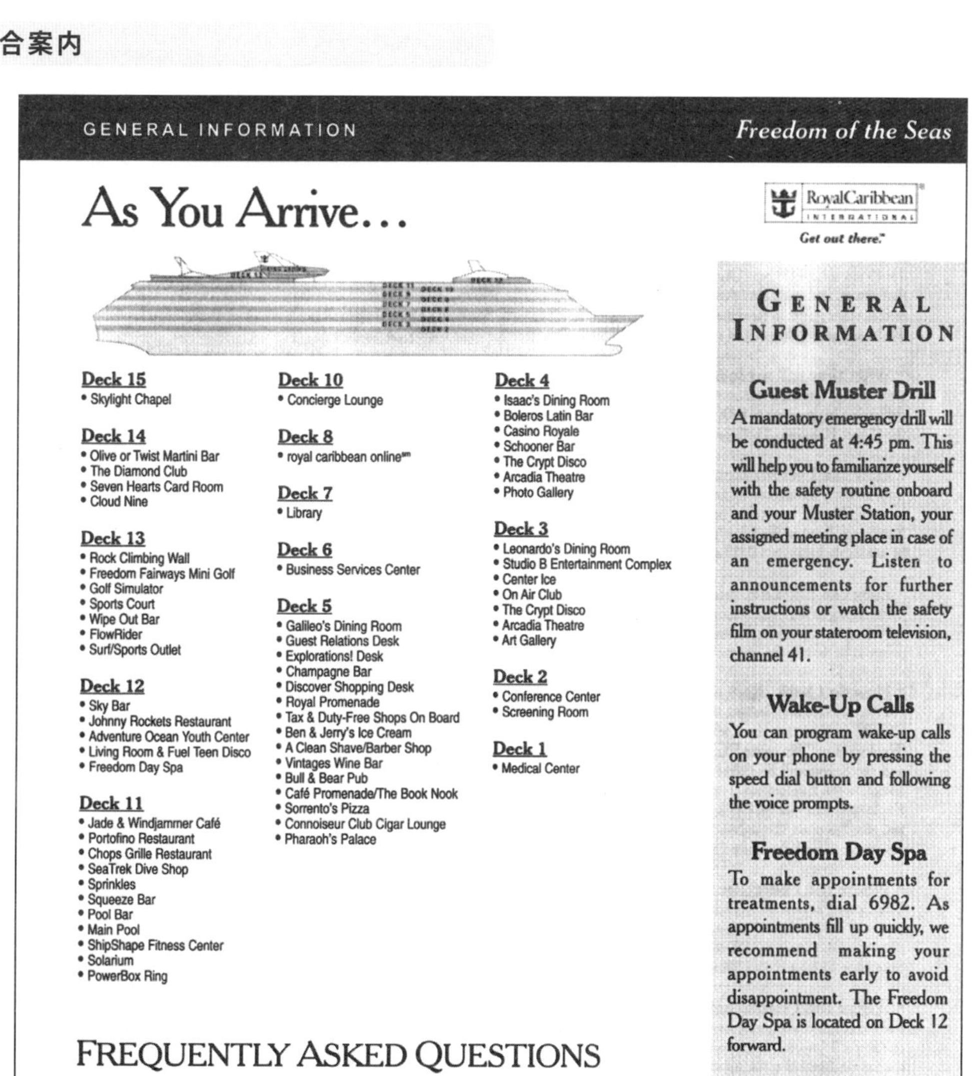

GENERAL INFORMATION　*Freedom of the Seas*

Royal Caribbean INTERNATIONAL
Get out there.

As You Arrive...

Deck 15
- Skylight Chapel

Deck 14
- Olive or Twist Martini Bar
- The Diamond Club
- Seven Hearts Card Room
- Cloud Nine

Deck 13
- Rock Climbing Wall
- Freedom Fairways Mini Golf
- Golf Simulator
- Sports Court
- Wipe Out Bar
- FlowRider
- Surf/Sports Outlet

Deck 12
- Sky Bar
- Johnny Rockets Restaurant
- Adventure Ocean Youth Center
- Living Room & Fuel Teen Disco
- Freedom Day Spa

Deck 11
- Jade & Windjammer Café
- Portofino Restaurant
- Chops Grille Restaurant
- SeaTrek Dive Shop
- Sprinkles
- Squeeze Bar
- Pool Bar
- Main Pool
- ShipShape Fitness Center
- Solarium
- PowerBox Ring

Deck 10
- Concierge Lounge

Deck 8
- royal caribbean online℠

Deck 7
- Library

Deck 6
- Business Services Center

Deck 5
- Galileo's Dining Room
- Guest Relations Desk
- Explorations! Desk
- Champagne Bar
- Discover Shopping Desk
- Royal Promenade
- Tax & Duty-Free Shops On Board
- Ben & Jerry's Ice Cream
- A Clean Shave/Barber Shop
- Vintages Wine Bar
- Bull & Bear Pub
- Café Promenade/The Book Nook
- Sorrento's Pizza
- Connoiseur Club Cigar Lounge
- Pharaoh's Palace

Deck 4
- Isaac's Dining Room
- Boleros Latin Bar
- Casino Royale
- Schooner Bar
- The Crypt Disco
- Arcadia Theatre
- Photo Gallery

Deck 3
- Leonardo's Dining Room
- Studio B Entertainment Complex
- Center Ice
- On Air Club
- The Crypt Disco
- Arcadia Theatre
- Art Gallery

Deck 2
- Conference Center
- Screening Room

Deck 1
- Medical Center

FREQUENTLY ASKED QUESTIONS

• **What is SeaPass and how do I activate it?**
The Freedom of the Seas works on a cashless system. Think of your SeaPass card as your onboard charge card. To activate your SeaPass card, we suggest using a credit card as it allows for automatic checkout at the end of your cruise vacation while also avoiding standing in long lines. Should you wish to establish a cash account, we encourage you to make a cash deposit at the Guest Relations Desk, Deck 5, to avoid having your SeaPass card de-activated should it exceed the pre-set $500 limit. Any remaining cash credits will be refunded to you on the last day of the cruise vacation.

• **When will my stateroom be ready?**
Our previous sailing guests have just departed this morning and we are still in the process of preparing the ship for your arrival. Staterooms will be ready for occupancy at approximately 1:00 pm.

GENERAL INFORMATION

Guest Muster Drill
A mandatory emergency drill will be conducted at 4:45 pm. This will help you to familiarize yourself with the safety routine onboard and your Muster Station, your assigned meeting place in case of an emergency. Listen to announcements for further instructions or watch the safety film on your stateroom television, channel 41.

Wake-Up Calls
You can program wake-up calls on your phone by pressing the speed dial button and following the voice prompts.

Freedom Day Spa
To make appointments for treatments, dial 6982. As appointments fill up quickly, we recommend making your appointments early to avoid disappointment. The Freedom Day Spa is located on Deck 12 forward.

Shops On Board
Our duty-free shops are located on the Royal Promenade on Deck 5 and will remain closed whenever the ship is in port. Once we set sail and have reached international waters, the shops will open. Check the daily opening hours in your Cruise Compass.

• **Where can I get something to eat?**
A delicious lunch buffet is available for all guests in the Windjammer Café, Deck 11, from 11:45 am until 4:00 pm. Once you have finished lunch in the Windjammer Café, we kindly ask that you go and enjoy the other public areas in the ship until your stateroom is available. This will enable other guests to have the opportunity to enjoy lunch as well. Sorrento's Pizza is open from 11:30 am until 3:00 am and the Café Promenade is also open for light snacks 24 hours a day.

• **What time will my luggage arrive to my stateroom?**
Due to the immense amount of luggage handled during the boarding process, we will be delivering luggage up to 8:00 pm. If your luggage has not arrived by this time, please contact the Guest Relations Desk on Deck 5.

• **How can I get my dinner seating changed?**
Your seating arrangements are printed on the front of your SeaPass card. Our Maitre d' will be available for any table or seating change requests between 1:00 pm and 3:45 pm on Sunday in front of the Isaac's Dining Room, Deck 4. On Monday, Dining Room Staff will be available between 10:00 am and 1:30 pm inside the Isaac's Dining Room, Deck 4.

• **How can I find out more about the onboard activities and services?**
Come along to the Freedom Expo! on the Royal Promenade on Deck 5 between 2:30 pm – 4:00 pm, meet our friendly staff and find out more about our onboard services. Over $800 in prizes to be won in the free raffle. Must be present to win!

• **How do I book shore excursions?**
For fast service and instant confirmation, we recommend ordering your tour tickets through RCTV on your stateroom television by pressing the 'menu' key on your remote and selecting 'Explorations!' By using RCTV, you will know immediately if the tour you want is available and the tickets will be delivered to your stateroom within 24 hours. Alternatively, you may complete the Explorations! order form in your stateroom and deposit it in the drop box at the Explorations! Desk on Deck 5 (part of the Guest Relations Desk). The Explorations! Desk will also be open throughout the week to assist you with your ticket purchases and answer questions. Check the daily Cruise Compass for opening hours.

• **Where can I purchase the Soda and Wine & Dine Packages?**
The Soda Package allows adults and children to enjoy unlimited fountain sodas for the week and may be purchased at any of the bars throughout the ship. The Wine & Dine Package allows guests to enjoy a different bottle of wine each evening of the week and may be purchased from your Dining Room Waiter or at Vintages on the Royal Promenade, Deck 5.

• **Where can I smoke onboard?**
For the comfort and enjoyment of our guests, smoking is prohibited onboard in most areas of our ships. However, to provide an onboard environment that also satisfies smokers, we have special designated smoking areas in many of our lounges and on all starboard side open-air decks.

• **What if I need medical treatment?**
The Freedom of the Seas' Medical Facility has two fully licensed doctors and three nurses and is located on Deck 1 aft. The Medical Facility can be contacted by dialing 51 or 911 in the case of an extreme emergency. The daily opening hours are listed in the daily Cruise Compass. Applicable charges are based upon the U.S. Medical Physician Fee Schedule as customarily charged for medical services rendered. We regret that we are unable to accept Medical or insurance assignments.

On behalf of the Captain, Hotel Director, Officers, Staff and Crew, we wish you a wonderful cruise vacation onboard the beautiful Freedom of the Seas.

Photographs

Our Photographers will be taking photographs throughout your cruise vacation from boarding to formal night and much more in between, even when you leave the ship in our port(s)-of-call. Leave it to us, you're under no obligation to purchase, all you need is a smile. All photos will be displayed in the Photo Gallery on Deck 4.

Specialty Restaurants

An intimate dinner awaits you. Enjoy fine Italian cuisine at Portofino or the perfect steak at Chops Grille. Reservations are recommended and a dining fee applies. For Chops reservations, dial 3055. For Portofino reservations, dial 3035.

Wi-Fi

This service allow guests to stay connected on the internet. If you already have wireless access on your laptop, proceed directly to **royal caribbean online**℠, Deck 8, and use one of the work stations to sign up.

Cell Phones

Did you know you can use your cell phone onboard? You can make and receive calls or text messages at sea! While in port or arriving/departing from port you can connect through the local service provider if available.

Ice Show Tickets

Due to seating limitations in Studio B, tickets for the upcoming FREEDOM-ICE.COM shows will be available later in the week. Please consult your Cruise Compass for the specific times and locations that tickets will be handed out. This is the only show you will require tickets for.

寄港地でのオプショナルツアー案内

Freedom of the Seas

June 11, 2006

Seven Night Western Caribbean

Best Way to Order:

1. Order through RCTV for fast service and instant confirmation on your stateroom television by pressing the 'menu' key on your remote and selecting 'Explorations!' By using RCTV, you will know immediately if the tour you want is available and the tickets will be delivered to your stateroom within 24 hours.

2. Choose a tour, circle a time on the order form, write the number of tickets you desire and indicate a second choice. Deposit in the Drop Off Box at Explorations!, Deck 5. Order forms are processed in the order received. Confirmation will be sent to your stateroom within 24 hours.

Book via RCTV – The fastest way to make a reservation. No need to wait in line.

Alternatively:

Purchase directly with *Explorations! Staff* at the Explorations! Desk, Deck 5 and *SeaTrek Divers* at the SeaTrek Dive and Water Activities Shop, Deck 11. Please note that you may have to wait in line.
(Refer to your Cruise Compass for opening hours.)

Name ____________ **Stateroom #** 6690

Signature ____________ **SeaPass #** 44578456

IMPORTANT TIPS AND INFORMATION

Explorations! Desk Hours: See your Compass for desk hours.

Explorations! Talk: Pre-recorded for your convenience and replayed continuously on channel 15. Explorations! information is also available on-demand through RCTV.

Weather Notice: All tours will go rain or shine unless considered unsafe by the tour operators. Bring sun protection.

Physically Challenged Guests: Please consult with our staff before booking a tour.

Exchanges or Cancellations: 24 hours notice required for exchanges or cancellations. No refunds will be given after the deadline.

Child Prices: Available for children 3 to 12 years.

Guests Under 18 years: Must be accompanied by an adult unless otherwise specified. All guests need to carry picture identification with them in all ports-of-call.

Towels: If you are on a water tour please be sure to bring a towel from your stateroom. Towels are not provided on the tour.

DVDs, Videos & Pictures: Purchased independently while on tour or otherwise are not part of Royal Caribbean International and the company can not be held responsible or liable for any problem or occurrence in regards to this.

Tour Duration: Please note that these are approximate times, and may vary.

LEVEL OF DIFFICULTY INDICATORS

MILD: These excursions may require leisurely walking for minimal distances over primarily flat terrain. There might be some steps.

MODERATE: These excursions may require more physical exertion and might involve uneven/steep terrain and/or water activity in a slight current.

STRENUOUS: These excursions are the most active and have been designed for participants in very good physical condition.

Space on tours is limited... so book early to avoid disappointment! All tours are sold on first come, first served basis.

Very Important: Please refer to your tour tickets for exact departure times and locations. Please use discretion.

Cozumel, Mexico – *Tuesday (7:00 am – 4:00 pm)* *Deadline for booking on RCTV is Monday at 3:00 pm*

EXPLORATIONS!	DURATION DEPARTURE (PLEASE REMAIN ON SHIP'S TIME IN THIS PORT)	NUMBER OF TICKETS
Tulum Mayan Ruins CZA1	**Duration: 6½ hrs. Departs: 7:00 am** Arrive at Playa Del Carmen after a 45 minute ferry ride and relax in the comfort of your air conditioned bus as you enjoy a scenic ride to Tulum. On arrival your guide will escort you through the walled Mayan city and share with you the secrets, facts, myths and legends of this long forgotten city. End your tour with a breathtaking view of the beautiful Caribbean sea and enjoy time to shop at the local market before returning on the ferry to the downtown area of Cozumel (**$6 return taxi to the ship at your own expense**). A light lunch and soft drinks are offered complimentary. **Please note: There is a fee of approximately $8–10 at Tulum for guests wishing to bring video cameras. Small cove available for swimming at your own risk.**	**$79 adult** ____ **$69 child** ____
Tulum/Xel-Ha Combination CZB1	**Duration: 8 hrs. Departs: 7:00 am** Excellent combination! Combine a visit to the Tulum Mayan ruins, as described above, with a couple of hours of relaxation at beautiful Xel-Ha. Xel-Ha is a collection of caves, coves, inlets, lagoons and cenotes filled with a combination of saltwater and freshwater. You will have approximately two hours of independent time to swim or snorkel in the cove (snorkel gear is available for rental) or navigate the river, sunbathe, stroll along the beach or walk through the jungle along paths. (Light snacks and complimentary sodas will be served.) At the end of your visit, you will return by bus to Playa del Carmen for the return ferry ride to the ship.	**$109 adult** ____ **$79 child** ____
Deluxe Reef Snorkeling & Beach Combo CZAY1	**Duration: 5 hrs. Departs: 8:15 am** First you will take a boat ride to two of the most beautiful dive sites of Cozumel's National Marine Park: Paraíso and Dzul Ha Reefs. After enjoy a scenic boat ride to Playa Mía beach club. At Playa Mía you'll enjoy a delicious buffet lunch. Afterwards, you will have plenty of time to relax and enjoy the sun or have fun with the many different activities which are included: the iceberg water mountain, kayaks, pedal boats, Hobiecats, floating mats, inflatable tubes, you may also visit the mini zoo or try your bargaining skills at the Mexican souvenir shopping center. After a fun time at Playa Mía , let the good times continue with your open bar and music while you enjoy your boat ride back to the cruise ship pier. **Please note: That this is a drift snorkel, so participants must know how to swim. Water depth ranges from to 10–40 feet. Minimum age 8 years of age.**	**$76 adult** ____ **$62 child** ____
Palancar Reef Snorkeling by Boat CZF1/F2	**Duration: 3½ hrs. Departs: 8:15 am** One of the best snorkel trips! The adventure begins as you board your snorkeling boat for the ride to one of the most popular reefs in Cozumel. This incredible coral formation is located in 15 to 20 feet of water and offers you a unique look at Cozumel's beautiful and unusual marine life. Mask, fins, snorkel and safety vests are provided. Complimentary soft drinks are offered after snorkeling. **Please note: That if weather and/or water conditions do not permit snorkeling at this reef, another site may be substituted closer to shore for the safety of guests. Minimum age 10 years old.**	**$44 each** ____

Beach Escape and Fiesta Cruise **CZI1**	**Duration: 4½ hrs. Departs: 11:15 am** **$45 each ____** Don't miss this lively Mexican Party! Join the Fiesta party boat for a fun-filled trip to Playa Mia, one of Cozumel's beautiful beaches. Unlimited margaritas, rum punch and a live band set the mood for fun. At Playa Mia, enjoy about one and a half hours of relaxing, swimming or working on your tan. The party continues after re-boarding the Fiesta with a pinata party, music and dancing. **Please note: Guests under the age of 21 must be accompanied by an adult.** Best tour for all party people.
Deluxe Beach Break **CZAX1/4**	**Duration: 5 hrs. Departs: 8:30 am / 10:30 am** **$62 adult ____ $54 child ____** Transportation to and from the Beach Club, full domestic open bar, an international buffet and fun attractions will be at your disposal. Upon arrival, you will be pampered by the friendly staff who will let you know of the club's amenities such as beach chairs, beach umbrellas, hammocks, non-motorized water toys like kayaks, Hobiecats, pedal boats, water trampolines, the incredible Water Iceberg and games such as ping pong tables and basketball. Of course, at any time, you'll be served your favorite soft drink or cocktail and when you feel a little hungry you'll enjoy a delicious buffet served under the Palapa. (Outdoor massage service, hair braiding, temporary tattoo parlor and a Mexican souvenir shop and motorized water sports are available at additional cost.)
Island Jeep Off Road & Snorkel Experience **CZN1**	**Duration: 4½ hrs. Departs: 8:00 am** **$92 each ____** Experience a unique visit to Cozumel island. After boarding your 4X4 vehicle you will travel to Pelicanos Beach to snorkel in the clear Caribbean waters. (Must be 8 years old to snorkel) Once you have experienced this natural wonder continue your drive through the different neighborhoods to the east side of the island. After a drive on the main road turn off road to Xpalbarco Beach to enjoy fresh tropical fruits and tasty Mexican snacks. Enjoy the beach, relax and unwind or take a siesta before your drive through town back to your cruise ship. Guests planning to drive must bring a valid driver's license and must be 21 years or over. All vehicles must follow the guide at all times. Not recommended for pregnant women, small children or those with neck or back problems. (Manual drive only). **Tour alteration: This tour will no longer make the full 45 minutes off road drive to Xpalbarco beach. Instead, the tour will go half way, stop for a short break and return along the same off-road. It will then take about a 20 minutes drive along the eastern coastal road to Punta Morena beach for the beach-break/snack portion of the tour.**
Playa Mia Beach Break **CZH1**	**Duration: At your leisure** **$28 adult ____ $18 child ____** Great for families! Your ticket includes: Unlimited domestic open bar, sun chairs, beach umbrella and the use of kayak, pedal boats, Hobiecats, water trampolines, water platforms, floating tubes, volleyball courts, ping-pong tables, hammocks and the incredible "Water Iceberg". Also, guided visits to the mini-Zoo. Available at a nominal fee: Wave runners, Parasailing, Banana boat, the climbing wall, Snorkel and Diving tours and the "Space walker" bungee trampoline. A four passenger taxi ride is $14.00 to $15.00 U.S. one way. Taxis are available at all times at the gate for your return to the pier. **Please note: Transportation and meals are not included. Park opens at 10:00 am.**
Mexican Cuisine Workshop & Tasting **CZAZ1**	**Duration: 5 hrs. Departs: 8:30 am** **$74 each ____** Enjoy a scenic ride to Playa Mia Grand Beach Park. Learn the secrets on how to prepare and serve some of the most delicious recipes from this wonderful culture. A full domestic open bar will be available for you to enjoy at any time. Once food preparation has been completed, all participants will be served their own creations accompanied by fine Mexican wines and a festive Mexican atmosphere. **Please note: Minimum age 18 years old.** Guest may enjoy the beach club after meals are completed. Please bring swimwear and towel.
Xel-Ha Snorkeling & National Aquarium **CZAU1**	**Duration: 8 hrs. Departs: 7:00 am** **$72 adult ____ $58 child ____** Visit the sacred Mayan paradise Xel-Ha. Xel-Ha offers a collection of caves, coves, inlets and lagoons. Rent snorkel equipment or just take a refreshing swim. Walk along the jungle paths or stroll on the beach. Xel-Ha has restrooms, changing facilities, lockers, showers, restaurants and shops. **Please note: Snorkel equipment is not provided but may be rented on arrival.**
Cozumel Off-Road Kart **CZBH1**	**Duration: 3½ hrs. Departs: 11:00 am** **$78 adult ____ $62 child ____** Get off the beaten path, after a short ride you will arrive to the Welcome Station, where your kart awaits. After a brief safety instruction, it's time to strap on your seat belt and helmet and drive this astonishing machine through winding narrow paths across the Mayan jungle towards a hidden beach. Once there you will have time to rest and relax for a while as you enjoy a light snack and refreshments. After your break, it's time to go back to the cruise ship pier. **Please note: Minimum age 16 years old with valid driver's license. Maximum combined weight 330 lbs. (2 per cart)**
Cozumel Power Snorkel **CZVT1**	**Duration: 2¾ hrs. Departs: 10:00 am** **$49 each ____** Come and enjoy the fun and freedom of flying underwater. Safe, fun and environmentally friendly, your hand held power motor will give you a new level of mobility and adventure, in depths between 3 and 20 feet. Enjoy the most spectacular views of coral formations and myriad of tropical fish that stretch to the shore line. After you enjoy this magnificent combination of technology and nature, you may enjoy the beach and relax before you get ready to take your taxi back to the pier. **Please note: Minimum age 12 years old.**
Bike and Snorkel Beach Break **CZAC1**	**Duration: 4 hrs. Departs: 8:00 am** **$62 each ____** Ride through the natural beauty and culture of Cozumel Island and snorkel a breathtaking coral reef. After a 30-minute bike ride you will arrive at a famous Cenote (underground river) named "Aerolito". You will then ride over to beautiful Pelicanos Beach. Cool down in the Caribbean waters with your complimentary snorkel equipment. When you get out of the water it's time to relax at the beach or if you prefer, start a volleyball game with your family and friends! Choose a shady spot under one of the palapas and take a siesta on the full service beach to prepare yourself for the ride back. **Please note: If you plan to swim or snorkel, please wear your bathing suit and bring a towel. A backpack is recommended, as well as sunscreen and adequate sun protection. Minimum age 12 years old.**
Jungle Hike **CZAH1**	**Duration: 3½ hrs. Departs: 7:15 am** **$57 each ____** Hike your way through a Cozumel jungle for a unique adventure and encounter the ancient holy grounds where the Mayans once dwelled thousands of years ago. On arrival at the Welcome Station you will receive an adventure pack complete with a compass, binoculars, granola bar, bottled water and an explorer booklet. Hike through winding, narrow paths that lead into the ancient past of the Mayans. Hike back along the spiraling path, keeping your eyes open for tropical birds and resident iguanas to enjoy a nice cold beverage. **Please note: Closed toe shoes are required. Minimum age 10 years old.**
Climbing Park & Snorkeling Beach Break Adventure **CZAV1**	**Duration: 3 hrs. Departs: 9:00 am** **$69 each ____** Join professional climbing guides for an exciting rock climbing adventure consisting of 6 spectacular state of the art climbing towers. After the climb, enjoy 1½ hours of free time at Playa Bella Vista to enjoy the snorkeling area, restaurant, and white sandy beach. **Please note: Wear comfortable shoes and clothing that allows for flexibility. Minimum age 8 years old.**
Dune Buggy Beach & Snorkel Combo **CZAO1**	**Duration: 4½ hrs. Departs: 8:00 am** **$86 each ____** The experience begins traveling around the island in a customized dune buggy, visiting Cozumel's most beautiful and rugged areas. Enjoy the opportunity to partake in a guided snorkel tour. (Children under 8 years will be unable to snorkel), then jump back into your buggy and head towards the other side of the island. After a brief stop, it's time to continue the caravan to a lovely secluded beach. You will have time to relax in the shade of an umbrella, walk along the white sands, or play games in the warm sun. Leaving the beach, you will head back to town on the west coast of the island. **Please note: Minimum age to drive a vehicle is 21 years old with a valid driver's license. Not recommended for pregnant women or those with back problems. Medical waiver required. Tour alteration: This tour will no longer drive all the way around the island but will drive 20 minutes on the eastern coastal road to Punta Morena beach. Tour will no longer make the photo stop at El Mirador. (Manual transmission only).**
All Terrain Truck **CZAN1**	**Duration: 4 hrs. Departs: 9:00 am** **$54 adult ____ $39 child ____** The journey begins at the pier by boarding our special All Terrain Truck along with your bilingual tour guide who will prepare you for the expedition. Start the experience by crossing between a freshwater lagoon and the ocean where you may see locals such as crocodiles and various types of tropical birds. Next it is on to a Mayan Ruin where we will share with you its magical history and cultural traditions. Continue your experience by crossing through mangroves and low tropical jungle. The next stop is a white, sandy, secluded beach where you will snorkel (must be 8 years old to snorkel). Afterwards, you can relax in the ocean breeze or stroll along the beach. **Please note: Not recommended for pregnant women or those with back problems. Tour alternation: This tour will no longer visit Punta Sur Park but will now include a Mexican snack buffet on the beach. Snorkeling will be done at Uvas Beach.** Beer is served on this tour after snorkeling. **The tour will also take you through San Miguel to the eastern coastal road for approximately 20 minutes to Punta Morena Beach.**

Hideaway Beach Boat Adventure
CZAD1

Duration: 4 hrs. **Departs: 9:15 am** **$94 each ____**
Get away from it all behind the wheel of your very own 2-person speedboat. After a short taxi ride, you arrive at a local beach where the boats await you. You'll cruise along the northern coastline in a sapphire and emerald sea until you come upon a hidden lagoon encircled by tropical mangroves. Dock your boat on the beach and experience its pure white sand and crystal clear waters. Enjoy a Mexican grilled buffet and bask in the sun, swim in the ocean or explore this isolated getaway. After soaking in all the beauties of Hideaway Beach, it's time to get back in your boat for the return trip to the local beach where there will be time to use the facilities to change and freshen up before returning to the ship. **Please note: For your safety, guests will be required to wear their lifejacket at all times while boating. Guests under the age of 18 must be accompanied by an adult. Guests must be 18 years of age or older to drive boats. Minimum age 12 years old.**

Clear Kayak Adventure & Snorkel
CZAS1

Duration: 3 hrs. **Departs: 8:15 am** **$64 adult ____** **$46 child ____**
Come see the Caribbean World from a totally unique view – gliding over the coral reef in your very own Clear Kayak on the ultimate dry snorkel! Your bilingual guide will greet you at the pier and take you to Cozumel's newest beach club – exclusive Uvas Beach. After your kayaking return to the beautiful white sandy beach to relax in the shade or soak up the sun. You will also have the opportunity to revisit the reef to get an even closer look at the marine life in an optional snorkel. **Please note: Due to safety reasons pregnant women are unable to participate. Not recommended for those with back problems. Minimum age 6 years old / to participate in snorkeling minimum of 8 years old.**

Catamaran Sail & Snorkel
CZL1

Duration: 3½ hrs. **Departs: 9:00 am** **$56 adult ____** **$39 child ____**
This luxurious 65-foot catamaran has it all: spacious sun decks, a shaded lounge and a friendly crew that will provide professional instruction on the use of snorkel equipment. Crew members accompany you into the crystal blue waters where you will discover stunning coral formations and a spectacular array of colorful marine life. After snorkeling, unlimited ice-cold beer, zesty margaritas, sodas and purified water are served as you cruise to the fabulous white sand beach. While there, relax in a palapa covered hammock or enjoy the complimentary kayaks, beach floats, volleyball, and of course, the drinks. This beach party provides everything for a perfect day of enjoyment.

Xcaret
Eco-Archeological Park
CZD1

Duration: 8 hrs. **Departs: 7:00 am** **$88 adult ____** **$66 child ____**
The adventure begins after a 45 minute ferry ride from Cozumel to Playa del Carmen, where you'll board buses for a 20-minute drive to Xcaret. After arrival at the park, a guide will give you an orientation of the facilities before you set off to explore the park. Take a Mayan swim, an unbelievable experience floating for 30 minutes on the surface of an underground river. Enjoy a dip in the Caribbean sea, in lagoons, natural pools, channels and cenotes. Browse through a botanical garden and nursery, visit the wild bird-breeding aviary or wonder at exhibits in the Museum of Mayan Archeological Sites. All too soon it will be time to re-board the bus. At Playa del Carmen, you'll board a ferry for a late afternoon return to the ship. **Lunch is not included.**

4x4 Cavern & Beach
CZAR1

Duration: 8 hrs. **Departs: 7:00 am** **$94 adult ____** **$72 child ____**
The adventure begins as you board your 4x4 vehicle, gear up and drive through the town of Playa del Carmen. Your first stop is to explore the mystical cavern. Discover natural formations such as stunning stalagmites and stalactites at every turn. You will also have the unique opportunity to swim in the crystal clear water of the blue "Cenote". After the cavern, you'll head to a charming restaurant on beautiful "Xpu-Ha" beach for a taste of genuine Mexican hospitality with lunch in Paradise. On the endless white sands you can enjoy playing a game of volley ball, relaxing in hammocks, sunbathing in the beach chairs, taking a romantic walk or swimming in the blue Caribbean Sea.
Please note: This tour takes place on the Mexican mainland, and involves taking a 45 minute ferry ride before and after the tour (included in the price). Minimum age to drive a vehicle is 21 years old with a valid driver's license. Not recommended for pregnant women or those with back problems. Children under 8 years will be unable to snorkel.

Muyil Ruins & Sian Kaán Jungle Explorer
CZBA1

Duration: 8 hrs. **Departs: 7:00 am** **$74 adult ____** **$54 child ____**
Explore beautiful Sian Kaán biosphere with its diverse wildlife and jungle paths, which will lead you straight into one of the oldest Mayan trading cities in the area, Muyil. Learn about the various flora and fauna in the region. Visit the Muyil lagoon, full of Mangrove and natural water channels. Once inside the ruins, you will discover the 17-meter high pyramid "El Castillo" and other partially excavated ruins. After the ruins, you will return to Playa del Carmen where your tender awaits to take you back to your ship. **Please note: There is a 45 minute ferry ride across to the mainland and back to the ship included in this trip. Transfer of 1½ hours from ferry point to the ruins.**

Catch the Wave Snorkel Safari
CZV1

Duration: 4 hrs. **Departs: 8:30 am** **$45 adult ____** **$31 child ____**
Whether you want snorkeling, sun or music, this trip has it all. On its three decks, there is an air-conditioned open bar serving unlimited ice-cold beer, tropical cocktails, along with frozen margaritas and daiquiris. There is a snorkel and water activity deck where all your snorkel equipment and any required instruction is provided by the friendly crew. There are sundecks, showers and full restrooms. Also included, is a glass bottom boat, available whilst on your tour on a first come first served basis. Delicious Mexican snacks, bottled water, soft drinks and alcoholic beverages (after completing snorkeling) are included as well as all snorkeling equipment. **Please note: Snorkel location is close to the ship. Minimum age 7 years old.**

Catch the Wave Undersea Bond Adventure
CZVA1

Duration: 4 hrs. **Departs: 8:30 am** **$78 each ____**
The highlight of this tour is a ride on your own underwater motorized scooter! Whether you want snorkeling, sun or music, this trip has it all. On its three decks, there is an air-conditioned open bar serving unlimited ice-cold Corona beer, tropical cocktails, along with frozen margaritas and daiquiris. There is a snorkel and water activity deck. Also included, is a glass bottom boat, available on a first come first serve basis. Delicious Mexican snacks, bottled water, soft drinks and alcoholic beverages (after completing snorkeling) are included as well as all snorkeling equipment. **Please note: Snorkel location is close to the ship. Minimum age 12 years old.**

Deep Sea Fishing
CZAW1

Duration: 4 hrs. **Departs: 7:15 am** **$139 adult ____** **$99 child ____**
Come and try your luck and skill at catching a big Blue Marlin, white Marlin, Sailfish, Tuna, Sierra, Dorado, Wahoo and Barracuda. Bring your enthusiasm and leave the rest to our Captain and crew to catch your prize. **All fish are catch and release.** You'll be briefed prior to departure regarding sea conditions and chair rotation. All the boats are rigged with downriggers and outriggers and all lines will be rotated throughout the tour. Your trip includes beverages (beer, soda or water), a snack, rods and reels, bait tackle. **Please note: Minimum age 8 years old.**

Cozumel Semi-Submersible
CZJ1

Duration: 1¼ hrs. **Departs: 11:00 am** **$46 adult ____** **$35 child ____**
Travel through the sea in a unique craft featuring seats that sit below water level with glass panels for viewing. Within minutes of leaving the dock, you will cruise over a sunken airplane, originally placed there during the filming of a movie. Then cruise over Paraiso Reef, where you will see colorful coral formations and fish through the glass panels in front of you. Don't miss this great opportunity to see Cozumel's underwater world without getting wet.

Atlantis Submarine
CZK1

Duration: 1¾ hrs. **Departs: 8:15 am** **$94 adult ____** **$58 child ____**
Experience why Cozumel is one of the top five dive destinations in the world. After a short ferry ride, board the high-tech, air-conditioned Atlantis submarine. At a depth of approximately 90 feet, glimpse the unique underwater paradise and marvel at the huge variety of tropical fish. **Please note: Guests must be at least 3 feet tall to participate. This is a 40-minute dive. Minimum age 4 years old.**

Horseback Riding Adventure
CZM1

Duration: 3½ hrs. **Departs: 8:15 am** **$89 each ____**
After a 30-minute drive to the southeast side of the island, you will arrive at Buena Vista Ranch. Here, horses are assigned and instructions are given. Then, with your guide, follow the trail through the jungle. Listen to the guide's explanation of Mayan history and customs while passing ruins and caves dating back 1,500 years. The last stop will be at the Cantina for a soft drink or beer. Happy trails! **Please note: Pregnant women are not able to participate. Long pants are recommended. Minimum age 12 years old. Maximum weight 220 lbs.**

Cozumel Country Club
CZZ1

Duration: 5 hrs. **Departs: 7:15 am** **$155 each ____**
We are proud to introduce Cozumel's 18-hole Championship Golf Course under management by ClubCorp International. The course totals 6,816 yards from the back tees. The course is built around Cozumel's native trees, red Mangroves and natural wetlands in order to preserve ecologically sensitive areas. Note: Price includes green fees, transportation and shared golf cart. Club rental $29.00 per person; shoe rental $14.00 per person - Prices are not part of the golf package and are subject to change at anytime. Payable at the golf course only. 18 holes/Par 72/6,816 yards/10 minute from port.

Mexican Folkloric Show
CZAA1

Duration: 2½ hrs. **Departs: 11:30 am** **$46 each ____**
The history, culture, and Mexican zest for life is explored through music and dance in this wonderful folkloric show. Experience the variety and color of the diverse types of dance and music expressing aspects of Mexican civilization which originated from the Indians, and combined with Spanish influence, evolved into their own unique and wonderful traditions.

Jungle ATV Adventure
CZBD1

Duration: 3½ hrs. **Departs: 7:15 am** **$89 adult ____**
Trek your way through the Cozumel jungle for a unique ATV adventure and encounter the ancient holy grounds where the Mayans once dwelled thousands of years ago. Your bus will arrive at the Welcome Station where you will find your All Terrain Vehicle and gear (helmets and goggles). You will arrive at the ancient Mayan burial grounds. Glance upon two original temples that once held the bodies of Mayan royalty. The ATV tour will end back at the Welcome Station where you will have a chance to use the rest rooms and enjoy a nice cold beverage.
Please note: Minimum age 18 years old. Maximum weight is 364 lbs. Must bring a valid driver's license. Wear closed toe shoes.

Tequila Tasting Seminar CZBC1	**Duration: 3¼ hrs. Departs: 11:00 am** **$82 each** ____	This is your chance to get a real taste of the history of Mexico. Delight yourself with a sampling of the very best in tequilas. Enjoy tasty Mexican snacks and as long as you moderate your sampling, you will return an expert! **Please note: Minimum age 18 years old.**
Temazcal Steamlodge Experience CZAF1	**Duration: 3 hrs. Departs: 7:30 am** **$98 each** ____	As one of the most significant ancient Mexican traditions, the Temazcal comes to the modern times as a relaxation therapy for the nervous system and reactivation of the skin cells. You will be led through the various phases of a process once reserved only for the elite classes of Mayan society. Once completed, you will then enter the Temazcal (Aztec for the house of steam) where you will experience the 4 phases of cleansing and the 5 point visualizations of your journey. You will emerge after approximately one hour feeling rejuvenated. Choose to shower, or rinse in the fresh water cenote before relaxing in a hammock to enjoy fresh fruits and juices. After 30 minutes, the option is yours if you wish to explore some of the nearby ruins before receiving your departing gift. (Minimum age: 15 years old. Medical waiver required and good physical health.)
Jungle Bike Adventure CZAM1	**Duration: 3½ hrs. Departs: 7:15 am** **$68 each** ____	Enjoy a unique guided biking adventure and visit ancient Mayan grounds. After being fitted with your bike and helmet, follow your guide along winding, narrow paths that lead into the ancient Mayan burial grounds that have not been excavated and are still in their natural state.
Cozumel Bike & Sail CZAK1	**Duration: 3 hrs. Departs: 12:30 pm** **$78 each** ____	Discover Cozumel by land and sea. Once fitted with your bicycle and safety gear enjoy an invigorating ride along the north hotel zone of Cozumel until you reach Barracuda Beach. Here, awaits your Hobie vessel. After receiving sailing instructions you will set sail and enjoy a refreshing swim. **Please note: Minimum age 10 years old. Minimum height 4 feet.**
Cozumel Island & Playa Mia Beach CZG1	**Duration: 4 hrs. Departs: 8:00 am** **$52 adult** ____ **$39 child** ____	After a short ride to the San Gervasio ruins, learn about the civilization and history of the Mayan culture. Back aboard your motor coach, continue your tour past white sand beaches and blue Caribbean waters on your way to Playa Mia. After a welcome drink, walk on the sandy beaches, relax in a hammock, play volleyball or try the water trampoline, kayaks or climb the water iceburg. Motorized watersports are available at an additional cost.
Unlimited Beach Snorkel CZU1	**Duration: 2 hrs. Departs: 8:15 am** **$29 each** ____	Sunset Beach has something for everyone. The depth varies from 2-25 feet with unlimited time to enjoy the beach, sun and undersea world. **Please note: Showers, restaurants/snack bar, bathrooms and lockers are available at an extra cost. Minimum age 6 years old.**
Dolphin Encounter in Puerto Aventuras CZBF1	**Duration: 8 hrs. Departs: 7:00 am** **$109 each** ____	Travel to the mainland of Mexico (Playa del Carmen) where a bus awaits you for a 30-minute drive to Dolphin Discovery. The Dolphin Encounter consists of: kiss, tail walk, song, bow jumps and spray! Enjoy time to relax on the beach or take in a complimentary snorkel tour. The Puerto Aventuras Marina also offers side walk cafés, restaurants, bars and few souvenir stores. Return back to the ship in time for sailing. **Please note: The actual Dolphin Program lasts approximately 30 minutes. This is a full day excursion. Minimum age 3 years old.**
Dolphin Swim in Puerto Aventuras CZBG1	**Duration: 8 hrs. Departs: 7:00 am** **$145 each** ____	Travel to the mainland of Mexico (Playa del Carmen) where a bus awaits you for a 30-minute drive to Dolphin Discovery. The Dolphin Swim consists of: a dolphin handshake, kiss, belly ride and at the end, enjoy moments of great spontaneity during your free time together! Guests will also have free time to enjoy Discovery beach and snorkeling at their leisure. Return back to the ship in time for sailing! **Please note: The actual Dolphin Program lasts approximately 30 minutes.** This is a full day excursion. **Minimum age 8 years old.**
Cozumel SNUBA Adventure CZQ1	**Duration: 2 hrs. Departs: 8:15 am** **$69 adult** ____ **$52 child** ____	After a short drive to a beach you'll have an orientation in Snuba. With your PADI certified dive master, you'll then enter the water from the beach where you'll become familiar and comfortable with the breathing equipment. Then enjoy approximately 25 minutes under the crystal-clear Mayan Caribbean waters exploring the beautiful Dzul-Ha reef. At the conclusion of your water experience, transportation will be available to take you back to the ship. **Please note: The Minimum age for this tour is 8 years old. Participants must be in good health. This tour is not available for women who are more than three months pregnant, participants will be required to complete and sign a waiver.**
Segway Adventure CZW1/CZW2	**Duration: 2½ hrs. Departs: 8:30 am / 12:15 pm** **$59 each** ____	After a safety briefing, with your Segway follow the guide on a scenic bike track that parallels the Caribbean Sea to Uva's Beach. Your guide will explain all the activities that are available at the beach such as: snorkeling, kayaking or simply relaxing on the beach. The tour provides guided snorkeling tour (gear included). You may stay and enjoy for as long as you like, a taxi coupon will be provided so you can return to the ship. **Please note: Minimum age: 12 years/maximum 65 years old. Minimum weight 100 lbs/maximum 250 lbs.**
Passion Island Escape CZO1	**Duration: 5 hrs. Departs: 8:15 am** **$69 adult** ____ **$35 child** ____	Our guide will offer a brief overview of Cozumel and Passion Island on the 35-minute air-conditioned bus ride to the pier . Here you may choose to cross to Passion Island on a four person canoe (15–20 minutes) or a motorized tender ferry (10 min.) In Passion Island you will find tropical palm trees, pine trees, hammocks, lounge beds, picnic tables, a variety of beach games including beach volleyball, beautiful shallow water and a gorgeous white sandy beach you will never forget. Buffet style lunch and drinks included.
Cozumel Highlights & Shopping CZBB1	**Duration: 4 hrs. Departs: 8:00 am** **$45 adult** ____ **$34 child** ____	Your journey begins as you board your air-conditioned bus at the ship's pier. Passing by the southern resort area of the island, soon you will come to the town of Cedral. Next, you will cruise the wilder side of the island for a look at the untamed beaches of Cozumel. Afterwards, you'll turn inland once again to experience the flavor of island life in the Caribbean in some residential areas. Upon arrival in the downtown waterfront, you will get an up close sight of living examples of the islands rich history and culture in Cozumels Museum. You will have the option to take your own tour of the Museums exhibits or to explore downtown and practice the art of shopping. Your guide will be nearby to answer any questions. After some time for shopping, you will be given the option to return to your ship aboard the bus or stay and return by taxi at your expense.
Helmet Dive CZT1	**Duration: 2 hrs. Departs: 8:15 am** **$92 adult** ____ **$78 child** ____	Explore the underwater world 20 feet below wearing a comfortable space age helmet. (Great for nonswimmers). **Please note: Minimum age 8 years old. Medical waiver required. (25 minute dive)**
Mini Jeep Outlander and Snorkel CZBE1	**Duration: 5 hrs. Departs: 8:00 am** **$98 each** ____	Ride the exciting and powerful Yamaha Rhino and explore the untouched! Head off to watch salt water crocodiles, visit an antique Mayan ruin and Punta Celarain Lighthouse. Follow your guide through the tropical jungle road, visit secret sand paths in Cozumel island and arrive to our private bay, to snorkel (snorkeling gear is included). **Please note: Guests must be 18 years old with a valid driver's license to drive. Pregnant women and people with back/neck injuries or recent surgery are not recommended to participate on the tour. People with diabetes, heart conditions and/or asthma must abstain snorkeling. Minimum age for snorkeling 8 years old.**

SECOND CHOICE FOR COZUMEL **Tour Name** **Time**

WE STRONGLY RECOMMEND THAT YOU INDICATE A SECOND CHOICE, AS TOURS ARE LIMITED AND DO SELL OUT QUICKLY!

Georgetown, Grand Cayman – *Wednesday (9:00 am – 5:30 pm) Deadline for booking on RCTV is Tuesday at 3:00 pm*

EXPLORATIONS!	DURATION DEPARTURE (PLEASE REMAIN ON SHIP'S TIME IN THIS PORT) NUMBER OF TICKETS	
Rays & Reef Combo GC01/GC02	**Duration: 3¼ hrs. Departs: 9:00 am / 12:30 pm** **$66 adult** ____ **$48 child** ____	This is a combination of two of the most popular excursions in the Caymans. Stop at the Coral Gardens, to go snorkeling at one of the most beautiful sites in the North Sound. Mask, fins, snorkel and floatation devices are provided. Depart the Coral Gardens and its on to the famous deep Stingray City site. Enter the water and swim or snorkel with these wondrous creatures. **Please note: That the average water depth is approximately 10–12 feet. Minimum age 5 years old. Guests 5–16 years of age must be accompanied by an adult.**

Rays & Reef Beach Combo with Lunch **GCW1**	**Duration: 6 hrs.** **Departs: 9:00 am** This tour has it all. Stop at the famous deep Stingray City site. Enter the water and swim or snorkel with these wondrous creatures. Stop at Coral Gardens to go snorkeling at one of the most beautiful sites in North Sound. Mask, fins, snorkel and floatation device are provided. Board a bus for a short transfer to the "Beach Club" Colony Resort, located on famous Seven Mile Beach. For lunch, there will be ample time to swim and relax before returning to the ship. Your tour includes transportation, snorkeling equipment, beverages, lunch and chaise lounge. **Please note: That the average water depth is approximately 10–12 feet. Minimum age 5 years old. Guest 5–16 years of age must be accompanied by an adult.**	**$94 adult ____ $68 child ____**
Land and Sea **GCK1**	**Duration: 3½ hrs.** **Departs: 10:00 am** A popular tour choice! Discover the wonders of Grand Cayman above and below the sea. Your trip in the air-conditioned Nautilus semi-submarine allows you to see the shipwreck and an abundance of marine life at Cheeseburger Reef. Back on land you will enjoy a visit to the Turtle Farm and Rum Cake Center as well as a stop in the "City of Hell" and the Governors House on 7 mile Beach.	**$62 each ____**
Stingray City Snorkeling **GCR1/GCR2**	**Duration: 3 hrs.** **Departs: 9:15 am / 1:00 pm** Swim and snorkel with the amazing stingrays! After a short bus ride, our boat will take you out to the famous sandbar shallow area where you will spend 45 minutes snorkeling and feeding the stingrays. This is a good tour choice if you want to shop or go to Seven Mile Beach after completing the tour. (Snorkel equipment is provided). No tropical fish or coral will be seen.	**$45 each ____**
Cockatoo Catamaran **GCN1**	**Duration: 3 hrs.** **Departs: 12:30 pm** Straight from the Cayman Island Yacht Club comes your 60 feet racing catamaran Cockatoo. Sail the waters of North Sound before stopping at world famous Stingray City for a swim with these incredible creatures. One of the best ways to swim with stingrays. **Please note: Minimum age 5 years old.**	**$59 each ____**
Aquaboats & Snorkel Adventure **GCAA1**	**Duration: 3 hrs.** **Departs: 1:00 pm** Enjoy the fun as you ride in your own 2-person, speedy, 12' inflatable motorboat! Skippering your own boat, you will arrive at Smith's Cove to swim and snorkel and then head north again, stopping at 'Cali' for a chance to see the ancient shipwreck! All equipment included. **Please note: Minimum age 10 years old. Must be over 18 to drive. Pregnant women may not participate.**	**$84 each ____**
Nautilus Undersea Tour & Reef Snorkel **GCL1**	**Duration: 1¾ hrs.** **Departs: 1:45 pm** Not only will you see the abundance of tropical fish as your semi-sub glides over shipwrecks and the famous Cheeseburger Reef, but you can snorkel right from the semi-sub if you wish! Feed the fish and keep an eye out for the amazing marine life. Great for all ages! Wear swimsuit under clothes if you wish to snorkel.	**$48 adult ____ $36 child ____**
Butterfly Farm & Nautilus Sub Sea Viewer **GCAC1**	**Duration: 2½ hrs.** **Departs: 11:45 am** After a short bus transfer, you will arrive at Grand Cayman's Butterfly Farm for an unforgettable encounter. Step into a tropical garden teeming with butterflies from around the globe and prepare to meet some of the world's most colorful creatures. Then it's on to the Nautilus, the most luxurious semi-submarine in the world, to see the spectacular undersea world that surrounds us. At the marine paradise of Cheeseburger Reef, watch in awe as our divers hand-feed thousands of tropical fish right at your window.	**$58 adult ____ $42 child ____**
Stingray City by Helicopter **GCAN1/AN2/AN3/AN4**	**Duration: 1 hr** **Departs: 10:30 am / 11:30 am / 2:00 pm / 3:00 pm** Departing the pier, take a quick trip by air-conditioned minibus to the airport. The tour takes you on a low flight over the beautiful turquoise waters to the reef. Hover over Stingray City. View not only the famous drop off wall - also the famous Seven Mile Beach. Next, view the grandeur of your cruise ship against the beautiful backdrop of George Town. Once you land, re-board your minibus for the short transfer back to the pier. **Please note: Flight time is approximately 15–20 minutes. The order of sights visited may vary. Participation on this tour is very limited. Maximum weight 300 pounds.**	**$139 adult ____ $124 child ____**
Seven Mile Beach Blast By Helicopter **GCF1/GCF2/GCF3**	**Duration: 45 min** **Departs: 1:00 pm / 1:15 pm / 1:30 pm** Departing from the pier, arrive by minibus to the airport to begin your adventure. You will take a low flight over the beautiful turquoise waters of the famous 7 mile beach. View the famous drop-off wall that makes diving in the Caymans famous before hovering over your cruise ship. Flying over Georgetown, the tour concludes where your minibus will transfer back to the pier. **Please note: Actual flight 7-10 minutes. Maximum weight 300 lbs.**	**$109 adult ____ $94 child ____**
Stingray City & Island Tour **GCB1/GCB2**	**Duration: 3½ hrs.** **Departs: 10:30 am / 1:30 pm** Combines viewing world-famous Stingray City through the windows of a semi-submersible, the Turtle Farm plus a tour to the city of Hell to view the unusual rock formations. You can even mail a postcard from Hell if you want!	**$59 adult ____ $36 child ____**
Grand Cayman Snorkeling **GCC1**	**Duration: 2½ hrs.** **Departs: 10:30 am** One of the finest Caribbean underwater experiences around. Enjoy snorkeling at both a shipwreck and a coral reef. **Please note: Minimum age 5 years old.**	**$34 each ____**
Reef & Wreck Snorkel **GCX1**	**Duration: 2½ hrs.** **Departs: 1:00 pm** Enjoy a short scenic cruise out to two snorkeling locations. The first is a majestic reef system, where you will have a perfect opportunity to swim with multitudes of colorful and tropical fish and marine life. The adventure continues with a snorkel stop over a shallow ship wreck, providing an up close encounter with what has become a superb artificial reef system in growth. Shade and snorkel equipment included. **Please note: Minimum age 5 years old.**	**$28 each ____**
Kayak & Snorkel Adventure **GCAE1**	**Duration: 4 hrs.** **Departs: 9:00 am** Good fun and adventure. Enjoy a little snorkeling and a leisurely kayak ride to Seven Mile Beach. These sit-on-top 2-person kayaks are fun and easy to operate. Spend as much time as you want at the resort or take the shuttle back. If you stay you will need your own transportation back to the ship. **Please note: Minimum age 10 years old.**	**$62 adult ____ $48 child ____**
Cayman 4x4 Safari **GCAB1**	**Duration: 5½ hrs.** **Departs: 10:30 am** After a brief orientation you will depart caravan style to explore Caymans natural mangroves. En route you will see the habitats of the Blue Iguanas and take in the beauty of the North Sound. You will have a refreshment stop at Spanish Bay Reef Resort for time to snorkel, swim or relax. Your convoy will reach its conclusion at Beach Club Colony located on Seven-mile beach. You will be returned to the pier via motor coach. **Please note 4 passengers per jeep. This is not an off road adventure. Minimum age to drive: 18 years old with a valid driver license. Lunch is not included. Order of stops may vary.**	**$67 each ____**
Grand Cayman SNUBA Adventure **GCAH1**	**Duration: 2 hrs.** **Departs: 12:30 pm** At last, there is a way to enjoy the beauty of the Caribbean waters without having to be a certified SCUBA Diver. After being transported to the SNUBA facility by bus, you will receive a full orientation and safety briefing. Enjoy the endless tropical fish and beautiful coral formations. A perfect adventure for the family! Pregnant women may not participate. **Please note: Minimum age 8 years old.**	**$54 each ____**
Ultimate Stingray SNUBA **GCM1**	**Duration: 2½ hrs.** **Departs: 1:15 pm** For those who love to snorkel and are not yet certified SCUBA divers... this is a fun way to visit the original DEEP water Stingray City site. You will be taken to the original site located in about 12-15 feet of water in the North Sound where your instructors will prepare you for a visit with the stingrays. **Please note: Minimum age 8 years old. Medical waiver required. Pregnant women may not participate.**	**$99 each ____**
Turtle Farm & Colony Beach Club **GCAF1**	**Duration: 3 hrs.** **Departs: 9:45 am** Visit the famous Turtle farm and the city of Hell then enjoy a relaxing 1½ hours on the famous 7 mile beach at the Beach Colony Hotel. (Lounge chair included.)	**$48 each ____**
Grand Cayman Highlights & Turtle Farm **GCA1/A2**	**Duration: 2 hrs.** **Departs: 9:30 am / 1:00 pm** You'll journey through charming George Town and up to the city of Hell, named for it's unusual rock formations. Next, enjoy the Hawksbill turtles at the sea turtle nursery before a photo stop of Seven Mile Beach.	**$30 adult ____ $22 child ____**
Seaworld Explorer Glass Bottom **GCG1**	**Duration: 1 hr** **Departs: 12:30 pm** This semi-submarine allows you to descend 6 feet under the surface as you glide over some of Cayman's most famous underwater sites including the wrecks of the Cali, Balboa and Cheeseburger Reef. Completely narrated.	**$37 adult ____ $28 child ____**
Seven Mile Beach Break **GCJ1** **GCH1 – With Lunch**	**Duration: At your leisure** Relax on the beautiful world famous Seven Mile Beach. This tour includes use of Beach Colony Hotel facilities (no pool), one beverage and a lounge chair. ($4 per person transportation each way) **Parasol rentals not included.** ($29 adult ___, $22 child ___)	**$19 adult ____ $12 child ____**

Trolley Roger **GCAJ1/AJ2**	**Duration: 1 hr** **Departs: 11:30 am / 2:00 pm** You will be filled in on the entertaining history of the island and it's unique past as you ride on this unique open air trolley. View some of the oldest buildings in Cayman nestled in between some of the 600 banks and admire central park and the statue of the local hero "Mr. Jim Bodden." After making a short stop at the local beach, ride along the coastal roads towards Hog's Bay. The tour concludes downtown in the shopping area.	**$24 adult** ____ **$14 child** ____
Save a Blue Iguana Tour **GCAM1**	**Duration: 3½ hrs.** **Departs: 10:00 am** The Grand Cayman blue iguana is a giant, blue, plant-eating lizard unique to the island of Grand Cayman. Profits for this tour go to help save the world's most endangered iguana from extinction! After a 45-minute drive, arrive at the QE II Botanical Park where you will see and learn the history of the Blue Iguana. You will observe them living in breeding pens reared to maturity before being released to the wild!	**$54 each** ____
Stingray City Swim & Snorkel **GCD1/GCD2**	**Duration: 3 hrs.** **Departs: 9:45 am / 1:00 pm** Spend approximately 45 minutes in shallow waters feeding, touching and playing the incredible Southern Stingrays. No tropical fish or coral will be seen. **Please note: Minimum age 5 years old. Snorkel equipment is provided.**	**$45 each** ____
Stingray City Adventure **GCV1**	**Duration: 2½ hrs.** **Departs: 10:00 am** Enjoy swimming with the stingrays at the shallow sandbar. Snorkel equipment is provided. No tropical fish of coral will be seen.	**$45 each** ____
Cayman Pirate Encounter **GCI1**	**Duration: 2 hrs.** **Departs: 11:00 am** Join the 80-foot square rigged galleon and swear on the oath to the Pirates Creed of Ethics. Watch in awe as pirates fire the cannon in a mock attack on your cruise ship. After the fun and pirate adventure enjoy a great swim stop. **Not a snorkel tour.**	**$39 adult** ____ **$31 child** ____
Butterfly Farm Express **GCAG1**	**Duration: 1 hr** **Departs: 11:30 am** Enjoy a guided tour through a tropical garden filled with over 1000 colorful butterflies from around the world. View the manicured gardens, ponds and waterfalls before being returned back to the pier. Do not forget your camera.	**$29 adult** ____ **$19 child** ____
The Butterfly Farm & Beach **GCAI1**	**Duration: 3 hrs.** **Departs: 11:45 am** Enjoy a guided tour through the tropical gardens, home to over 1000 colorful butterflies from all over the world. Next, it is on to "Sea Grape Beach," located on the world famous 7 Mile Beach. (Lounge chair provided, beach bar and grill available at an additional cost.)	**$49 adult** ____ **$34 child** ____
Sea Trek of Grand Cayman (Helmet Dive) **GCAQ1/GCAQ2**	**Duration: 1 hr** **Departs: 12:15 pm / 2:15 pm** The SeaTrek experience is a zero gravity stroll underwater wearing the specially designed SeaTrek underwater helmet. Air is pumped to the helmets from the state-of-the-art compressor system. This tour is fun and interactive! **Please note: Minimum age 8 years old to participate. Medical waiver form required.**	**$89 each** ____
Cruise, Sightsee & Seven Mile Beach Break **GCAO1**	**Duration: 5 hrs.** **Departs: 11:00 am** Enjoy a scenic cruise to the world-famous 7 Mile Beach, then at your leisure enjoy the azure waters and powdery white sands. (Watersports, restaurant and bar service are available at an additional cost.) Guests will be collected by mini bus for their transfer back to the pier. (Beach chair and welcome drink included).	**$42 each** ____
Thriller Boat Sea & Sand **GCAP1**	**Duration: 2¾ hrs.** **Departs: 11:00 am** This offshore racing boat accelerates you towards a day of fun! Fly across the water at speeds of 50-60 m.p.h. as you journey south along Grand Cayman's shoreline. Stop at Sea Grape Beach on the famous Seven Mile Beach. Enjoy a beach chair for approximately 1 hour of beachtime before it's time to race back to Georgetown on the Thriller!	**$78 adult** ____ **$62 child** ____
Atlantis Submarine **GCE1/GCE2**	**Duration: 1½ hrs.** **Departs: 11:30 am / 1:30 pm** Experience Cayman's underwater wonders onboard the ultramodern 48-guest Atlantis XI Submarine. Dive to depths of 100 ft. in air-conditioned comfort through colorful coral and a rich diversity of marine life. **Please note: Minimum age 4 years old. Minimum height 3 feet.**	**$94 adult** ____ **$49 child** ____
Cayman Bicycle Adventure **GCP1**	**Duration: 3 hrs.** **Departs: 12:30 pm** Together with your tour guide and mountain bike you'll see spectacular coastline and Seven Mile Beach. You'll also travel inland and to the north to Spanish Bay Reef Hotel for refreshments on the beach. Next, head south down to the Colony Hotel. Take the shuttle back or stay and pay your own transportation back to the pier. **Please note: Minimum age 12 years old.**	**$69 each** ____
Spanish Bay Reef Resort **GCY1**	**Duration: 5 hrs.** **Departs: 10:00 am** Enjoy this secluded oceanfront resort facing the undersea world's renowned North Wall snorkeling paradise! Swim in the pool, snorkel all day soaking up the sun while taking full advantage of the open bar and buffet lunch. Snorkeling equipment is included.	**$84 adult** ____ **$59 child** ____
Island Tour & Turtle Farm **GCU1**	**Duration: 2½ hrs.** **Departs: 10:00 am** Visit the famous Seven Mile Beach and enjoy viewing the entire stretch of white sand beach. Next stop is the Turtle Farm where Hawksbill turtles are raised. Afterwards visit the town of Hell!	**$36 each** ____
Catamaran Sail & Snorkel **GCAL1**	**Duration: 3 hrs.** **Departs: 1:00 pm** This tour is a 2½ hour combination sail with two snorkel stops on a shallow reef or wreck along spectacular Seven Mile Beach. Both snorkel stops will be in shallow water with great visibility and lots of marine life. Snorkel equipment supplied. **Please note: Minimum age 5 years old.**	**$52 each** ____
2-man Bubble Sub **GCT1/GCT2/GCT3**	**Duration: 1½ hrs.** **Departs: 12:30 pm / 1:30 pm / 2:30 pm** The Bubble Sub has a unique design where 2 people sit comfortably inside a giant non-claustrophobic glass bubble. You have a 360 degree view so you can see all of the underwater wonders. Travel beside schools of fish, watch as an eagle ray glides overhead, follow along side a turtle or sneak up on a sleeping nurse shark. **Please note: Minimum age 4 years old. Maximum weight 500 lbs. Maximum height 6'8".**	**$198 each** ____

SECOND CHOICE FOR GRAND CAYMAN **Tour Name** **Time**

WE STRONGLY RECOMMEND THAT YOU INDICATE A SECOND CHOICE, AS TOURS ARE LIMITED AND DO SELL OUT QUICKLY!

Montego Bay, Jamaica – *Thursday (7:00 am – 4:00 pm) Deadline for booking on RCTV is Wednesday at 3:00 pm*

EXPLORATIONS!	DURATION / DEPARTURE (PLEASE REMAIN ON SHIP'S TIME IN THIS PORT)	NUMBER OF TICKETS
The Best Of Jamaica **MJJ1**	**Duration: 7 hrs.** **Departs: 7:30 am** Your tour guide will take you on a scenic 2 hour, narrated drive down to the Jamaica Grande Hotel where a luxury Yacht awaits. Sail around the Ocho Rios coastline down to Dunn's River Falls where you will have the opportunity to climb the Falls, then cruise round to Dolphin Cove where lunch is served and you can watch the Dolphins at play. Buses will be waiting at Dolphin Cove for your return to the cruise ship pier in Montego Bay.	**$119 each** ____
Catamaran Sail & Snorkel **MJN1**	**Duration: 3 hrs.** **Departs: 8:00 am** Upon departing the cruise ship pier, your crew will take you on a scenic 60 minute sail around the environs of Montego Bay before anchoring at a snorkel location where guests will have approximately 60 minutes to enjoy swimming and snorkeling. The anchors are then raised, and your crew sails you back to Montego Bay and the cruise ship pier.	**$42 adult** ____ **$35 child** ____
Dunn's River Falls Tour **MJC1**	**Duration: 7½ hrs.** **Departs: 7:30 am** A scenic 2 hour drive, a little lively history, a visit to Jamaica's most famous waterfall, and a sumptuous lunch are just some elements of this package. Journey through the busy city of Montego Bay, and then along the coastal road with its exclusive hotels edged by sandy beaches. Our first stop will be the Yow Travel Stop in Rio Bueno, for those who wish to browse for island hand-crafted souvenirs or use the rest room facilities. Then it's off to Dunn's River Falls. From Dunn's River, we take you to the outskirts of Ocho Rios for a sumptuous Jamaican buffet luncheon prepared for you at a renowned Margaritaville restaurant. After lunch, we begin our return journey to Montego Bay, pausing at Columbus Park. There you get a glimpse of Jamaica's history as you wander through this lovely, outdoor museum, set high above a beautiful bay.	**$74 adult** ____ **$63 child** ____

Essence of Montego Highlights MJAB1	**Duration: 4 hrs. Departs: 8:00 am** **$29 adult____ $25 child____** Enjoy a scenic drive through the countryside of Montego Bay, stopping at Rose Hall Great House, White Witch Golf Course and Half Moon Hotel where guests can enjoy duty-free shopping, before returning to the cruise ship pier via Aqua Sol Beach where guests relax with a complimentary welcome drink.
Catamaran Sail, Beach & Party MJAD1	**Duration: 3 hrs. Departs: Noon** **$52 each____** You will be provided with a wonderful opportunity to experience sailing/motoring aboard a catamaran for approximately two hours at the beach (public beach) before a fun party sail back to the cruise ship pier across Montego Bay Harbor.
Appleton Estate Rum Tour MJM1	**Duration: 6½–7 hrs. Departs: 7:30 am** **$65 each____** Appleton Estate Rum Tour is a "Journey Through Time". The tour begins with a demonstration by "PAZ" resident Donkey who shows how the juice from the sugar cane was extracted before steam was invented. Historical artifacts are on display, and the tours take in the Distillery and ageing house. Guests can then "grind" their own sugar cane, sample the juice, make special "wishes" at the Aqueduct Water Wheel, boil wet sugar and visit the V/X museum. A tasting of the different blends and ages of Rum is done inside John Wray's Tavern, and a complimentary bottle of Appleton Rum is given as a gift for each visitor.
Beach Comber Beach Break MJAF1	**Duration: 6½–7 hrs. Departs: 8:30 am** **$82 adult____ $76 child____** Enjoy a scenic 50-minute drive along the seashore to Negril .On arrival at the Beach Comber Resort you will receive a welcome drink and chaise lounge. After a wonderful day of relaxation enjoy the return ride to the pier. Snorkel equipment and buffet lunch included.
Calico Pirate Cruise MJF1	**Duration: 3½ hrs. Departs: 8:00 am** **$32 each____** Depart cruise ship terminal for a 45 minute sail, stop for snorkeling in the Montego Bay Marine Park. After enjoy a 45 minute stop at Jimmy Buffet's Margaritaville, during return sail alcoholic beverages are served.
Croydon Plantation Tour MJV1	**Duration: 6 hrs. Departs: 8:00 am** **$56 adult____ $26 child____** Croydon in the Mountains is a fascinating tour to the interior of the island, highlighted by a conducted tour of our famous pineapple and coffee plantation. You will be able to sample exotic fresh fruit and drinks made from them. Unusual fruits like carambola, oneca, otahiti apple and passion fruit are available in season. Our 'pineapple stop', where you will taste different varieties of chilled pineapple slices, is an unforgettable and unique experience and one of the highlights of the tour. You will enjoy a sumptuous Jamaican barbecued lunch served with world famous 'Blue Mountain Coffee' as the final treat of this excellent excursion.
Rose Hall Great House MJA1	**Duration: 3¾ hrs. Departs: 9:00 am** **$43 adult____ $33 child____** Who could resist a beautiful woman with an evil heart, whose wicked ways finally brought her to destruction in her own magnificent home? Jamaica's 'White Witch' was all this and more. The stage for this true to life melodrama was the Rose Hall Great House. We don't guarantee that you will see the White Witch walking the halls in broad daylight, but you are sure to enjoy this retreat into time. On your way back to Montego Bay, you can see some of the hotels which have played host to our visitors over the years offering the best of sea and sun as well as warm hospitality. We then take you into the center of Montego Bay, which residents call the 'Friendly City', to browse and shop before returning to the quayside.
Horseback Ride N' Swim MJE1/MJE2/MJE3	**Duration: 3 hrs. Departs: 8:00 am / 9:00 am / 11:30 am** **$86 each____** Depart from the terminal in air-conditioned comfort enjoying a scenic bus ride along the coast to Chukka Blue, which is located 20 minutes west of Montego Bay. Upon arriving at Chukka Blue Guests will be given a brief orientation before mounting their horses. The ride departs from The Blue Hole Estate stables and passes the ruins of the Great House. The tour continues along a cliff with a panoramic view over the bay with the beautiful mountain back drop and leads to the Caribbean Sea to the cove where you will swim with the horses in the sea. **Please note: Minimum age 6 years old. Maximum weight 250 lbs.**
4X4 Safari Adventure MJH1/MJH2	**Duration: 4½ hrs. Departs: 7:30 am / 11:30 am** **$82 each____** The tour takes guests into the interior of Jamaica, but not before enjoying a majestic view of the city, cutting through the Mountains by a mixture of rural roadways, farm access roads, and at times tracks made only by the very jeeps of the tour. Cut through the jungles and even enjoy the thrill of driving through the river (subject to river level). There is the stop at the river where you may get out in bathing suites and walk up a mineral spring to a hidden waterfall. You will be completely wet when you are ready to finally get back to the jeeps to Four wheel drive out of the Jungle back to the farm roads. The Flora and fauna is breathtaking. **Please note: Minimum age 6 years old. Water shoes rental $4.00 each pair.**
Jungle River Tubing MJI1/MJI2	**Duration: 4 hrs. Departs: 7:30 am / 11:30 am** **$68 each____** You'll sit in a comfortable tube as you float through an unspoilt river course in Jamaica. Depending on the amount of rain fall there are some very small rapids that will be encountered during the float at other times it's just a peaceful float down river with excellent guides. The river tubing ends at a point on the river where guests must be carried out of the jungle by themed 4 wheel drive vehicles. Money, dry clothes, towel, etc. will be kept in the bus and supervised during the tour. Pax must wear a bathing suit and bring a towel since they will get wet during this tour. **Please note: Minimum age 6 years old.**
Helmet Diving MJU1/MJU2	**Duration: 3 hrs. Departs: 8:00 am / 12:30 pm** **$94 each____** A space age helmet is set comfortably on your shoulders while a continuous flow of air lets you breath with ease. For the duration of 25 minutes, your guide will show you all there is to see at the bottom of the Sea - Amazing marine life and spectacular live coral. Sea trek is great for the non swimmer. After your underwater adventure you have 1¾ hours to enjoy the beautiful surroundings at Doctor's Cave Beach. You can purchase a snack and refreshments at the Groovy Grouper Restaurant, snorkel or just relax on the beach. Once you have soaked up the Sun, Sea and Sand we will transfer you back to your ship. **Please note: Minimum age 12 years old. Medical waiver required.**
Black River Safari MJB1	**Duration: 6¼ hrs. Departs: 7:30 am** **$62 adult____ $58 child____** This trip takes you from Montego Bay along the Southern Coast; you will pass through quaint seaside villages where colorful fishing boats line the shore. As you continue along the coastline toward Black River, rustic roadside stands beckon you to stop and sample escoveitched fish, peppered shrimp and bammy, a great Jamaican delight. Journey by boat up the Black River, where you will see exotic birds, plants, mangroves and wildlife such as the endangered American crocodiles. Look out for the resident crocs' Freddi and Allo at the dockside. We travel from Black River to Bridge House Inn for a scrumptious lunch before returning to the ship. This is a tour you will not want to miss.
Mobay Highlight & Shopping Extravaganza MJG1	**Duration: 3½ hrs. Departs: 9:30 am** **$30 adult____ $21 child____** From the quayside we wind our way through the hustle and bustle for a drive-by of the world famous Rose Hall Great House. From this taste of Jamaica's ghostly past we will take you to the Half Moon Shopping Centre and the new retail and entertainment centre known as the "Bob Marley Experience". There you have the comfort of an air-conditioned, wide-screen theatre where you view a film on the life of the reggae legend, Bob Marley. We take a short climb to Richmond Hill Hotel, with its panoramic view of the city and harbor for a complimentary drink. While there, visit the Little Gallery and see the paintings of one of Jamaica's best-known artists, Barrington Watson. As we wind our way down into the centre of town, we pause for shopping at the local craft's market.
Negril Splendor MJL1	**Duration: 5¾ hrs. Departs: 7:30 am** **$68 adult____ $65 child____** The leisurely ride from the Pier takes you through the South Western side of the island with breathtaking views of the coast. Along the route you will see the magnificent landscape that Jamaica is famous for as well as exquisite homes and farmland. Our first stop is at the Royal Palm Reserve, a tranquil forest located within the Negril Great Morass. Here, nature walks, bird watching and relaxation awaits. Complete the day with a sumptuous Buffet Luncheon at Jimmy Buffet's Margaritaville, and enjoy the seven miles of white sand beach in gorgeous Negril as well as the many games and activities that your host will offer.
Johns Hall Plantation Tour MJO1	**Duration: 4 hrs. Departs: 9:00 am** **$58 adult____ $28 child____** We will take the sun, the rum and jamming to our reggae music into the glittering hillside, while we update you on our region and history at our Johns Hall Estate. View the scenic downtown of Montego Bay before embarking on a 1½ half hour drive through the green countryside to a local school. After you will reach Johns Hall Estate. Enjoy a presentation of the different fruits, vegetation, authentic birds and animals exhibited on the plantation, sample also the coffee grown and processed on the plantation. At the end of the tour buffet lunch is served.
Jamaica Wonders MJP1	**Duration: 3½ hrs. Departs: 9:30 am** **$58 adult____ $48 child____** Our journey will take us into the interior of Jamaica to a tiny village named Lethe, where the beautiful Great River meanders gently through the mountains. We invite you to the estate of the Tullochs in an area called "the Green" for a cool drink of local fruit punch. From there we go on a Jitney ride into the banana plantation. You will have an opportunity to taste different liqueurs made from coconut, banana and our Jamaican special rum. The tour continues to Rockland bird sanctuary where you may view over 300 species of birds found in Jamaica. Continue the journey of discovery along the picturesque landscape of St. James and Hanover. Our next stop is the Waterwheel at Tryall. Sample Jamaican grown fruits at the base of the wheel before you are returned to the pier where the journey ends.

Dunn's River Express
MJW1

Duration: 5 hrs. **Departs: 8:00 am** **$56 each ___**
Enjoy a scenic 2 hour drive from Montego Bay to Ocho Rios, through the historical town of Falmouth, Discovery Bay and Runaway Bay, before arriving at Dunn's River Falls. Climb and swim at the Fall, before returning to Montego Bay. **Please note: Minimum age to climb the falls is 4 years old. Lockers/ watershoes available at an additional cost.**

ATV Adventure
MJK1/MJK2

Duration: 3 hrs. **Departs: 7:30 am / 11:00 am** **$89 each ___**
This adventure offers a scenic ride through the rugged terrain of the Blue Hole Estate to the beautiful community of Cascade. Taste Jamaica's rich culture and heritage as you journey through farming villages and learn of the history of Pumpkin Bottom, Smithfield and Cascade. Visit one of the oldest Presbyterian Churches that was built by slaves in 1840. See nature's natural flora and fauna and learn of their multiple uses by our expert guides. After the tour, guests return to base, where they will get the chance to cool off and relax while soaking up nature's unspoilt natural beauty, before being transferred to their ship. **Please note: Minimum age 16 years old. Maximum weight 250 lbs.**

Canopy Tour
MJS1/MJS2/MJS3

Duration: 4 hrs. **Departs: 8:00 am / 9:00 am / 11:00 am** **$89 each ___**
This unique tour consists of a series of decks and platforms mounted in the cliffs and trees of the forest. Your Canopy Tour Guide will instruct you in using the high-angle equipment and techniques to move from platform to platform. After the first traverse participants will take a short bus trip or a brisk 10-minute walk at Diving Board where the traversing adventure continues with five additional traverses coupled with intermittent nature walks. Hear and see the rushing waters below as you traverse up to 35 miles an hour across springs, the Great River and a 150 year old Dam. **Please note: Minimum age 12 years old. Maximum weight 220 lbs. Closed toe shoes required.**

Mountain Valley Rafting
MJD1/MJD2

Duration: 4¼ hrs. **Departs: 9:00 am** **$52 adult ___** **$42 child ___**
On arrival at the little village called Lethe you will board a bamboo raft built for two persons. Hear the stories regarding the mystery of the river as can only be told by your rafts captain. Stop along the river bank and learn about the different trees and foliage. At the recreation area at the end of your journey we invite you for a cool drink of local fruit punch. You will have an opportunity to taste different liqueurs made from the coconut, banana and/or Jamaican special rum. Our final stop before returning to the ship is for shopping at the City Centre Shopping Complex. Here you can select your purchases from a variety of duty-free, souvenir and specialty shops.

Coral Semi-sub
MJT1

Duration: 2 hrs. **Departs: 9:00 am** **$34 adult ___** **$19 child ___**
View the fascinating underwater ecosystem of Jamaica's only marine park in a comfortable air-conditioned underwater semi-sub. Learn about the colorful corals, exotic tropical fish and other sea life from marine experts. A stop for 30 minutes or longer is provided for those who wish to snorkel. The semi-sub continues its sail along the coastline passing Sandals Inn, Jack Tar Hotel and the famous Doctor's Cave Beach and Margaritaville before returning to the cruise ship pier.

Beach Break
MJY1/MJY2

Duration: 3¾ hrs. **Departs: 9:00 am** **$34 adult ___** **$21 child ___**
The Caribbean Beach Park is the perfect place for family to spend a most enjoyable day in Jamaica. Located just 20 minutes drive from the port; you can enjoy a secure and private environment offering a day of relaxation and sun. On arrival you will be welcomed to the rhythms of island music and a complimentary drink of rum punch, draft beer, soda or fruit juice. On the beach you can relax in the sun or partake in an organized volleyball or basketball game, or play a challenging game of table tennis or darts with a fellow patron. There is an extra charge for motorized water sports and a game of pool. Snacks and additional drinks are available for purchase at the Pavilion. Lounge chairs are included in your package.

Rafting On The Martha Brae
MJAC1

Duration: 4¼ hrs. **Departs: 9:00 am** **$45 each ___**
Treat yourself and your family to an exhilarating ride on a 30 feet bamboo raft, over three miles of the legendary Martha Brae River. Visit Miss Martha's Herb Garden, a presentation of Jamaican herbs famous for their many medicinal properties. "Rafting on the Martha Brae" is a unique Jamaican experience not to be missed by any visitor to the island. Raft ride approximately 1¼ hours.

Cinnamon Hill Golf Club
MJZ1

Duration: 5 hrs. **Departs: 8:00 am** **$115 each ___**
The Ocean Golf Course at Rose Hall will encompass 18 holes and 7,000 yards of lush tropical splendor. Note: Price includes green fees, transportation, shared cart and caddy. Club rental $30.00 per person; shoe rental not available. Prices are not part of the golf package and are subject to change at anytime. Payable at the golf course only. 18 holes/par 71/6,637 yards.

SECOND CHOICE FOR MONTEGO BAY **Tour Name** **Time**

WE STRONGLY RECOMMEND THAT YOU INDICATE A SECOND CHOICE, AS TOURS ARE LIMITED AND DO SELL OUT QUICKLY!

Labadee, Haiti – *Friday (9:00 am – 5:00 pm)* *For more details for Labadee, please see SeaTrek Tour Order Form*

Miami, Florida – *Sunday (Arrival 6:30 am)* *Deadline for booking on RCTV is Saturday at noon*

EXPLORATIONS! | DURATION | DEPARTURE | NUMBER OF TICKETS

Miami Highlights Tour
MIA1

Duration: 2 hrs. **Departure: 7:00 am** **$32 adult ___** **$18 child ___**
Depart aboard an air-conditioned coach through portions of downtown Miami, heading to the world-famous Art Deco District of South Beach. After a colorful tour of this historic and trendy area, return to Miami along Brickell Avenue, the financial heart of the city. Your tour then takes you to the lush bohemian village of Coconut Grove. A drive through Coral Gables will take you past the Biltmore Hotel, the Venetian Pool and Miracle Mile. Passing through parts of Little Havana your tour concludes at Miami International Airport. **Please note: This tour is only available to guests with flights departing from Miami airport at 3:00 pm or later for domestic flights/4:00 pm or later for international flights.**

Everglades Safari Tour
MIB1
MIBA1 (FLL)

Duration: 4 hrs. **Departure: 7:00 am** **$49 adult ___** **$39 child ___**
Experience the unique natural beauty of the Florida Everglades in a once-in-a-lifetime opportunity. Leaving the Port of Miami aboard an air-conditioned coach, you'll drive through downtown Miami as your guide explains the city's rich history. Information about the ecology and history of the River of Grass will also be provided. Delight in a 40-minute airboat tour in this unspoiled wilderness! See fish, turtles, alligators, birds and other wildlife in their natural habitat. Enjoy a wildlife show and alligator wrestling. Your tour concludes at Miami International Airport. **Please note: This tour is only available to guests with flights departing from Miami airport at 3:00 pm or later for domestic flights/4:00 pm or later for international flights.** **(FLL) $58 adult___ / $47 child___**

South Beach and Island Queen Cruise
MIC1
MICA1 (FLL)

Duration: 4 hrs. **Departure: 7:00 am** **$48 adult ___** **$39 child ___**
Depart Port of Miami aboard an air-conditioned coach heading to the world-famous South Beach Art Deco District. After a tour of this colorful area your destination is Bayside Marketplace, where you will board the 92-foot Island Queen for a one-and-a-half hour cruise. See the homes of the rich and famous Gloria Estefan, the legendary Al Capone, and the honeymoon cottage of Elizabeth Taylor and Eddie Fisher – as you cruise the calm waters of Biscayne Bay and pass Star, Palm and Hibiscus Islands. Enjoy splendid views of downtown Miami before concluding your tour and Miami International Airport. **Please note: This tour is only available to guests with flights departing from Miami airport at 3:00 pm or later for domestic flights/4:00 pm or later for international flights.** **(FLL) $57 adult___ / $48 child___**

Parrot Jungle Island & City Tour
MID1

Duration: 4½ hrs. **Departure: 7:00 am** **$52 adult ___** **$38 child ___**
View the metropolitan area of Miami taking in some of the historic sights along downtown Miami, South Beach and the Art Deco District. Then visit Parrot Jungle Island Miami's new attraction. Enjoy the tropical landscape and entertaining wildlife shows as well as the beautiful birds, playful baby apes, tigers, alligators and the 20ft crocodile. The Everglades Habitat is a must see before your tour concludes at the Miami International Airport. **Please note: This tour is only available to guests with flights departing from Miami airport at 3:00 pm or later for domestic flights/4:00 pm or later for international flights.**

Parrot Jungle Island On Your Own
MIE1

Duration: At your leisure **Departure: At your leisure** **$26 adult ___** **$21 child ___**
Experience how wild Miami really is! Come and see the new Parrot Jungle Island. You may stroll the lush tropical landscapes and elegant gardens while encountering more than 3,000 animals and over 500 species of plants and trees. Enjoy the world famous parrot show, waterfront Serpertarium, ape and monkey exhibits, reptile exhibits and open aviaries Come and have a junglerific day. **Please note: This tour is at guests leisure and does not include transportation to or from the venue.**

船内新聞「CRUISE COMPASS」

〔1日目〕マイアミ港出港

CRUISE COMPASS

DAY 1

*Freedom of the Seas*SM
DEPART MIAMI, FLORIDA
Sunday, June 11, 2006

TODAY'S FORECAST Thundershowers, 85°F

SUNRISE 6:29 am
SUNSET 8:10 pm

TOMORROW At Sea

WELCOME TO THE FIRST DAY OF THE MOST AMAZING VACATION OF YOUR LIFE.
This week will be filled with boatloads of activities, wonderful adventures and unforgettable experiences. In other words, there is a lot to do and see. The list below highlights the things you won't want to miss. Each night you will receive a Cruise Compass and Daily Planner in your stateroom so you can get a jump-start planning your adventures for the next day. Remember that our crew is here to ensure you receive the best service and have the best vacation, so feel free to ask any questions you may have. Get out there!

TODAY'S TIPS
- Please note that luggage delivery is a lengthy process. Thank you for your patience and understanding.
- Familiarize yourself with the safety routine onboard and your Assembly Station, your assigned meeting place in case of emergency. View our safety film on your stateroom television, channel 41. There will be a compulsory mustering at 4:45 pm.
- A nibble of this, a bite of that, try it all at the Windjammer buffet on Deck 11 aft from 11:30 am to 3:30 pm.
- Once we set sail, stop by and browse our duty-free shops and boutiques.
- Tonight's the perfect night to dine at Chops Grille or Portofino.

TONIGHT'S ENTERTAINMENT

Your Cruise Director, **Ken Rush**, proudly presents the

WELCOME ABOARD SHOW
Starring
The Comedy of
Steve Bruner

7:00 pm (Pre-Dinner Show) for Second Seating Guests
9:00 pm for Main Seating Guests
Arcadia Theatre, Decks 3 & 4

Please be reminded that the saving of seats and video taping of shows is strictly prohibited. Also, children in the first three rows of the theatre must be accompanied by a parent or guardian.

DON'T MISS THE EXCITING FREEDOM EXPO!
Join us on the Royal Promenade from 2:30 pm. Let us show you what we have in store for you this cruise vacation. Over $800 in prizes to be won in the free raffle. The raffle will be drawn at 4:00 pm. (You must be present to win).

FAMILY WELCOME ABOARD SHOW
We invite all families to meet our Youth Staff for a special welcome aboard show and introduction to our award-winning Adventure Ocean Youth Program.
Doors open at 7:45 pm, Showtime at 8:15 pm, Studio B, Deck 3

PARTY AROUND THE WORLD PARADE
Don't miss the first of the two beautiful parades! Bring your cameras!
11:15 pm, Royal Promenade, Deck 5

FREEDOM-ICE.COM
Don't miss this incredible ice show. Due to seating limitations, complimentary tickets are needed for this spectacular event. Ticket distribution is tomorrow from 9:30 am – 11:00 am in the On Air Club. Shows will take place on Monday, Wednesday and Thursday.

DINNER IS SERVED — **Casual Attire**
Main Seating – 6:15 pm in Leonardo's, Isaac's & Galileo's Dining Rooms, Decks 3, 4 & 5
Second Seating – 8:45 pm in Leonardo's, Isaac's & Galileo's Dining Rooms, Decks 3, 4 & 5

THE TOP 10 THINGS TO DO ON A ROYAL CARIBBEAN CRUISE VACATION
1. Bring your cruise vacation to new heights – literally. Soaring 200-ft above the sea, the Rock Climbing Wall is something to see but, more importantly, conquer.
2. Indulge yourself, you deserve it. Book an appointment today at our Freedom Day Spa.
3. Let our entertainers amaze you in the Arcadia Theatre.
4. With so many shore excursions like SCUBA diving and parasailing, we have just the thing for everyone seeking new adventures. Book today.
5. Whether you like salsa, rock or ballroom dance, one of our many bars and lounges will surely suit your dancing needs.
6. Napa Valley in the middle of the Caribbean? You bet! Be sure to visit Vintages, our wine bar at sea.
7. The Royal Promenade is lined with boutiques and specialty shops with everything from designer duds to exquisite jewelry. So get out there and shop 'til you drop.
8. Practice your figure 8's on one of the only ice rinks at sea.
9. An intimate dinner awaits you. Enjoy fine Italian cuisine at Portofino or the perfect steak at Chops Grille. Call and make your dinner reservations now.
10. Top off your evening in Olive or Twist or one of our other many exciting bars and lounges.

Get ready for fun! Check out the Daily Planner for each day's activities.

EXPLORATIONS!℠ Shore Excursions and Port Information – We have a wide variety of shore excursions to offer in Cozumel, Grand Cayman, Montego Bay, Labadee and Miami. Want to book right now? Just grab your remote control. You can book a reservation for any shore excursion through our state-of-the-art interactive TV (RCTV) and receive instant confirmation. You can also complete the order form found in your stateroom and deposit it in our drop-box located at the Explorations! Desk.

DISCOVER SHOPPING ASHORE

Tune in to channel 22 today and discover why tomorrow's Discover Shopping Show is the must see red carpet event of the week. All who attend receive free jewelry.

CROWN & ANCHOR SOCIETY AND LOYALTY AMBASSADOR

Welcome back Crown & Anchor Society members! Enroll now to be a Crown & Anchor Society member, enrollment is free. Members earn great benefits and the more you cruise, the bigger they get. Pick up your forms at the Loyalty Ambassador's desk on Deck 6 or at Guest Relations on Deck 5, Centrum.

CHECK YOUR DAILY PLANNER FOR EVENT/DINING TIMES AND LOCATIONS

FOOD & DRINK

Drink of the Day – C.S.I. Miami for $5.95 and keep the glass. Get in to the swing of the cruise with this popular cocktail, a blend of Vodka, Blue Curacao, Gin, Chambord, Sweet & Sour topped with Sprite. Also a souvenir sloozie glass for $8.95 with any frozen drink.

Soda Package Special – Unlimited fountain soda and juices for the week for only $42 for guests 18 years and up and $28 for guests 17 years and younger. Ask any Bar Staff for details.

Wine & Dine Package – Seven chances to be a true connoisseur. Buy a Wine & Dine Package and enjoy a different bottle each evening with dinner. You may purchase a Wine & Dine Package in the main dining room or at Vintages.

Specialty Restaurants – Onboard you can enjoy various upscale dining options with impeccable service and an intimate atmosphere. When your taste buds desire Italian with a romantic flair, then head towards Portofino or if you want a succulent steak, visit Chops Grille. Both specialty restaurants are located on Deck 11 aft. Remember, these are specialty restaurants, so be sure to make a reservation.

Dining Room Questions – For all of your culinary queries, the Maitre d' will be available to answer your questions. Just remember, stick to the dining schedule and you'll receive the best service. And please note that bare feet, shorts, tank tops and t-shirts are not permitted in the Dining Room at dinner.

Casual Dinner Tonight – Come as you are! Enjoy selections from the Dining Room menu in a casual setting in the Windjammer Café. If you love Asian cuisine, then Jade is the place for you. Located within the Windjammer Café, this restaurant features a sushi bar and Asian-fusion dishes. Deck 11.

ACTIVITIES

Freedom Expo! – Tour virtually everything we have to offer at today's Freedom Expo! Enter the free raffle to win a variety of prizes just for visiting. Excitement is at your fingertips.

Studio B Open House – Come see our famous ice-skating rink. This is where our Ice Show FREEDOM-ICE.COM show takes place and where you can also go ice skating and learn how to skate. Please note: The ice rink is open for viewing only. Studio B, Deck 3 aft.

First-Time Cruiser's Club – What to do, where to do it, when to do it. A member of the Cruise Director's Staff will give you the inside scoop and make sure you don't miss a thing on your cruise.

Bon Voyage Sailaway Party – The party is starting as we depart for Cozumel and you're invited! Come join us poolside for live music, some dancing and a whole lot of fun as we depart Miami.

Adventure Ocean Open House – Come explore our youth facilities and get a youth program overview of all the fun, games and adventures we'll offer during this cruise vacation.

Party Around The World Parade – There's nothing more fun than a parade. Come experience the festivities as we officially open our Royal Promenade.

Family Disco – This is an hour that all ages are welcome to dance in our adult disco. Kids, take your parents for a spin around the dance floor.

Community Bulletin Board – Looking for like minded interest? Stop by the Library on Deck 7 and fill out a card.

royal caribbean online℠ – No matter where your adventure takes you, you can always stay in touch. Check e-mail and stocks, or e-mail a friend. This service is available 24 hours a day, at a nominal charge. Also check out Wireless Internet Access for a weekly rate at the Guest Relations Desk.

SPORTS & FITNESS

Specialty Fitness Classes – There is a wide variety of specialty fitness classes such as Yoga, Pilates and Spinning onboard. Check your Daily Planner each day for a complete list of the ShipShape specialty classes offered. They also offer other personal training and full metabolism testing.

Waverunner Fun – Rent a waverunner and explore the beautiful waters and coastline of Labadee. A very cool way to spend your day. Space is limited so sign up today!

SeaTrek Dive Program – Get certified on this cruise! We offer PADI scuba diver certifications, imagine learning to dive in two of the world's top diving destinations. Sign up now and ask about our fantastic equipment packages.

FlowRider – Join us for the ride of your life! The FlowRider simulates a real surfing and body boarding experience. 30,000 gallons of water per minute RUSH underneath the rider at 30 mph creating a five-foot ocean-like wave. Riders must be at least 58" to surf.

SPA & BEAUTY

The Freedom Day Spa is located on Deck 12. Dial 6982 for an appointment.

Freedom Day Spa Tours – Discover the largest and newest spa and fitness center at sea. Our caring team of internationally-trained therapists will introduce you to our wide variety of exclusive spa treatments. Here we can meet all your hair, makeup, spa, and beauty needs. We recommend making your appointments early to avoid disappointment.

Sail-Away Special – Start your cruise vacation off with a little pampering. Receive $20 per session off all signature spa treatments - tonight only.

Gentle Touch Teeth Whitening – Whiten your teeth and brighten your smile with our Gentle Touch Tooth Whitening Treatment. This is the safest, most effective system available today, which uses an FDA-approved non-toxic gel; your teeth will be visibly whiter after just one session. Contact or visit the spa for more information. Ask about our special introductory offer.

ONBOARD SHOPPING

Shops On Board – "Never Undersold, Always Guaranteed." Shops On Board guarantees to match the price on any identical item from our port(s)-of-call. Ask a Retail Associate for more details on the Royal Promenade, Deck 5.

Liquor – Lift a glass to our amazing deals on liquor. No matter what's your favorite, we have a huge selection of wine and spirits to choose from. Today's specials include 2 Absolut for $19.95, 2 Crown Royal for $32.95 + 2 Baileys for $33.95.

Get Out There – The latest edition to our Shops On Board, this store brings a brand new selection of sport & logo apparel never seen before.

Art Auction – Stop by the Art Gallery tonight and get a glimpse of some of the artwork available at tomorrow's live art auction. Free raffle tickets for the $2000 art raffle just for stopping by.

Photo Gallery – Relax, you're on vacation. Let us handle taking the photographs. During your cruise vacation we will be photographing most of the events onboard. But remember, you're under no obligation to purchase the photographs. All photos will be on display in our Photo Gallery.

Photo Gallery – Do you own a digital camera? If not, stop by the Photo Gallery and see how easy and cheap it is to enter the world of digital photography.

Gifts & Gear, Get it Here – Celebrate this cruise vacation in style. Choose from a wide assortment of edible treats, chocolate-covered strawberries, stateroom decorations, terry robes, towels and so much more. Contact the Guest Relations Desk on Deck 5 for more information.

Australian Airbrushed Tattoos – Meet Tracy, our airbrushed tattoo artist on the Royal Promenade and see all of our unique designs. Our airbrushed tattoos are henna free, safe, FDA approved and best of all, temporary!

GAMING

Gaming Lessons – We'll teach you to win cash from us. Come and enjoy complimentary gaming lessons at 6:00 pm today. We offer Blackjack, Roulette, Caribbean Stud Poker, 3 Card Poker, Craps, Let It Ride, Texas Hold'em plus 306 Slots.

Did You Know – You can bet from $50 up to $1000 table maximum upon request at our Blackjack tables at Casino Royale, Deck 4. Good Luck!!!!

Cash Prize Bingo – Thousands of dollars and many raffle prizes to be won at each fun session of Bingo. First session begins tomorrow.

Fisher-Price If you have a child under the age of 3 years, check out our **Aqua babies** and **Aqua Tots** programs, developed specially for Royal Caribbean International by early childhood experts at Fisher-Price. All classes feature wonderfully-fun age-appropriate storytelling, creative arts, play, music, movement and more. check your Daily Planner for times and location.

OPEN HOURS

A Clean Shave/Barber	11:30 am – 8:00 pm	Deck 5, Dial 2981
Art Gallery	7:00 pm – 10:00 pm	Deck 3
Australian Airbrushed Tattoo Artist	2:30 pm – 4:00 pm & 7:30 pm – 11:00 pm Royal Promenade, Deck 5	
Casino Royale	6:30 pm – late	Deck 4, Dial 3170
Discover Shopping Guide	2:30 pm – 4:00 pm	Deck 5, Dial 3830
Doctors' Hours	5:00 pm – 7:00 pm	Deck 1,
Emergency Only	*(In case of)*	Dial 911
Explorations! Desk (Tour Sales)	Noon – 4:00 pm & 5:00 pm – 9:00 pm Royal Promenade, Deck 5	
Floral Cart	5:00 pm – 9:00 pm Royal Promenade, Deck 5	
Freedom Day Spa	11:30 am – 10:00 pm	Deck 12, Dial 6982
Freedom Fitness Center	11:30 am – 10:00 pm	Deck 11
Guest Relations	24 hours	Deck 5, Dial 0
H2O Zone	8:00 am – 8:00 pm	Deck 11

Per United States Public Health, pull ups/swimmer's or diapers are not permitted in the swimming pools including the H2O Zone. All children must be potty trained.

Loyalty Ambassador	2:00 pm – 7:00 pm	Deck 6 aft
Medical Facility	5:00 pm – 7:00 pm	Deck 1 aft, Dial 51
Operator	24 hours	Dial 0
Photo Gallery	6:00 pm – 11:00 pm	Deck 4, Dial 2973
Photo Shop	6:30 pm – 11:00 pm	Deck 4
SeaTrek Dive Shop	Noon – 6:30 pm	Deck 11
Shops On Board	6:30 pm – 11:00 pm	Deck 5
Swimming Pool	24 hours	

Selected Pools and Hot Tubs are open 24 hours for your enjoyment. Guests are reminded there is no Lifeguard on duty. The Solarium Pool, Hot Tubs and the Solarium area are reserved for guests 16 and older only. Children must be supervised in the two Main Pools and in the H2O Zone at all times. As a courtesy to other guests, reserving of deck chairs is not permitted. Chairs left vacant for 30 minutes or longer may be reassigned by the Pool Attendant to another waiting guest. Royal Caribbean International is not responsible for theft or loss of property left on deck chairs or pool areas.

Tuxedo Rentals	2:30 pm – 4:00 pm	Royal Promenade, Deck 5

SPORTS DECK ACTIVITIES & OPEN HOURS

(Adult sporting events are for guests 18 and older.)

Basketball Court - Open Play	24 hours	Deck 13
FlowRider *(Demonstration)*	Noon – 1:00 pm	
(Boogie Boarding)	1:00 pm – 4:00 pm 5:30 pm – 7:30 pm	

Please wear a t-shirt. Please remove all jewelry items. A FlowRider waiver must be completed every day of the cruise.

Freedom Fairways Golf	24 hours	Deck 13
Ping Pong	24 hours	Deck 14
Rock Climbing Wall		Deck 13
(Open Climbing)	Noon – 3:00 pm	
(Introduction to Rock Climbing)	3:00 pm – 4:00 pm	
(Open Climbing)	5:30 pm – 7:30 pm	

Please bring socks, t-shirt & shorts/pants. A Rock Climbing Wall waiver must be completed once a cruise. Parents must be present to sign a waiver for children under the age of 18 and must stay to supervise children under the age of 13 years. Children must be 6 years of age or older to climb. Adults must be able to fit into the XXL harness to climb.

Shuffleboard	24 hours	Deck 4
Sports Pool	Noon – 6:00 pm *(Open swimming)* Starboard side	Deck 11

Please note: All bars and venues will be closed from 4:15 pm until the Guest Emergency Assembly Drill is completed.

For further information and descriptions of onboard services and policies, please read Important Information on the back of your Day 1 Cruise News.

✲ – Denotes ShipShape Dollars awarded for this activity.

FOLD HERE ↑

DAY 1

Freedom of the Seas SM
DEPART MIAMI, FLORIDA
Sunday, June 11, 2006
Departure: 5:30 pm

DINING SCHEDULE & AFTERNOON SNACKS

LUNCH

Noon – 4:00 pm	Windjammer Café	Deck 11

DINNER IS SERVED — **CASUAL**

6:15 pm	Main Seating, All Dining Rooms	Decks 3, 4, 5
8:45 pm	Second Seating, All Dining Rooms	Decks 3, 4, 5

Please note that dinner for second seating will be served at 8:30 pm for the remainder of the cruise vacation.
Please also note that bare feet, short pants and tank tops are not permitted at dinner.

6:30 pm – 9:00 pm	Casual Dinner, Windjammer Café	Deck 11
6:30 pm – 9:00 pm	Casual Dinner, Jade	Deck 11

Casual Dinner in Jade and the Windjammer Café is buffet style.
Please note that bare feet and tank tops are not permitted in these venues at dinner.

SPECIALTY DINING

6:00 pm – 9:45 pm	**Chops Grille,** The best steak on the high seas $20 dining fee per person applies, gratuities included (Reservations recommended, Dial 3055)	Deck 11
6:00 pm – 9:45 pm	**Portofino,** Intimate Italian Dining $20 dining fee per person applies, gratuities included (Reservations recommended, Dial 3035)	Deck 11

Dress Suggestion: Smart Casual. No short pants please. Guests 13 years and older are welcome. Please allow approximately two hours for dinner. Cancellations with less than 24 hours notice will be charged a fee of $10 per person.

SNACKS

24 hours	Café Promenade Please note the Café Promenade will be temporarily closed from 4:15 pm until the Assembly Drill is completed.	Deck 5
11:30 am – 3:00 am	Sorrento's Pizza	Deck 5
Noon – 11:00 pm	Ben & Jerry's Ice Cream (Nominal fee)	Deck 5
5:30 pm – 1:00 am	Johnny Rockets, ($3.95 cover charge per person, gratuities included).	Deck 12
Midnight – 1:00 am	Midnight Delights, Boleros & Casino Royale	Deck 4

AFTERNOON ACTIVITIES

24 hours	Community Bulletin Board, Library	Deck 7
24 hours	royal caribbean online℠	Deck 8
Continuous	Safety Video on your stateroom TV	Channel 41
Continuous	Explorations! Information	Channel 15
Continuous	Discover Shopping Information	Channel 22
11:30 am – 4:00 pm	Studio B Open for viewing	Deck 3
11:30 am – 3:45 pm	Dining Questions, (Maitre d' is available), Isaac's Dining Room	Deck 4
Noon – 4:00 pm	Adventure Ocean Youth Program Open House	Deck 12
12:30 pm	Library Opens	Deck 7
1:00 pm – 4:00 pm	Spa Tours with Free Raffle, Freedom Day Spa	Deck 12
2:30 pm – 4:00 pm	Freedom Expo!, Royal Promenade	Deck 5
4:45 pm	**Assembly Drill, Listen for further instructions**	Decks 4,5
5:00 pm – (approx.)	**Freedom of the Seas sets sail**	

AFTERNOON ACTIVITIES

5:00 pm – 6:00 pm	Bon Voyage Sailaway Party, Poolside	Deck 11
5:15 pm – 6:00 pm	Sail Away Social, (Ages 12-17) Pool Deck	Deck 12
5:30 pm – 6:00 pm	Singles Get-Together, Olive or Twist	Deck 14

EVENING ACTIVITIES

6:00 pm – 6:30 pm	Complimentary Gaming Lessons, Casino Royale	Deck 4
7:00 pm	Welcome Aboard Show, (Pre-Dinner Second Seating Guests), Arcadia Theatre	Decks 3,4
7:00 pm – 10:00 pm	"Sneak Peek" Exhibition, Art Gallery	Deck 3
8:00 pm	Freedom Mix 'n Mingle, (Ages 12-14), Living Room	Deck 12
8:00 pm	First Time Cruisers' Club, On Air Club	Deck 3
8:15 pm	Adventure Ocean Family Welcome Aboard Show, (Doors open at 7:45 pm), Studio B	Deck 3
8:30 pm – 10:00 pm	Adventure Ocean Registration & Family Fun Expo! Adventure Ocean	Deck 12
9:00 pm	NBA Finals - Game 2, *Miami Heat vs. Dallas Mavericks*, Pharaoh's Palace	Deck 5
9:00 pm	Teen Mix 'n Mingle, (Ages 15-17), Fuel	Deck 12
9:00 pm	Welcome Aboard Show, (Main Seating Guests), Arcadia Theatre	Decks 3,4
9:30 pm – 10:30 pm	Introduction to Wine Tasting, (Nominal fee), Vintages Wine Bar	Deck 5
10:00 pm – 11:00 pm	Family Disco, The Crypt	Deck 3,4
10:30 pm – midnight	Karaoke Kick Off, On Air Club	Deck 3
11:15 pm	Party Around The World Parade, Royal Promenade	Deck 5

FREEDOM FACTS

Occupancy
- Full Guest Occupancy – 4,375
- Double Guest Occupancy – 3,634
- Crew Onboard – 1,400

Statistics
- 160,000 GRT
- 1,112 Feet Long
- 184 Feet Wide
- 28 Feet Draft
- 21.6 Knots Cruising Speed
- 15 Guest Decks
- 14 Guest Elevators

Staterooms
- Total: 1,817
- Ocean View: 1,084 (842 with Balconies)
- Interior: 733 (172 Promenade View)
- Presidential Family Suite: 14 Guests, 1,215 Square Feet
- All staterooms convert to queen-size bed configuration and have private bath, phone, interactive flat-screen plasma TV, mini-bar, hair dryers and individually controlled air conditioning.

Safety
- Life Boats: 30 – Capacity 150 = 4,500
- Life Rafts: 78 – Capacity 35 = 2,930

Dining Rooms
- ***Leonardo's**, Deck 3*
  Named after the great scientist, inventor and artist Leonardo DaVinci, most well known for his painting of the mysterious Mona Lisa.
- ***Isaac's**, Deck 4*
  Named after Sir Isaac Newton - mathematician and scientist - who developed the Laws of Motions after seeing an apple fall in his orchard.
- ***Galileo's**, Deck 5*
  Named after the scientist and mathematician Galileo Galilei whose improvements to the telescope allowed for important astronomical discoveries.

MUSIC & DANCING

Live Bands & DJ's

Piano Melodies	*With Tara Davis* Noon – 2:00 pm, Boleros, Deck 4
Island Music	*With our Island Band, Caribbean Fusion* 2:00 pm – 3:00 pm & 3:15 pm – 4:00 pm Poolside, Deck 11
Sailaway Party	*With our Island Band Caribbean Fusion* 5:00 pm – 6:00 pm, Poolside, Deck 11
Dining Room Music	*With Tara Davis* 6:15 pm – 7:15 pm & 8:45 pm – 9:45 pm All Dining Rooms
Guitar Music	*With Leandro* 7:45 pm – 8:45 pm, Royal Promenade, Deck 5
Welcome Aboard Rock Party	*The Stingrays* 10:00 pm – 11:00 pm & 11:30 pm – 12:30 am *(with breaks)*, Olive or Twist, Deck 14
Cocktail Music	*With Viva Expressia* 10:00 pm – 11:00 pm *and Leandro* 11:30 pm – 12:30 am, Schooner Bar, Deck 4
Pub Entertainment	*Vocal Guitarist Jimmy Blakemore Plays Requests* 10:00 pm – 11:00 pm & 11:30 pm – 1:00 am *(with breaks)*, Bull & Bear Pub, Deck 5
Resident DJ	*With DJ Jamie* 11:00 pm – late The Crypt, Decks 3 & 4 Guests must be 18 years or older for admission and photo identification will be required if you appear to be under 18.

Please note, the Viking Crown Lounge which includes Olive or Twist is for adults 18 and older only after 10:00 pm.

BAR SERVICE HOURS

***Proof of Age** – Guests eighteen to twenty (18-20) years of age are welcome to enjoy beer and wine. Guests twenty-one (21) years of age and older are welcome to enjoy all alcoholic beverage. ** Applicable regulatory age restrictions apply while the ship is in port and until the vessel enters international waters. Picture Identification required.*

Café Promenade Bar	7:00 am – 3:00 am	Deck 5
Solarium Bar	11:30 am – 6:00 pm	Deck 11
Wipe Out Bar	11:30 am – 7:00 pm	Deck 13
Squeeze Bar	11:30 am – 7:00 pm	Deck 11
Plaza Bar	11:30 am – 11:30 pm	Deck 11
Schooner Bar	11:30 am – 1:00 am	Deck 4
Vintages Wine Bar	11:30 am – 1:00 am	Deck 5
Sky Bar	11:30 am – 7:00 pm	Deck 12
Pool Bar	11:30 am – 7:00 pm	Deck 11
Bull & Bear Pub	11:30 am – 2:00 am	Deck 5
Boleros	11:30 am – 2:00 am	Deck 4
Sorrento's Bar	11:30 am – 3:00 am	Deck 5
Olive or Twist	1:00 pm – 1:00 am	Deck 14
Champagne Bar	1:00 pm – 1:00 am	Deck 5
On Air Club	3:00 pm – 1:00 am	Deck 3
Casino Bar	6:00 pm – late	Deck 4
Arcadia Theatre	6:30 pm – 10:00 pm	Decks 3,4
Studio B	7:45 pm – 8:45 pm	Deck 3
Connoisseur Club Cigar Lounge	8:00 pm – 1:00 am	Deck 5
Fuel Bar (Guests ages 15-17)	8:00 pm – 1:00 am	Deck 12
Pharaoh's Palace	10:00 pm – 1:00 am	Deck 5
The Crypt (Adult only)	11:00 pm – late	Decks 3,4
DRINK OF THE DAY	**C.S.I. MIAMI**	**$5.95**

〔2日目〕終日航海日

CRUISE COMPASS

TEAR HERE

DAY 2

*Freedom of the Seas*SM
AT SEA
Monday, June 12, 2006

TODAY'S FORECAST
Scattered Thunderstorms, 89°F

SUNRISE 6:37 am
SUNSET 8:23 pm

NEXT PORT-OF-CALL
Cozumel, Mexico

Today is your day on the open sea. So, it's the perfect time to take advantage of all there is to do onboard *Freedom of the Seas*. The Rock Climbing Wall, miniature golf, basketball, a spa treatment-you're going to have a full day of adventure. Look inside for a sampling of today's highlights.

Explorations!
Don't miss this opportunity to explore the best secrets of the Caribbean. Tulum Mayan Ruins, pristine Mexican beaches, snorkel Mexican reefs or swim with dolphins. Come by the Explorations! Desk and SeaTrek Dive and Water Activities Shop to sign up for your shore excursions and dive tours in our upcoming port(s)-of-call. ***Get out there*** and make the most of your time ashore. Use the RCTV interactive television system in your stateroom to book your tour tickets.

DISCOVER SHOPPING ASHORE
Discover Shopping Show – A one time only red carpet event. Everyone in the live TV audience will receive a FREE Caribbean charm bracelet and could win FREE diamond jewelry!
Discover Style After Party – Join Discover Shopping Team for a sneak peak at Hollywood's jewelry and watch favorites. Fashion expert Tasha will consult on how to build the ultimate jewelry wardrobe.

TONIGHT'S ENTERTAINMENT

FREEDOM-ICE.COM
In Search of Freedom
Starring Our International Ice Cast
& Special Guest Stars **George & Anna**
Showtime: 5:00 pm, Studio B, Center Ice, Deck 3
Doors open 30 minutes prior to showtime.

CELEBRITY SHOWTIME
Starring Singer/Comedy Impressionist
Scott Record

Showtime for Main Seating Guests 9:00 pm
Showtime for Second Seating Guests 10:45 pm
Arcadia Theatre, Decks 3 & 4

Please be reminded that the saving of seats and video taping of shows is strictly prohibited. Also, children in the first three rows of the theatre must be accompanied by a parent or guardian. We also kindly ask that you refrain from using your cell phones in the theatre.

CAPTAIN'S WELCOME ABOARD RECEPTION
The Captain invites all guests to a welcome reception held in your honor. Captain Bill Wright will be located in Pharaoh's Palace for guests who wish to have their photograph taken with him from 7:30 pm – 8:00 pm. The Captain will give his welcome speech from the Flying Bridge at 8:15 pm.
Formal Attire, 7:30 pm – 8:30 pm, Royal Promenade and Pharaoh's Palace, Deck 5

ROCK-A-ROKIE
It's new, it's hot and it's here. Perform with a LIVE Rock Band and live out you fantasies.
It's your turn to be a Rock Star! (Adult participants only.)
11:15 pm – 12:30 am, Pharaoh's Palace, Deck 5

LATE NIGHT COMEDY
Starring
Steve Bruner
Adults Only. 12:15 am, Arcadia Theatre, Decks 3 and 4

CAPTAIN'S GALA DINNER **Formal Attire**
Main Seating – 6:00 pm in Leonardo's, Isaac's & Galileo's Dining Rooms, Decks 3, 4 & 5
Second Seating – 8:30 pm in Leonardo's, Isaac's & Galileo's Dining Rooms, Decks 3, 4 & 5

TODAY'S TIPS

- We know you want to get out into the beautiful sunshine. But the Caribbean sun is stronger than you think, so don't forget that sunscreen. If you did, buy it at the General Store on the Royal Promenade.
- **FREEDOM-ICE.COM** – One person per party can pick up a ticket for our upcoming ice shows for every person in their party today from 9:30 am – 11:00 am in the On Air Club, Deck 3 aft. Any remaining tickets will be available at the Guest Relations Desk after 11:30 am today.
- **Recommended Ice Show for Main Seating Guests:** Wednesday's showtime is at 9:00 pm. Thursday's showtime is at 5:00 pm. Please note there will not be a main show in the Arcadia Theatre on Wednesday evening.
- **Recommended Ice Show for Second Seating Guests:** Wednesday's showtime is at 7:00 pm. Thursday's showtime is at 7:00 pm. Please note on Wednesday evening there will not be a main show in the Arcadia Theatre.
- **Recommended Ice Show for either seating** – Tonight at 5:00 pm. Be one of the first guests this cruise vacation to enjoy this incredible show.
- **Blackjack Tournament** – Join in the fun at Casino Royale, sign up today. See the Casino Cashier for details.
- **Tonight is Formal Night** – Have your formal portrait taken tonight by our professional onboard Photographers at the studio set ups located on Decks 3, 4 and 5.

Get ready for fun! Check out the Daily Planner for each day's activities.

CROWN & ANCHOR SOCIETY AND LOYALTY AMBASSADOR

By the end of this amazing cruise vacation everyone will be ready to book another! Make your appointment with the Future Cruise Consultant now because appointments fill up quickly and we want you to come back and visit us again. Sign up in the binder located in the Business Services Center, Deck 6 aft. Crown & Anchor guests, don't miss our Wine Tasting at 3:00 pm in Leonardo's Dining Room.

CHECK YOUR DAILY PLANNER FOR EVENT/DINING TIMES AND LOCATIONS

FOOD & DRINK

Drink of the Day – Drink a *Wipeout* for $5.95 and keep the souvenir glass. A blend of Midori, Banana Liquor Malibu, Pineapple Juice and Orange Juice. Also available for $8.95, a souvenir sloozie glass with any frozen drink.

Wine Appreciation Hour – Sample fine wine & cheese. Only time you can redeem your Crown & Anchor Complimentary voucher.

Premium Wine Tasting – Sample some of our premium wines and gain more knowledge from our expert Winetenders. Nominal fee applies. Dial 3044 for more information. Vintages Wine Bar.

Squeeze Bar – This is the best place to rejuvenate after your gym workout. There is a large selection of freshly squeezed juices, shakes and protein drinks. Which is your favorite?

Champagne Bar – You're with loved ones, making new friends, on an amazing vacation on a state-of-the-art ship; what's not to celebrate? Pop open a bottle of bubbly in our Champagne Bar and toast an amazing vacation.

Cigar Aficionados – Enjoy a first-class cigar as you sip fine Cognac with friends in the Connoisseur Club. Minimum age 21.

ACTIVITIES

Community Bulletin Board – Looking for like minded interest? Stop by the Library on Deck 7 and fill out a card.

First Time Cruisers' Club – Become a Royal Caribbean International expert. Join us at the First Time Cruisers' Club to learn all the tricks to make your cruise vacation even more enjoyable.

Service Club Meeting – Member of service clubs (Rotary, Kiwanis, etc...) are invited for a meeting.

Singles Mingle – Hang out with old friends, make some new ones, enjoy your favorite beverage and dance the night away at this friendly singles mixer.

Men's International Belly Flop Competition – Oh, the prestige, the glamour, the recognition of the belly flop! Come and make a splash with your Cruise Director's Staff in this unique event.

Battle of the Sexes – Which is the stronger sex, men or women? Yes. That answer will never get us in trouble. Join your Cruise Director's Staff for all the crazy competion of men vs. women.

Honeymooners' Get-Together – How much happiness can we squeeze into one room? Join your Cruise Director's Staff to meet your fellow newlyweds and share wedding stories.

Cry Baby Cinema – Parents, tired of missing the latest movies because of your infant or toddler? Welcome to *Cry Baby Cinema* - a new movie experience designed with you and your baby in mind, where cry babies are welcome! Bring your family, gather the baby and enjoy the film.

Family Karaoke – All ages welcome however children under 12 years of age must be accompanied by an adult. Great for the whole family.

Mr. Sexy Legs Competition – Ooh, la, la... Show the women what you've got at the Men's Sexy Legs Contest. This will be one of the most funniest contest you'll ever see.

Scrapbook Workshop – What's the hottest trend in preserving your cruise vacation memories? Scrapbooking! This complimentary hands-on class hosted by your Cruise Director's Staff will give you a jump start at creating a scrapbook page. Pictures are not required. Maximum of 50 guests.

EXPLORER ACADEMY

Learn to Skate Sign Up – Sign up today for our *Learn to Skate* program on Saturday in Studio B.

SPORTS & FITNESS

Polar BodyAge System – Is your body older than you? Do you know your body age can be older than your actual age? So how old is yours? Find out with our Polar BodyAge system, as seen on The Today Show, CNN and Oprah.

Specialty Fitness Classes – There is a wide variety of specialty fitness classes such as Yoga, Pilates and stretching onboard.

SeaTrek Dive Shop – Scuba diving in these beautiful waters is indescribable. It'll open up a whole new world to you, but first, you need to be certified. Our classes are fun, safe and easy. Best of all, they won't take much of your time. Check at the SeaTrek Dive Shop for more information.

Waverunner Fun – Rent a waverunner and explore the beautiful waters and coastline of Labadee. A very cool way to spend your day. Space is limited so sign up today!

Rock Climbing Wall – It's challenging, it's fun and it's on the top deck. Be your own king of the mountain on the top of the Rock Climbing Wall.

Ice Skating – Show off your best figure eight or pirouette on one of the only ice-skating rinks at sea. Long pants and socks are required. Children under age of 18 must have a parent or legal guardian present to sign a waiver for them.

SPA & BEAUTY

The Freedom of the Seas is located on Deck 12. Dial 6982 for an appointment.

Secret to a Flatter Stomach – Ionithermie Have you ever wondered why you diet and exercise but still can't get the flat stomach you always wanted? Join our Body Specialist to find out how Ionithermie figure-correcting treatment can work for you. After just one treatment, you'll see instant results.

Acupuncture Seminar – Find out how acupuncture can help with pain management, weight loss, stress, detox, and can help you quit smoking. James, our Acupuncturist, will explain how he can help release your blocked energy, restore balance and improve your well-being today.

Bad Hair Days No More – Say "goodbye" to bad-hair days forever. Our international salon stylists will show you the latest color trends and techniques designed to make you look amazing, no matter what the occasion.

ONBOARD SHOPPING

Make-Over Demonstration – Learn all the tricks-of-the-trade by having a professional make-up consultation with Emma, our cosmetic specialist. 2:45 pm, Perfume Shop.

$10 Super Sale – Watches, scarves, belts, necklaces and more - all for just $10 each! Don't miss all the incredible savings at our sidewalk sale.

Champagne Art Auction – You'll be amazed at the phenomenal collection onboard the Freedom of the Seas. Your exciting opportunity to bid on works by Max, Picasso, Dali, Tarkay and hundreds of other artists, many works 40-80% below retail gallery prices. Complimentary champagne, and over $2000 worth of art to be given away!

Digital Camera Seminar – Learn how simple, fun and interesting digital photography can be at this morning's seminar. Seminar starts at 11:30 am, On Air Club, Deck 3.

Photo Shop – Stock up on all your photographic needs from batteries, film, underwater cameras, one-time-use cameras and memory cards. All available onboard.

Australian Airbrushed Tattoos – Visit Tracy poolside today and get a temporary tattoo. These airbrushed tattoos are henna free, safe, FDA approved and fun for all ages.

GAMING

Explorations! & Discover Shopping Prize Bingo – Not one but two great sessions today. Don't miss the fun and the cash! Buy the Bingo Special and have a chance at prizes from our Explorations! Staff, Bar Division and Discover Shopping Guide.

Blackjack Tournament – Feeling lucky today? Enter our Blackjack tournament at the Casino Cashier. You could leave as Freedom of the Seas tournament champion.

Blackjack Fact – Did you know a face card and ace are dealt approximately every 20 hands in Blackjack? That's why it's so popular! Try your luck today.

EVERYTHING ELSE

Stay Connected – Your cell phone may work! What's happening in the office? Well, find out by using your cell phone onboard. Check out the Cellular Service brochure in your stateroom or at the Guest Relations Desk. Service provided for AT&T/ Cingular Wireless, Sprint and T-Mobile users.

Smoking Policy – The Freedom of the Seas has designated smoking areas. Guests are asked to only smoke in specific areas.

Katherine Louise Calder

Godmother
Freedom of the Seas
Read more about her in Boleros, Deck 4

Royal Caribbean INTERNATIONAL
Get out there.

DAILY PLANNER

FOLD HERE ↑

DAY 2

Freedom of the Seas SM
AT SEA
Monday, June 12, 2006

DINING SCHEDULE & AFTERNOON SNACKS

BREAKFAST

6:00 am – 9:00 am	Grab and Go Breakfast, Plaza Bar	Deck 11
7:30 am – 11:00 am	Breakfast, Jade & Windjammer Café	Deck 11
8:00 am – 10:00 am	Leonardo's Dining Room	Deck 3

LUNCH

11:30 am – 3:00 pm	Jade & Windjammer Café	Deck 11
Noon – 2:00 pm	Leonardo's Dining Room	Deck 3

DINNER	CAPTAIN'S GALA DINNER	FORMAL
6:00 pm	Main Seating, All Dining Rooms	Decks 3, 4, 5
8:30 pm	Second Seating, All Dining Rooms	Decks 3, 4, 5

Please also note that bare feet, short pants and tank tops are not permitted at dinner.

6:30 pm – 9:00 pm	Casual Dinner, Windjammer Café	Deck 11
6:30 pm – 9:00 pm	Casual Dinner, Jade	Deck 11

Casual Dinner in Jade and the Windjammer Café is buffet style.
Please note that bare feet and tank tops are not permitted in these venues at dinner.

SPECIALTY DINING

6:00 pm – 9:45 pm	**Chops Grille,** The best steak on the high seas $20 dining fee per person applies, gratuities included (Reservations recommended, Dial 3055)	Deck 11
6:00 pm – 9:45 pm	**Portofino,** Intimate Italian Dining $20 dining fee per person applies, gratuities included (Reservations recommended, Dial 3035)	Deck 11

Dress Suggestion: Smart Casual. No short pants please. Guests 13 years and older are welcome. Please allow approximately two hours for dinner. Cancellations with less than 24 hours notice will be charged a fee of $10 per person.

FOLD HERE ↑

SNACKS

24 hours	Café Promenade	Deck 5
11:00 am – 11:00 pm	Ben & Jerry's Ice Cream (Nominal fee)	Deck 5
11:30 am – 1:00 am	Johnny Rockets ($3.95 cover charge per person, gratuities included.) Johnny Rockets is reserved for Adventure Ocean Kids' dinner 5:30 pm – 7:00 pm	Deck 12
11:30 am – 3:00 am	Sorrento's Pizza	Deck 5
3:00 pm – 5:00 pm	Afternoon Tea, Windjammer Café	Deck 11
Midnight – 1:00 am	Midnight Delights, Boleros & Casino Royale	Deck 4

Please Note: Ben & Jerry's, Café Promenade and Sorrento's Pizza will be closed during the Captain's Welcome Aboard Reception from 7:30 pm – 8:30 pm.

MORNING ACTIVITIES

24 hours	Community Bulletin Board, Library	Deck 7
24 hours	Cards & Games available, Seven Hearts Card Room	Deck 14
24 hours	Discover Shopping Information	Channel 22
7:30 am	Balance in Motion ✻, ShipShape Center Center	Deck 11
8:00 am	Salsamania ✻, ShipShape Center	Deck 11
8:30 am	New Balance Walk-a-Mile ✻, Sky Bar	Deck 12
9:00 am	Pathway to Yoga ✻, ($10 fee), ShipShape Center	Deck 11
9:00 am	Daily Trivia Sheet available, Library	Deck 7
9:00 am – 5:00 pm	Quiet Zone, Library	Deck 7
9:30 am	First Time Cruisers' Club, Schooner Bar	Deck 4
9:30 am – 11:00 am	FREEDOM-ICE.COM ticket distribution, On Air Club	Deck 3
10:00 am	Adult Dodgeball Tournament ✻, Basketball Court	Deck 13
10:00 am	Cry Baby Cinema: *Must Love Dogs*, Screening Room	Deck 2
10:00 am	Discover Shopping Show, Studio B	Deck 3
10:00 am	Seminar: *Burn Fat Fast*, ShipShape Center	Deck 11
10:45 am – 11:15 am	Cash Prize Bingo: *Cards on Sale*, Pharaoh's Palace	Deck 5
11:00 am	Seminar: *Bad Hair Day*, Freedom Day Spa	Deck 12
11:00 am	PowerBox Conditioning Class ✻, ($10 fee), ShipShape Center	Deck 11
11:00 am	Seminar: *Detox for Weight Loss*, ShipShape Center	Deck 11
11:15 am	Cash Prize Bingo Games Begin, Pharaoh's Palace	Deck 5
11:15 am – noon	Sign up for our *Learn to Skate* Program, Studio B	Deck 3

FOLD HERE ↑

OPEN HOURS

A Clean Shave/Barber	8:00 am – 8:00 pm	Deck 5	
Adventure Ocean Back Deck	24 hours (Ages 12-17 only)		Deck 12
Art Gallery	24 hours	Deck 3	
Australian Airbrushed Tattoos	10:30 am – 5:30 pm	Deck 11,	Poolside
Casino Royale Slots	9:00 am – late		
Tables	1:00 pm – late	Deck 4	Dial 3171
Discover Shopping Desk	5:00 pm – 6:00 pm		
	7:30 pm – 8:30 pm	Deck 5	
Doctor's Hours	9:00 am – 11:00 am		
	5:00 pm – 7:00 pm	Deck 1 aft	Dial 51
Emergency	***(In case of)***		Dial 911
Explorations! Desk	10:00 am – noon		
	2:00 pm – 4:00 pm		
	7:00 pm – 9:00 pm	Deck 5	Dial 3936
Floral Cart	9:00 am – noon		
	5:00 pm – 9:00 pm	Deck 5, Royal Promenade	
Freedom Day Spa	8:00 am – 8:00 pm	Deck 12	Dial 6982
Hair & Beauty Salon	8:00 am – 8:00 pm		
Accupuncture	8:00 am – 8:00 pm		
Teeth Whitening	8:00 am – 8:00 pm		
Guest Relations	24 hours	Deck 5	Dial 0
H2O Zone	8:00 am – 8:00 pm	Deck 11	

Per United States Public Health, pull ups/swimmer's or diapers are not permitted in the swimming pools including the H2O Zone. All children must be potty trained.

Language Assistance	10:00 am – 11:45 am		
	6:00 pm – 7:30 pm	Deck 6	
Library	24 hours	Deck 7	
Loyalty Ambassador	9:00 am – 5:30 am	Deck 6	
Medical Facility	9:00 am – 11:00 am		
	5:00 pm – 7:00 pm	Deck 1 aft	Dial 51
Operator	24 hours		Dial 0
Assistance for Overseas calls	8:00 am – 11:00 pm		Dial 0
	(All outside phone calls are $7.95 per minute)		
Photo Gallery and Shop	9:00 am – 11:00 pm	Deck 3	Dial 3870
royal caribbean onlineSM	24 hrs	Deck 8	
SeaTrek Dive Shop	9:00 am – 6:00 pm	Deck 11	Dial 3889
ShipShape Center	6:30 am – 10:00 pm	Deck 11	
Shops On Board	9:00 am – 11:00 pm	Deck 5	Dial 3867
Sports Outlet	9:00 am – 5:00 pm	Deck 13	
Swimming Pool	24 hours		

Selected Pools and Hot Tubs are open 24 hours for your enjoyment. Guests are reminded there is no Lifeguard on duty. The Solarium Pool, Hot Tubs and the Solarium area are reserved for guests 16 and older only. Children must be supervised in the two Main Pools and the H2O Zone at all times. As a courtesy to other guests, reserving of deck chairs is not permitted. Chairs left vacant for 30 minutes or longer may be reassigned by the Pool Attendant to another waiting guest. Royal Caribbean International is not responsible for theft or loss of property left on deck chairs or pool areas. Reserved for Navigators (*ages 12-14*) from 10:30 pm – 11:30 pm.

Tuxedo Rentals	9:00 am – 11:00 am	Royal Promenade, Deck 5

SPORTS DECK ACTIVITIES & OPEN HOURS

(Adult sporting events are for guests 18 and older.)

Basketball Court - Open Play	22 hours (Reserved 10:00 am & 5:00 pm)	Deck 13
FlowRider		Deck 13
Advanced Stand-Up Surfing	8:15 am – 9:00 am	
Boogie Boarding	9:00 am – 1:00 pm	
Stand-Up Surfing	1:00 pm – 3:00 pm	
Boogie Boarding	3:00 pm – 5:00 pm	

Please wear a t-shirt. Please remove all items of jewelry. A FlowRider waiver must be completed every day of the cruise.

Freedom Fairways Golf	23 hours (Reserved 5:00 pm)	Deck 13
Ping Pong Tables	24 hours	Deck 14
Rock Climbing Wall		Deck 13
Beginner Climbing Session	9:00 am – 10:00 am	
Open Sessions	10:00 am, 11:00 am, 1:00 pm, 2:00 pm 3:00 pm, 4:00 pm & 5:00 pm	

Sign-ups for the session will begin on the hour and last for 15 minutes. The session will last one hour. Please bring socks, t-shirt & short/pants.

Shuffleboard	24 hours	Deck 4
Sports Pool		Deck 11
Lap Swimming	6:00 am – 10:00 am	
Adult Belly Flop Contest	12:30 pm	
Adult Water Aerobics	3:00 pm	

For further information and descriptions of onboard services and policies, please read important information on the back of your Day 1 Cruise News.

MORNING ACTIVITIES		
11:15 am – noon	Advanced Ice Skating, (Must have own skates), Studio B	Deck 3
11:30 am	Seminar: *Accupuncture, The Point of Wellbeing,* Freedom Day Spa	Deck 12
11:30 am	Seminar: *Digital Photo,* On Air Club	Deck 3

AFTERNOON ACTIVITIES		
Noon	Captain's Report, Public Address System	
Noon	Adult Speed Climb Challenge ✵, Rock Climbing Wall	Deck 13
Noon	World Cup, *USA vs. Czech Republic,* Schooner Bar	Deck 4
Noon	Pilates Reformer Group Class ✵, (15 fee), ShipShape Center	Deck 11
Noon	Blackjack Tournament, Casino Royale	Deck 4
Noon – 12:30 pm	Krooze Komics Entertain, Royal Promenade	Deck 5
Noon – 12:45 pm	Ice Skating, Studio B	Deck 3
12:15 pm	Service Club Meeting, Boleros	Deck 4
12:30 pm	Belly Flop Competition, Poolside	Deck 11
12:45 pm – 1:30 pm	Ice Skating, Studio B	Deck 3
12:45 pm – 1:30 pm	Aqua Babies Play Time, (Ages 6-18 months), On Air Club	Deck 3
1:00 pm	Family Basketball Tournament ✵, Basketball Court	Deck 13
1:00 pm	Screening Room Movie: *Must Love Dogs*	Deck 2
1:00 pm	Scrapbook Workshop, (Maximum of 50 guests), Olive or Twist	Deck 14
1:00 pm	Champagne Art Auction Preview, Pharoah's Palace	Deck 5
1:00 pm – 2:00 pm	Making Sense of Wine, (Nominal fee), Vintages Wine Bar	Deck 5
1:30 pm	Mr. Sexy Legs Contest, Poolside	Deck 11
1:45 pm	Champagne Art Auction, Pharoah's Palace	Deck 5
1:45 pm – 2:30 pm	Aqua Tots Play Time, (Ages 18-36 months), On Air Club	Deck 3
2:00 pm	Pub Trivia, Bull & Bear	Deck 5
2:00 pm	Seminar: *Eat More To Weigh Less,* ShipShape Center	Deck 11
2:00 pm	Informal Bridge Play, Seven Hearts Card Room	Deck 14
2:15 pm – 2:45 pm	Krooze Komics Entertain, Poolside	Deck 11
3:00 pm	Adult Water Aerobics ✵, Sports Pool	Deck 11
3:00 pm	The After the Party Jewelry Show, On Air Club	Deck 3
3:00 pm	Seminar: *Secrets to a Flatter Stomach,* Freedom Day Spa	Deck 12
3:00 pm – 4:00 pm	Crown & Anchor Wine Tasting, Leonardo's Dining Room	Deck 3
3:30 pm	Seminar: *Live Longer, Look Younger,* ShipShape Center	Deck 11
3:30 pm	Cash Prize Bingo: *Cards on Sale,* Arcadia Theatre	Deck 3
4:00 pm	Screening Room Movie: *Must Love Dogs*	Deck 2
4:00 pm	Cash Prize Bingo: *Games Begin,* Arcadia Theatre	Deck 3
4:00 pm	Wheels in Motion ✵, ($10 fee), ShipShape Center	Deck 11
5:00 pm	Teen Dodgeball Tournament ✵, Basketball Court	Deck 13
5:00 pm	Boot Camp Circuit Class ✵, ShipShape Center	Deck 11
5:00 pm	FREEDOM-ICE.COM, (Tickets required), Studio B	Deck 3
5:00 pm	Friends of Bill W. Meeting, Skylight Chapel	Deck 15
5:00 pm – 6:30 pm	Formal Portraits, Centrum	Decks 3,4,5

EVENING ACTIVITIES		
7:00 pm – 1:00 am	Cigar Aficionados, Connoisseur Club	Deck 5
7:30 pm – 8:00 pm	Meet Captain Bill Wright, Pharaoh's Palace	Deck 5
7:30 pm – 8:30 pm	Captain's Welcome Aboard Reception, (All Guests), Royal Promenade	Deck 5
7:30 pm – 8:45 pm	Formal Portraits, Centrum	Decks 3,4,5
8:00 pm	NHL Hockey Game: *Edmonton vs. Carolina,* Studio B	Deck 3
8:15 pm	Captain's Welcome Speech, Flying Bridge	Deck 5
9:00 pm	Celebrity Showtime, (Main Seating Guests), Arcadia Theatre	Decks 3,4
10:15 pm	Battle of the Sexes, Pharaoh's Palace	Deck 5
10:15 pm	Honeymooners' Get-Together, Poolside	Deck 11
10:00 pm – 11:00 pm	Formal Portraits, Centrum	Decks 3,4,5
10:00 pm – 11:00 pm	Family Karaoke, On Air Club	Deck 3
10:45 pm	Celebrity Showtime, (Second Seating Guests), Arcadia Theatre	Decks 3,4
11:00 pm – midnight	Jazz Under the Stars, Poolside	Deck 11
11:15 pm – 12:30 am	Rock-a-Rokie, (Ages 18 & over), Pharaoh's Palace	Deck 5
11:45 pm	Search for Singles, The Crypt	Decks 3,4
12:15 am	Late Night Comedy, (Adults only), Arcadia Theatre	Decks 3,4

✵ – Denotes ShipShape Dollars awarded for this activity.

MUSIC AND DANCING

Live Bands & DJs

Island Band	*With Caribean Fusion* 12:45 pm – 1:30 pm, 3:15 pm – 4:00 pm & 4:30 pm – 5:30 pm Poolside, Deck 11 (Weather permitting)
Classical Melodies	*With Viva Expressia* 2:00 pm – 3:00 pm, Royal Promenade, Deck 5
Piano Melodies	*With Tara Davis* 5:00 pm – 5:45 pm, Schooner Bar, Deck 4
Guitar Music	*With Leandro* 5:00 pm – 6:00 pm & 9:00 pm – 9:45 pm Olive or Twist, Deck 14 6:45 pm – 7:30 pm, Vintages Wine Bar
Live Dining Room Music	*With Tara Davis* 6:00 pm – 7:00 pm & 8:30 pm – 9:30 pm All Dining Rooms
Captain's Welcome Aboard Reception	*With Musical Director Greg Carger & The Freedom of the Seas Orchestra* 7:30 pm – 8:30 pm, Royal Promenade, Deck 5
Rock-a-Rokie	*With our Showband The Stingrays* 11:15 pm – 12:30 am, Pharaoh's Palace, Deck 5
Easy Listening Music	*With Viva Expressia* 10:00 pm – 12:30 am, Olive or Twist, Deck 14
Pub Entertainment	*Vocal Guitarist Jimmy Blakemore Plays Requests* 10:00 pm – 1:00 am, Bull & Bear, Deck 5
Latin Dance Music	*With Sol y Arena* 7:30 pm – 8:15 pm & 10:00 pm – 1:00 am Boleros, Deck 4
Piano Bar Entertainment	*With Peter Ritz* 10:00 pm – 1:00 am, Schooner Bar, Deck 4
Jazz Under The Stars	*With the Australian Jazz Quartet* 11:00 pm – 12:30 am, Poolside, Deck 11
Nightclub Dancing	*With DJ Jamie* 10:30 pm – midnight & 2:00 am – late The Crypt, Decks 3 & 4 Guests must be 18 years or older for admission and photo identification will be required if you appear to be under 18.
Celebrity DJ	*With DJ Smoke from the Scratch DJ Academy* Midnight – 2:00 am, The Crypt, Decks 3 & 4 Guests must be 18 years or older for admission and photo identification will be required if you appear to be under 18.

Please note, the Viking Crown Lounge which includes Olive or Twist is for adults 18 and older only after 10:00 pm.

BAR SERVICE HOURS

***Proof of Age** – Guests eitghteen to twenty (18-20) years of age are welcome to enjoy beer and wine. Guests twenty-one (21) years of age and older are welcome to enjoy all alcoholic beverage. ** Applicable regulatory age restrictions apply while the ship is in port and until the vessel enters international waters. Picture Identification required.*

Café Promenade Bar	7:00 am – 3:00 am	Deck 5
Wipe Out Bar	9:00 am – 7:00 pm	Deck 13
Squeeze Bar	9:00 am – 7:00 pm	Deck 11
Pool Bar (Reservered from 10:00 pm – 11:00 pm for Honeymooners)	9:00 am – 7:00 pm 10:00 pm – midnight	Deck 11
Sky Bar	11:00 am – 7:00 pm	Deck 12
Solarium Bar	11:00 am – 7:00 pm	Deck 11
Plaza Bar	11:30 am – 11:30 pm	Deck 11
Sorrento's Bar	11:30 am – 3:00 am	Deck 5
Boleros	Noon – 2:00 am	Deck 4
Bull & Bear Pub	Noon – 2:00 am	Deck 5
Casino Bar	Noon – late	Deck 4
Vintages Wine Bar	1:00 pm – 1:00 am	Deck 5
Olive or Twist	1:00 pm – 1:30 am	Deck 14
Schooner Bar	1:00 pm – 2:00 am	Deck 4
On Air Club	3:00 pm – 1:00 am	Deck 3
Champagne Bar	3:00 pm – 1:00 am	Deck 5
Studio B	4:30 pm – 6:00 pm 8:00 pm – 10:30 pm	Deck 3
Connoisseur Club Cigar Lounge	7:00 pm – 1:00 am	Deck 5
Fuel Bar (15-17 years)	8:00 pm – 1:00 am	Deck 12
Arcadia Theatre	8:15 pm – midnight	Deck 3
Pharaoh's Palace	10:00 pm – 1:00 am	Deck 5
The Crypt (Adults only)	11:00 pm – late	Decks 3,4
DRINK OF THE DAY	**WIPEOUT**	**$5.95**

〔3日目〕メキシコのコスメル寄港日

CRUISE COMPASS

TEAR HERE ↑

DAY 3

Freedom of the Seas℠
COZUMEL, MEXICO
Tuesday, June 13, 2006

TODAY'S FORECAST
Scattered Thunderstorms, 86°F

SUNRISE 6:43 am
SUNSET 8:26 pm

NEXT PORT-OF-CALL
Grand Cayman

Welcome to Cozumel, Mexico, one of the most famous dive and snorkeling destinations in the world. Other attractions include amazing beaches, unique Mexican markets and handicrafts, underwater caves and caverns and ancient Mayan ruins. Enjoy all your adventures ashore today!

IMPORTANT TENDERING INFORMATION
Freedom of the Seas will be anchoring off Cozumel, and tendering guests ashore. In order to ensure a safe and successful tendering operation, please read the departure information below.

If your tour meets in Arcadia Theatre – You should arrive there at least 10 minutes before the time printed on your ticket.

If your tour meets on the pier in Cozumel – You must be on the tender at least 45 minutes before the time printed on your ticket. If you have a tour ticket then a tender ticket is not required.

Please remember that the last tender back to the ship leaves the dock at 3:30 pm. You must be on the pier by 3:15 pm in order to return to the Freedom of the Seas prior to sailing. Guests returning to the ship do not require tender tickets.

EXPLORATIONS! Shore Excursions and Port Information
There are still many exciting tours available in our upcoming port(s)-of-call. Stop by the Explorations! Desk for more information.

TEAR HERE ↑

DISCOVER SHOPPING ASHORE
See Tasha & Karl on the gangway today from 8:00 am – 10:00 am with diamond VIP Cards, free gemstone stickers and so much more!

TONIGHT'S ENTERTAINMENT

Your Cruise Director, Ken Rush, proudly presents

MARQUEE
A tribute to shows that LIGHT up Broadway
Starring the Royal Caribbean Singers and Dancers
Showtime for Second Seating Guests 7:00 pm (Pre-Dinner Show)
Showtime for Main Seating Guests 9:00 pm
Arcadia Theatre, Decks 3 & 4

Please be reminded that the saving of seats and video taping of shows is strictly prohibited. Also, children in the first three rows of the theatre must be accompanied by a parent or guardian.

DANCIN' IN THE STREET – 70's STYLE
Music from the 70's with the Cruise Director's Staff. Join the fun and dance on the Royal Promenade Street.
10:15 pm, Royal Promenade, Deck 5

70's SONG TRIBUTE
Royal Caribbean Singers starring **Melissa** and **Craig**
10:45 pm, Royal Promenade, Deck 5

THE QUEST!
Are you ready? Don't miss the craziest game on the high seas. Doors open at 10:45 pm, Studio B, Deck 3 aft. The fun begins at 11:15 pm.

CROWN & ANCHOR DINNER — **Casual or 70's Attire**
Main Seating – **6:00 pm** in Leonardo's, Isaac's & Galileo's Dining Rooms, Decks 3, 4 & 5
Second Seating – **8:30 pm** in Leonardo's, Isaac's & Galileo's Dining Rooms, Decks 3, 4 & 5

TEAR HERE ↑

ARRIVAL: 7:00 am (Ship's time)
FIRST TENDER: 7:30 am (Ship's time)
LAST TENDER: 3:30 pm (Ship's time)
DEPARTURE: 4:00 pm (Ship's time)
GANGWAY: Deck 1

TODAY'S TIPS
- **Please note: Cozumel is one hour behind ship's time.**
- **Exchange Rate** – The approximate rate of exchange is $1 U.S.D. to 11 Mexican pesos.
- **SeaPass Boarding Card** – Don't forget to bring your SeaPass Boarding Card with you today when you leave for Cozumel. You will not be allowed off or on the ship without it. A security checkpoint will be set up at the gangway for your safety.
- **Moped & Motor Scooter Advisory** – Guests are strongly discouraged against renting mopeds or motor scooters in port(s)-of-call. Unfamiliar surroundings, unusual traffic and weather conditions, etc. may result in an accident. You may wish to consider an organized tour or taxi instead.
- **Important Departure Information** – In order to ensure a smooth departure on Sunday, we kindly request all guests to please complete the Departure Information sheet in your stateroom and return to the Guest Relations Desk before 6:00 pm on Thursday evening.
- **Pre-Paid Gratuities** – The suggested guideline for gratuities for the Stateroom Attendant and Dining Room Staff may be charged to your SeaPass account if you choose. Please contact the Guest Relations Desk, Deck 5.

Get ready for fun! Check out the Daily Planner for each day's activities.

CROWN & ANCHOR SOCIETY AND LOYALTY AMBASSADOR
Today is our Welcome Back Party for all those guests who have sailed with us before. Get together to share your favorite stories with other Crown & Anchor members. Please bring your invitation. If you've sailed with Royal Caribbean International before but haven't received an invitation, please contact the Guest Relations Desk. (Consult your invitation for time and location.)

CHECK YOUR DAILY PLANNER FOR EVENT/DINING TIMES AND LOCATIONS

FOOD & DRINK

Drink of the Day – Drink a *Margarita Azul* for $5.95 and keep the glass. This is a cool drink for those hot Mexican days, a blend of Tequila, Triple Sec and Blue Curacao. Also a souvenir sloozie glass for $8.95 with any frozen drink.

Champagne Bar – A little bubbly and a view. Grab your old and new friends and enjoy a drink in refined elegance. Get your Compass signed by your Cruise Director Ken and get 20% off in the Champagne Bar only.

Schooner Bar – Classic, dark teaks; oiled ropes; the smell of leather. The image that Royal Caribbean International had in mind when designing our original bar, the Schooner Bar. Create timeless memories as you have a drink and hum along to the piano melodies of Peter Ritz, Deck 4.

Cigar Aficionados – Enjoy a first-class cigar as you sip fine Cognac with friends in the Connoisseur Club. Munimum age 21.

Wine Trivia – Who is the famous Hollywood producer, who also make fine wine? For answers, visit Vintages.

ACTIVITIES

Napkin Folding Class – Learn all about it from your Cruise Director's Staff.

Family Parent Trap – Calling all families! Join the fun.

Family Rock & Roll Night – For an hour, we open up the adult nightclub to families with kids of all ages for music, fun and games.

Team Trivia – Test your knowledge with your Cruise Director's Staff and fellow guests.

Adult Karaoke – Awaken your inner rock or pop star as you step up to the mic tonight at Karaoke. Don't forget to bring a camera with you.

Screening Room Movie: *Batman Begins.* Action adventure starring Christian Bale and Katie Holmes. Limited seating.

SPORTS & FITNESS

Polar BodyAge System – Rewind the clock! Discover how to take up to 10 years off your body's age with our Polar BodyAge System.

Rock Climbing Wall – Feel a rush of adrenaline as you climb hand over hand up our 44-foot rock climbing wall. Instructional sessions start on the hour. Please bring socks. Children under the age of 18 must have a parent or legal guardian sign a waiver. No reservations required.

FlowRider – Join us for the ride of your life! The FlowRider simulates a real surfing and body boarding experience. 30,000 gallons of water per minute RUSH underneath the rider at 30 mph creating a five-foot ocean-like wave. Riders must be at least 58" to surf.

SPA & BEAUTY

The Freedom Day Spa located on Deck 12 is open in port. Dial 6982 for an appointment.

Ionithermie Weight Loss – Discover a whole new you with Ionithermie, a revolutionary treatment that can remove up to eight inches off your hips, thighs and abdomen. We use algae therapy to remove toxins from your body, leaving you with firmer skin. Call for more information.

Next Generation Skin Care – It's called Glycolic, and it's the best resurfacing skin care for aging, sun-damaged, and acne-prone skin. Radiant skin is just around the corner with this advanced skin care solution.

Therapy Facial – This new, effective, non-invasive face therapy will soften fine lines, skin discoloration, enlarged pores, sun and age spots, and scars from chicken pox or acne. The result after just one treatment is youthful, more vibrant skin. Who doesn't need a miracle?

SHOPPING

Jewelry – Visit our onboard jewelry experts tonight on the Royal Promenade. This beautiful layered gold jewelry comes with a lifetime guarantee. Prices start from just $1.00 per inch. Special offer: Purchase a 21" necklace and get a matching 7" bracelet FREE.

Photo Gallery – Just arrived! Brand new Silk Folios for your portraits that come with a FREE photo of the ship. Come and see the great offers available.

Crystal Memories – Keep your formal portraits for life! Have them inscribed into crystal.

Art Gallery – An evening with Picasso. Picasso is unquestionably regarded as the most influential artist of the 20th Century. Stop by the Art Gallery this evening and see the amazing collection of several hand signed masterworks from this Spanish genius.

Australian Airbrushed Tattoos – Temporary tattoo designs include: Chinese symbols for passion, love and happiness – Only $7.95

GAMING

Earn The Nickname "Lucky" – Flip on RCTV in your stateroom to learn a new game. Then head down to the casino on Deck 4 to test your skills. Blackjack, Craps, Roulette, Texas Hold'em, Let It Ride, Caribbean Stud, and 3-Card Poker. Try the latest in Vegas-style slots including the very popular "Wheel of Fortune." Casino Royale has them all. So come and enjoy the fun and exciting atmosphere. Who knows – you could go home a big winner. It's that easy.

Cash Prize Bingo! – Jackpot stands at over $2000. Join the fun and good luck.

EVERYTHING ELSE

Someone is Missing You – Even though you've only been gone a few days, let them know you miss them, too, with one of our many ways to access your e-mail. Get online with royal caribbean online℠ workstations, or our new Wi-Fi wireless access. To help in choosing the best option for you, pick up a brochure in the Internet Café.

Quiet Zone – Looking for a nice quiet place onboard? Try the Library on Deck 7. All guests are requested to respect silence in this area.

Smoking Policy – The Freedom of the Seas has designated smoking areas. Guests are asked to only smoke in specific areas.

Nostalgic Bite – What's not to love about the 50's? Great music, great style, but most of all, great burgers and shakes. Yearning for some classic American fare? Then swing by our own Royal Caribbean Johnny Rockets.® Friendly, old-fashioned service awaits you.

OPEN HOURS

A Clean Shave/Barber *(Appointment advisable)*	8:00 am – 8:00 pm	Deck 5	Dial 2981
Adventure Ocean Back Deck	24 hours(Ages 12-17 only)	Deck 12	
Art Gallery	7:00 pm – 10:00 pm	Deck 3	
Australian Airbrushed Tattoos	3:00 pm – 5:00 pm	Deck 11	Poolside
	8:00 pm – 11:00 pm	Deck 5, Royal Promenade	
Casino Royale	4:15 pm – late	Deck 4	Dial 3170
Discover Shopping Guide	3:00 pm – 4:00 pm	Deck 5	Dial 3830
	7:30 pm – 8:30 pm		
Doctor's Hours	9:00 am – 11:00 am		
	5:00 pm – 7:00 pm	Deck 1 aft	Dial 51
Emergency	*(In case of emergency)*		Dial 911
Explorations! Desk	6:30 am – 9:30 am	Deck 5	Dial 3936
	6:30 pm – 9:00 pm		
Floral Cart	5:00 pm – 9:00 pm	Deck 5, Royal Promenade	
Freedom Day Spa	8:00 am – 10:00 pm	Deck 12	Dial 6982
Hair & Beauty Salon	8:00 am – 10:00 pm		
Teeth Whitening	8:00 am – 10:00 pm		
Guest Relations	24 hours	Deck 5	Dial 0
H2O Zone	8:00 am – 8:00 pm	Deck 11	

Per United States Public Health, pull ups/swimmer's or diapers are not permitted in the swimming pools including the H2O Zone. All children must be potty trained.

Language Assistance	10:00 am – 11:45 am		
	6:00 pm – 7:30 pm	Deck 6	
Library	24 hours	Deck 7	
Loyalty Ambassador	9:00 am – 11:00 am		
	3:00 pm – 7:00 pm	Deck 6	
Medical Facility	9:00 am – 11:00 am		
	5:00 pm – 7:00 pm	Deck 1 aft	Dial 51
Operator	24 hours		Dial 0
Assistance for Overseas calls	8:00 am – 11:00 pm		Dial 0
	(All outside phone calls are $7.95 per minute)		
Photo Gallery	4:00 pm – 11:00 pm	Deck 4	Dial 2973
Photo Shop	4:30 pm – 11:00 pm	Deck 4	
royal caribbean onlineSM	24 hrs	Deck 8	
SeaTrek Dive Shop	7:00 am – 9:00 am	Deck 11	
	4:00 pm – 6:00 pm		
ShipShape Center	6:30 am – 10:00 pm	Deck 11	
Shops On Board	4:00 pm – 11:00 pm	Deck 5	Dial 3867
Sports Outlet	4:00 pm – 7:00 pm & 10:00 pm – 11:30 pm	Deck 13	
Swimming Pools	24 hours		

Selected Pools and Hot Tubs are open 24 hours for your enjoyment. Guests are reminded there is no Lifeguard on duty. The Solarium Pool, Hot Tubs and the Solarium area are reserved for guests 16 and older only. Children must be supervised in the two Main Pools and the H2O Zone at all times. As a courtesy to other guests, reserving of deck chairs is not permitted. Chairs left vacant for 30 minutes or longer may be reassigned by the Pool Attendant to another waiting guest. Royal Caribbean International is not responsible for theft or loss of property left on deck chairs or pool areas. Reserved for Teens *(ages 15-17)* from 11:45 pm – late.

Tuxedo Rentals			Dial 7094

SPORTS DECK ACTIVITIES & OPEN HOURS

(Adult sporting events are for guests 18 and older.)

Basketball Court - Open Play	21 hours (Reserved 11:00 am, 4:00 pm & 5:00 pm)	Deck 13
FlowRider		Deck 13
Advanced Stand-Up Surfing	8:15 am – 9:00 am	
Stand-Up Surfing	9:00 am – 10:00 am	
Boogie Boarding	10:00 am – 11:00 am	
Teen (12-14) Boogie Boarding	4:00 pm – 5:00 pm	
Boogie Boarding	5:00 pm – 6:00 pm	
Teen (15-17) Boogie Boarding	10:00 pm – 11:30 pm	

Please wear a t-shirt. Please remove all items of jewelry. A FlowRider waiver must be completed every day of the cruise.

Freedom Fairways Golf	24 hours	Deck 13
Ping Pong Tables	24 hours	Deck 14
Rock Climbing Wall		Deck 13
Open Climbing	9:00 am – 10:00 am	
Kids (6-12 years) Session	10:00 am – 11:00 am	
Open Climbing	4:00 am – 7:00 pm	

Please bring dry socks, t-shirt & shorts/pants. A Rock Climbing Wall waiver must be completed once a cruise. Children must be 6 years of age or older to climb. Adults must be able to fit into the XXL harness to climb.

Shuffleboard	24 hours	Deck 4
Sports Pool		Deck 11
Lap Swimming	6:00 am – 10:00 am	
Volleyball Court – Open Play	4:00 pm – 6:00 pm	Deck 13

For further information and descriptions of onboard services and policies, please read important information on the back of your Day 1 Cruise News.

✖ – Denotes ShipShape Dollars awarded for this activity.

DAILY PLANNER

FOLD HERE ↑

DAY 3

Freedom of the SeasSM
COZUMEL, MEXICO
Tuesday, June 13, 2006

Arrival:	7:00 am (Ship's time)
First Tender:	7:30 am (Ship's time)
Last Tender:	3:30 pm (Ship's time)
Departure:	4:00 pm (Ship's time)
Gangway	Deck 1

DINING SCHEDULE & AFTERNOON SNACKS

BREAKFAST

6:00 am – 8:00 am	Grab & Go Breakfast, Plaza Bar	Deck 11
6:30 am – 11:00 am	Breakfast, Jade & Windjammer Café	Deck 11
7:00 am – 9:00 am	Leonardo's Dining Room	Deck 3

LUNCH

11:30 am – 3:30 pm	Jade/Windjammer Café	Deck 11

CROWN & ANCHOR DINNER — **CASUAL ATTIRE**

6:00 pm	Main Seating, All Dining Rooms	Decks 3, 4, 5
8:30 pm	Second Seating, All Dining Rooms	Decks 3, 4, 5

Please also note that bare feet, short pants and tank tops are not permitted at dinner.

6:30 pm – 9:00 pm	Casual Dinner, Windjammer Café	Deck 11
6:30 pm – 9:00 pm	Casual Dinner, Jade	Deck 11

Casual Dinner in Jade and the Windjammer Café is buffet style.
Please note that bare feet and tank tops are not permitted in these venues at dinner.

FOLD HERE ↑

SPECIALTY DINING

6:00 pm – 9:45 pm	**Chops Grille,** The best steak on the high seas $20 dining fee per person applies, gratuities included (Reservations recommended, Dial 3055)	Deck 11
6:00 pm – 9:45 pm	**Portofino,** Intimate Italian Dining $20 dining fee per person applies, gratuities included (Reservations recommended, Dial 3035)	Deck 11

Dress Suggestion: Smart Casual. No short pants please. Guests 13 years and older are welcome. Please allow approximately two hours for dinner. Cancellations with less than 24 hours notice will be charged a fee of $10 per person.

SNACKS

24 hours	Café Promenade	Deck 5
11:30 am – 3:00 am	Sorrento's Pizza	Deck 5
1:30 pm – 1:00 am	Johnny Rockets, ($3.95 cover charge per person, gratuities included)	Deck 12
2:00 pm – midnight	Ben & Jerry's Ice Cream (Nominal fee)	Deck 5
3:30 pm – 5:00 pm	Afternoon Tea, Windjammer Café	Deck 11
Midnight – 1:00 am	Midnight Delights, Boleros & Casino Royale	Deck 4

MORNING ACTIVITIES

24 hours	Community Bulletin Board, Library	Deck 7
24 hours	royal caribbean online℠	Deck 8
24 hours	Cards & Games available, Seven Hearts Card Room	Deck 14
24 hours	Discover Shopping Information	Channel 22
7:00 am	**Estimated Time of Arrival in Cozumel**	
7:30 am (approximate)	**First Tender Leaves**	
7:30 am	Balance in Motion ✖, ShipShape Center	Deck 11
8:00 am	Fab Abs✖, ShipShape Center	Deck 11
8:00 am – 8:45 am	Aqua Babies Play Time, (6-18 months), On Air Club	Deck 3
8:00 am – 10:00 am	Discover Shopping Guide Available, Gangway	Deck 1
8:30 am	New Balance Walk-a-Mile ✖, Sky Bar	Deck 12
9:00 am	Daily Trivia Sheet available, Library	Deck 7

FOLD HERE ↑

MORNING ACTIVITIES

9:00 am – 9:45 am	Aqua Tots Play Time, (18-36 months), On Air Club	Deck 3
9:00 am – 5:00 pm	Quiet Zone, Library	Deck 7
11:15 am	Family Trivia, Schooner Bar	Deck 4

AFTERNOON ACTIVITIES

2:15 pm	Freedom Challenge, Schooner Bar	Deck 4
3:30 pm	**Last Tender Departs from Cozumel**	
3:30 pm	Napkin Folding Class, Schooner Bar	Deck 4
3:15 pm	Cash Prize Bingo Cards on Sale, Pharaoh's Palace	Deck 5
3:45 pm	Cash Prize Bingo, Pharaoh's Palace	Deck 5
4:00 pm	**Freedom of the Seas sails for Grand Cayman (334 nautical miles)**	
4:00 pm	Screening Room Movie: *Batman Begins*	Deck 2
4:00 pm	Aerobics in Motion ✵, ShipShape Center	Deck 11
4:00 pm	Seminar: Three Minute Make Over, Freedom Day Spa	Deck 12
4:00 pm	Adult Water Polo ✵, Sports Pool	Deck 11
4:00 pm – 5:00 pm	Latin Sailaway, Poolside	Deck 11
4:30 pm	Seminar: Next Generation Skin Care, Freedom Day Spa	Deck 12
5:00 pm	Friends of Bill W. Meeting, Skylight Chapel	Deck 15
5:00 pm – 6:30 pm	Casual Portraits will be taken, Centrum	Decks 3,4
5:15 pm	Family Parent Trap, On Air Club	Deck 3

EVENING ACTIVITIES

7:00 pm	Marquee Production Show, (Second Seating Guests), (Pre-Dinner show), Arcadia Theatre	Decks 3,4
7:00 pm – 10:00 pm	An Evening with Picasso, Art Gallery	Deck 3
7:00 pm – 1:00 am	Cigar Aficionados, Connoisseur Club	Deck 5
7:30 pm – 8:45 pm	Casual Portraits will be taken, Centrum	Decks 3,4
7:45 pm	Wine Tenders Perform: Heard It Through The Grape Vine, Royal Promenade	Deck 5
9:00 pm	Marquee Production Show, (Main Seating Guests), Arcadia Theatre	Decks 3,4
9:00 pm	NBA Playoffs - Game 3, *Miami Heat vs. Dallas Mavericks*, Pharaoh's Palace	Deck 5
10:00 pm – 11:00 pm	Family Rock & Roll Night, The Crypt	Decks 3,4
10:00 pm – 11:00 pm	Casual Portraits will be taken, Centrum	Decks 3,4
10:15 pm	Dancin' in the Street - 70's Style, Royal Promenade	Deck 5
10:15 pm – 11:00 pm	Dancing Margarita & Mojito Madness, Royal Promenade	Deck 5
10:45 pm	70's Song Tribute, Royal Promenade	Deck 5
10:45 pm	The Quest! Game Show Doors Open, Studio B	Deck 3
11:15 pm	The Quest! Game Show, (Adults only), Studio B	Deck 3
11:45 pm – late	Teen Disco Pool Party, (15-17 years), Poolside	Deck 11
Midnight – 1:00 am	Tequila Karaoke, (Adults only), On Air Club	Deck 3

✵ – *Denotes ShipShape Dollars awarded for this activity.*

Ship's Agent in Cozumel, Mexico:
Agencia Consignataria del Sureste, Telephone #
011 - 529 - 872 - 1508

MUSIC AND DANCING
Live Bands & DJs

Latin Sailaway Party	*With Sol y Arena* 4:00 pm – 5:30 pm Poolside, Deck 11
Guitar Music	*With Leandro* 5:00 pm – 6:00 pm & 7:30 pm – 8:30 pm Olive or Twist, Deck 14
Live Dining Room Music	*With Tara Davis* 6:00 pm – 7:00 pm & 8:30 pm – 9:30 pm All Dining Rooms
Latin Music in the Sky	*With Sol y Arena* 7:45 pm – 9:00 pm Flying Bridge, Royal Promenade, Deck 5
Piano Bar Entertainment	*With Peter Ritz* 10:00 pm – 1:00 am, Schooner Bar, Deck 4
Standards Dancing & Jazz	*With the Australian Jazz Quartet* 10:30 pm – 1:00 am Olive or Twist, Deck 14
Latin Dance Music	*With Sol y Arena* 11:00 pm – 12:30 am Boleros, Deck 4
Nightclub Dancing	*With DJ Jamie* 11:00 pm – late, The Crypt, Decks 3 & 4 Guests must be 18 years or older for admission and photo identification will be required if you appear to be under 18.
Pub Entertainment	*Vocal Guitarist Jimmy Blakemore Plays Requests* 11:15 pm – 1:00 am, Bull & Bear, Deck 5

Please note, the Viking Crown Lounge which includes Olive or Twist is for adults 18 and older only after 10:00 pm.

BAR SERVICE HOURS

*Proof of Age – Guests eighteen to twenty (18-20) years of age are welcome to enjoy beer and wine. Guests twenty-one (21) years of age and older are welcome to enjoy all alcoholic beverage. ** Applicable regulatory age restrictions apply while the ship is in port and until the vessel enters international waters. Picture Identification required.*

DRINK OF THE DAY	MARGARITA AZUL	$5.95
Café Promenade Bar	7:00 am – 3:00 am	Deck 5
Wipe Out Bar	9:00 am – 11:00 am 2:30 pm – 7:00 pm	Deck 13
Squeeze Bar	9:00 am – 7:00 pm	Deck 11
Pool Bar	9:00 am – 7:00 pm	Deck 11
Plaza Bar	11:30 am – 11:30 pm	Deck 11
Sorrento's Bar	11:30 am – 3:00 am	Deck 5
Solarium Bar	1:00 pm – 7:00 pm	Deck 11
Sky Bar	1:00 pm – 9:00 pm	Deck 12
Bull & Bear Pub	1:00 pm – 2:00 am	Deck 5
Boleros	2:00 pm – 2:00 am	Deck 4
Schooner Bar	2:00 pm – 2:00 am	Deck 4
Vintages Wine Bar	3:00 pm – 1:00 am	Deck 5
On Air Club	3:00 pm – 1:30 am	Deck 3
Olive or Twist	3:00 pm – 1:30 am	Deck 14
Pharaoh's Palace	3:45 pm – 5:30 pm 9:00 pm – 1:00 am	Deck 5
Champagne Bar	4:00 pm – 1:00 am	Deck 5
Casino Bar	5:00 pm – late	Deck 4
Arcadia Theatre	6:45 pm – 10:00 pm	Deck 3
Connoisseur Club Cigar Lounge	7:00 pm – 1:00 am	Deck 5
The Crypt (Family hour) **The Crypt** (Adults only)	10:00 pm – 11:00 pm 11:00 pm – late	Decks 3, 4
Studio B	10:15 pm – 11:30 pm	Deck 3

〔4日目〕グランドケイマン島寄港日

CRUISE COMPASS

TEAR HERE ↑

DAY 4

Freedom of the Seas℠
GEORGETOWN, GRAND CAYMAN
Wednesday, June 14, 2006

TODAY'S FORECAST
Isolated Thunderstorms, 88°F

SUNRISE 6:47 am
SUNSET 8:02 pm

NEXT PORT-OF-CALL
Montego Bay, Jamaica

If it's day four, it must be Grand Cayman – Welcome! Get ready to do and see things you only imagined in your dreams. Like swimming with and hand-feeding stingrays, seeing a house made entirely of shells or touring a turtle farm. This tropical paradise is perfect for any outdoor adventure. Be sure to check with the Explorations! Desk for remaining availability on today's shore excursions. And if you've already booked a shore excursion, make sure you check your tickets for departure times and locations. And, as always, don't forget your SeaPass card and photo ID when going ashore. If you're already thinking about tomorrow's adventures in Montego Bay, Jamaica, feel free to sign up for those excursions today!!

IMPORTANT TENDERING INFORMATION – Freedom of the Seas will be anchoring off in Grand Cayman, and tendering guests ashore. In order to ensure a safe and successful tendering operation, please read the departure information below.
If your tour meets in Arcadia Theatre – You should arrive there at least 10 minutes before the time printed on your ticket.
If your tour meets on the pier in Grand Cayman – You must be on the tender at least 45 minutes before the time printed on your ticket. If you have a tour ticket then a tender ticket is not required. Please remember that the last tender back to the ship leaves the dock at 5:00 pm. You must be on the pier by 4:45 pm in order to return to the Freedom of the Seas prior to sailing. Guests returning to the ship do not require tender tickets.
EXPLORATIONS!℠ Shore Excursions and Port Information – Be sure to check with our Explorations! Desk & SeaTrek Dive Shop for remaining availability on our shore excursions for Grand Cayman, Labadee and Miami. If you've already booked a shore excursion, make sure you check your tickets for departure times and locations.
DISCOVER SHOPPING ASHORE – See your Discover Shopping Team on the gangway today from 9:00 am – 11:00 am with VIP cards and discounts on liquor! Plus, don't miss their shopping desk hours tonight for last minute tips on Montego Bay.

TEAR HERE ↑

TONIGHT'S ENTERTAINMENT

FREEDOM-ICE.COM

In Search of Freedom
Starring Our International Ice Cast & Special Guest Stars
George & Anna
Showtimes: 7:00 pm and 9:00 pm, Studio B, Center Ice, Deck 3
Doors open 30 minutes prior to showtime.
(Tickets are required for admission. If tickets are still available, they can be obtained at the Guest Relations Desk on Deck 5. Guests without tickets will be permitted entrance once the show begins if seats are available.)

SUMMER BREEZE SONGS
With Featured Singers
Karina and Evan
Songs of the Summer
10:00 pm – 10:30 pm
Royal Promenade, Deck 5

SURFING & CLIMBING UNDER THE STARS
Join your Sports Staff and do some night time surfing, body boarding and rock climbing. Please note the Sports Deck is an alcohol-free zone.
10:00 pm – 11:15 pm, Deck 13
Don't forget to join us afterwards for our Dance Party by the pool starting at 11:15 pm.

THE LOVE & MARRIAGE GAME SHOW
Join your Cruise Director, Ken, as he puts your fellow shipmates through the matrimonial test of time. Don't miss the laughs! Recommended for Adults Only.
Showtime is 10:30 pm, Arcadia Theatre, Decks 3 & 4

SURFIN' UNDER THE STARS DANCE PARTY
Tonight it's time for some socializing and dancing. Come to the pool and experience our dance party featuring our Showband, *The Stingrays*, and your Cruise Director's Staff.
11:15 pm – 12:45 am, Poolside, Deck 11

SURF DINNER Casual or Surf Attire
Main Seating – 6:00 pm in Leonardo's, Isaac's & Galileo's Dining Rooms, Decks 3, 4 & 5
Second Seating – 8:30 pm in Leonardo's, Isaac's & Galileo's Dining Rooms, Decks 3, 4 & 5

TEAR HERE ↑

ARRIVAL	9:00 am (Ship's time)
FIRST TENDER	9:30 am (Ship's time)
LAST TENDER	5:00 pm (Ship's time)
DEPARTURE	5:30 pm (Ship's time)
GANGWAY	Deck 1

TODAY'S TIPS

- **Please note: Grand Cayman is one hour behind ship's time.**
- **SeaPass Boarding Card** – Don't forget to bring your SeaPass Boarding Card with you today when leaving for Grand Cayman. You will not be allowed off or on the ship without it. A security checkpoint will be set up at the gangway for your safety.
- **Moped & Motor Scooter Advisory** – Guests are strongly discouraged against renting mopeds or motor scooters in port(s)-of-call. Unfamiliar surroundings, unusual traffic and weather conditions, etc. may result in an accident. You may wish to consider an organized tour or taxi instead.
- **Customs Allowance** – Your Customs exemption allows you duty-free status on $800 in merchandise from any of our ports or purchased onboard. One carton of 200 cigarettes - must be 18 years or older. Excess U.S.–manufactured cigarettes made for export only will be seized. Foreign-manufactured tobacco products will be subject to duty and internal revenue taxes. One hundred cigars - must be 18 years or older. No Cuban cigars are permitted into the United States. One liter of alcohol - must be 21 years or older. One additional liter is permitted if it was manufactured in Jamaica. Applicable internal revenue taxes and duties will be assessed on overages.
- **Pre-Paid Gratuities** – The suggested guideline for gratuities for the Stateroom Attendant and Dining Room Staff may be charged to your SeaPass account if you choose. Please contact the Guest Relations Desk before midnight on Friday.
- **Important Departure Information** – In order to ensure a smooth departure on Sunday, we kindly request all guests to please complete the Departure Information sheet in your stateroom and return to the Guest Relations Desk before 6:00 pm on Thursday.

You Can Earn Royal Points – Imagine if all your everyday purchases added up to a Royal Caribbean International cruise vacation. They can with the Royal Caribbean International Visa credit card. Apply today and start earning points toward onboard credits, stateroom upgrades and even free cruise vacations. Stop by the Future Cruise Consultant's desk on Deck 6 for an application form or look for one in the Cruise Services Directory in your stateroom.

Receive Up To $200 Per Stateroom – Already planning your next Royal Caribbean International cruise vacation? Reserve your stateroom while you're still onboard and receive an onboard credit of up to $200 per stateroom – a special offer available only while you're onboard this cruise vacation. Just stop by the Future Cruise Consultant's desk on Deck 6 and schedule your appointment.

CHECK YOUR DAILY PLANNER FOR EVENT/DINING TIMES AND LOCATIONS

FOOD & DRINK

Drink of the Day – Drink a *Pineapple Rum Cake* for $5.95 and keep the souvenir glass. Enjoy the taste of this beautiful island, a blend of Bacardi, vanilla, pineapple juice and a touch of Grenadine. Mmmmm.... Also available for $8.95, a souvenir sloozie glass with any frozen drink.

Wine Trivia – Where is Maipau Valley. Find out at Vintages, Deck 5.

Boleros – This is the place to let go. Enjoy the hot Latin rhythms and the ice-cold mojitos. It's always a fiesta at Boleros.

Chops Grille – If you're a meat lover, have we got the place for you! Choice meats are our specialty. So order up one of your favorite cuts and savor every tender bite. Don't miss out – make your reservation today.

Cook Books – Love the food? Well now you can eat like you're always on a Royal Caribbean International cruise vacation with our exclusive cook book. Available at the Guest Relations Desk or from your Dining Room Waiter.

ACTIVITIES

Family Karaoke – All ages welcome however children under 12 years of age must be accompanied by an adult. Great for the whole family.

Crayola® – Coloring outside the lines - we encourage it! Adventure Art by Crayola® is a fun, creative and imaginative way for your kids to learn about all the fantastic places they're going to visit. And with five different supervised groups (ages 3-5, 6-8, 9-11, 12-14 and 15-17), your kids are sure to make some friends their own age.

Majority Rules – What's the correct answer? You decide. Join your Cruise Director's Staff for all the fun.

Adult Karaoke – It's your chance to be a star. Warm up your voice and practice your performance for tonight's Karaoke.

Movie – Staying onboard today? Catch a movie in our very own Screening Room. Today's movie: *Red Eye.* Drama thriller starring Gillian Murphy and Rachael McAdams. Limited seating.

Brain Teasing Trivia – Test your knowledge with your Cruise Director's Staff and fellow guests. Win prizes.

SPORTS & FITNESS

Polar BodyAge System – Take the BodyAge Challenge! Are you in your 40's but feel more like 60? Or maybe you're 70 and feel like a kid! Find out your body's true age with our unique system.

Walk-a-Mile – Start out your morning with fresh air and breath-taking ocean views during a one-mile walk on deck.

SPA & BEAUTY

The Freedom Day Spa is located on Deck 12. Dial 6982 for an appointment.

Hot Stone Massage – If you've never experienced a hot stone massage, run, don't walk, to make your appointment at the Day Spa. We will drench your body in the sensual aromas of the Orient and massage you with hot volcanic stones on key energy points of your body to balance the spirit. This experience is nothing short of divine.

HydraLift Facial – This anti-aging, deep-cleansing facial is customized to suit your individual needs, and the results will make your skin glow. Book today and we will include a free collagen eye treatment.

Color Me Beautiful Seminar – Here's your chance to find out the perfect colors for you – learn the secrets from a hair and makeup specialist.

Body Blitz – Our lime and ginger exfoliation, followed by a 25 minute massage, is the ultimate way to rejuvenate your body for the adventures to come. Your body will be polished to perfection, leaving you ready for anything! Only $89 per session.

ONBOARD SHOPPING

Designer Watch Sale – These prices are definitely worth watching. Check out the sale on our designer watches. Name brands such as Seiko, Citizen, Anne Klein and Fossil, to name just a few. Sale starts today.

Yummy, Yummy – Taste the delicious Tortuga rum cake straight from Grand Cayman. Lowest price guaranteed. Tasting begins at 5:30 pm at our Shops On Board.

Lifestyle Portraits – It's fun, it's fresh, it's new. Stop by the lifestyle Studio on Deck 4 and let us capture your personality.

Photo Art Exhibition – Tonight only! This is your chance to view our onboard collection of Photo Art. All photographs for sale are original prints from National Geographic, AP, Time Life, and many more. Champagne reception starts at midnight.

Art Seminar – "The Movers and Shakers in Art History." Ever wonder how a work of art by Picasso can sell for over $100 million? Do you ever see works of art by the great masters and think "Hey - my kid could do that?" Join us this evening to learn more about the top four artists of the 20th century and why their artworks are highly sought after by collectors the world over.

Australian Airbrushed Tattoos – Any grandparent that brings a photo of their grandchildren will get a tattoo for only $7.95 and a free color shimmer upgrade.

GAMING

Slot Machines – Cha-ching! The grand jackpots on our slot machines are growing bigger and bigger. We've already paid out thousands of dollars on this cruise vacation alone. It could be your turn!

Cash & Spa Prize Bingo – The jackpot has not been won yet, the winner could be you! Win big cash prizes plus Spa treatments in our Royal Raffle drawing.

EVERYTHING ELSE

Community Bulletin Board – Looking for like minded interest? Stop by the Library on Deck 7 and fill out a card.

Quiet Zone – Looking for a nice quiet place onboard? Try the Library on Deck 7. All guests are requested to respect silence in this area.

royal caribbean online℠ – No matter where your adventure takes you, you can always stay in touch. Check e-mail and stocks, or e-mail a friend. This service is available 24 hours a day, at a nominal charge.

Smoking Policy – The Freedom of the Seas has designated smoking areas. Guests are asked to only smoke in specific areas.

Nostalgic Bite – Grab your leather jackets and poodle skirts; we're going back to the '50s! Juicy burgers, thick shakes, tasty fries and friendly service are all served up '50s style at Johnny Rockets.® It's the perfect place to grab a quick, casual bite.

DAY 4

Freedom of the Seas SM
GRAND CAYMAN
Wednesday, June 14, 2006
Arrival: 9:00 am (Ship's Time)
First Tender: 9:30 am (Ship's Time)
Last Tender: 5:00 pm (Ship's Time)
Departure: 5:30 pm (Ship's Time)
Gangway: Deck 1

FOLD HERE ↑

OPEN HOURS

A Clean Shave/Barber 8:00 am – 8:00 pm Deck 5
(Appointments advisable)
Adventure Ocean Back Deck 24 hours(Ages 12-17 only) Deck 12
Art Gallery 7:00 pm – 10:00 pm Deck 3
Australian Airbrushed Tattoos 7:30 pm – 11:00 pm Deck 5 Royal Promenade
Casino Royale 5:45 pm – late Deck 4 Dial 3171
Discover Shopping 5:00 pm – 6:00 pm
7:30 pm – 8:30 pm Deck 5
Doctor's Hours 9:00 am – 11:00 am
5:00 pm – 7:00 pm Deck 1 aft Dial 51
Emergency *(In case of)* Dial 911
Explorations! Desk 8:30 am – 10:30 am
6:30 pm – 9:00 pm Deck 5 Dial 3936
Floral Cart 5:00 pm – 9:00 pm Deck 5 Royal Promenade
Freedom Day Spa 8:00 am – 10:00 pm Deck 12 Dial 6982
Hair & Beauty Salon 8:00 am – 10:00 pm
Teeth Whitening 8:00 am – 10:00 pm
Guest Relations 24 hours Deck 5 Dial 0
H2O Zone 8:00 am – 8:00 pm Deck 11
Per United States Public Health, pull ups/swimmer's or diapers are not permitted in the swimming pools including the H2O Zone. All children must be potty trained.
Language Assistance 10:00 am – 11:45 am
6:00 pm – 7:30 pm Deck 6
Library 24 hours Deck 7
Loyalty Ambassador 9:00 am – 11:00 am Deck 6
3:00 pm – 7:00 pm
Medical Facility 9:00 am – 11:00 am
5:00 pm – 7:00 pm Deck 1 aft Dial 51
Operator 24 hours Dial 0
Assistance for Overseas calls
8:00 am – 11:00 pm Dial 0
(All outside phone calls are $7.95 per minute)
Photo Gallery 5:00 pm – 11:00 pm Deck 3 Dial 3870
Photo Shop 5:30 pm – 11:00 pm Deck 3 Dial 3870
royal caribbean onlineSM 24 hrs Deck 8
SeaTrek Dive Shop 8:00 am – 10:00 am Deck 11 Dial 3889
4:00 pm – 6:00 pm
ShipShape Center 6:30 am – 10:00 pm Deck 11
Shops On Board 5:30 pm – 11:00 pm Deck 5 Dial 3867
Sports Outlet 5:30 pm – 7:00 pm Deck 13
10:00 pm – 11:15 pm
Swimming Pools 24 hours
Selected Pools and Hot Tubs are open 24 hours for your enjoyment. Guests are reminded there is no Lifeguard on duty. The Solarium Pool, Hot Tubs and the Solarium area are reserved for guests 16 and older only. Children must be supervised in the two Main Pools and the H2O Zone at all times. As a courtesy to other guests, reserving of deck chairs is not permitted. Chairs left vacant for 30 minutes or longer may be reassigned by the Pool Attendant to another waiting guest. Royal Caribbean International is not responsible for theft or loss of property left on deck chairs or pool areas.
Tuxedo Rentals Dial 7094

SPORTS DECK ACTIVITIES & OPEN HOURS

(Adult sporting events are for guests 18 and older.)
Basketball Court - Open Play 23 hours Deck 13
(Reserved 5:00 pm)
FlowRider Deck 13
Advanced Stand-Up Surfing 8:15 am – 9:00 am
Surf Camp 9:00 am – 10:00 am
Boogie Boarding 10:00 am – 11:00 am & 6:00 pm – 7:00 pm
Stand-up Surfing 5:00 pm – 6:00 pm
Please wear a t-shirt. Please remove all items of jewelry. A FlowRider waiver must be completed every day of the cruise.
Freedom Fairways Golf 24 hours Deck 13
Golf Simulator
Reservations 9:00 am – 11:00 am Deck 13
5:00 pm – 7:00 pm
(Reservation required, $25 fee)
You may reserve a time with one of your Sports Team at the Rock Climbing Wall Desk.
Ping Pong Tables 24 hours Deck 14
Rock Climbing Wall Deck 13
Open Climbing Session 9:00 am – 11:00 am & 5:00 pm – 6:00 pm
Bouldering Sessions 6:00 pm – 7:00 pm
Please bring socks, t-shirt and shorts/pants. Clothing must be dry. No wet clothing please. A Rock Climbing Wall waiver must be completed once a cruise. Children must be 6 years of age or older to climb. Adults must be able to fit into XXL harness to climb.
Shuffleboard 24 hours Deck 4
Sports Pool Deck 11
Lap Swimming 6:00 am – 10:00 am
Family Floating Golf 3:00 pm

For further information and descriptions of onboard services and policies, please read important information on the back of your Day 1 Cruise News.

DINING SCHEDULE & AFTERNOON SNACKS

BREAKFAST

6:00 am – 8:00 am	Grab and Go Breakfast, Plaza Bar	Deck 11
7:00 am – 11:00 am	Breakfast, Jade & Windjammer Café	Deck 11
7:30 am – 9:30 am	Leonardo's Dining Room	Deck 3

LUNCH

11:30 am – 4:00 pm	Jade & Windjammer Café	Deck 11

DINNER	SURF DINNER	CASUAL
6:00 pm	Main Seating, All Dining Rooms	Decks 3, 4, 5
8:30 pm	Second Seating, All Dining Rooms	Decks 3, 4, 5

Please also note that bare feet and tank tops are not permitted at dinner.

6:30 pm – 9:00 pm	Casual Dinner, Windjammer Café	Deck 11
6:30 pm – 9:00 pm	Casual Dinner, Jade	Deck 11

Casual Dinner in Jade and the Windjammer Café is buffet style.

Please note that bare feet and tank tops are not permitted in these venues at dinner.

FOLD HERE ↑

SPECIALTY DINING

6:00 pm – 9:45 pm	**Chops Grille**, The best steak on the high seas $20 dining fee per person applies, gratuities included (Reservations recommended, Dial 3055)	Deck 11
6:00 pm – 9:45 pm	**Portofino**, Intimate Italian Dining $20 dining fee per person applies, gratuities included (Reservations recommended, Dial 3035)	Deck 11

Dress Suggestion: Smart Casual. No short pants please. Guests 13 years and older are welcome. Please allow approximately two hours for dinner. Cancellations with less than 24 hours notice will be charged a fee of $10 per person.

SNACKS

24 hours	Café Promenade	Deck 5
11:30 am – 3:00 am	Sorrento's Pizza	Deck 5
1:30 pm – 1:00 am	Johnny Rockets ($3.95 cover charge per person, gratuities included.) Johnny Rockets is reserved for Adventure Ocean Kids' dinner 5:30 pm – 7:00 pm	Deck 12
2:00 pm – midnight	Ben & Jerry's Ice Cream (Nominal fee)	Deck 5
4:00 pm – 5:00 pm	Afternoon Tea, Windjammer Café	Deck 11
11:30 pm – 12:30 am	Surfing Midnight Delights, Poolside	Deck 11

MORNING ACTIVITIES

24 hours	Community Bulletin Board, Library	Deck7
24 hours	royal caribbean online sm	Deck 8
24 hours	Discover Shopping Talk, Continuous	Channel 22
24 hours	Cards & Games available, Seven Hearts Card Room	Deck 14
7:30 am	Balance in Motion ★, ShipShape Center	Deck 11
8:00 am	Pilates Reformer Group Class ★ ($15 fee), ShipShape Center	Deck 11
8:00 am	Wheels in Motion ★,($10 fee), ShipShape Center	Deck 11
8:00 am – 8:45 am	Aqua Babies Play Time, (Ages 6-18 months) On Air Club	Deck 3
8:30 am	New Balance Walk-a-Mile ★, Sky Bar	Deck 12
9:00 am	**Estimated Time of Arrival in Grand Cayman**	
9:00 am	Pathway to Yoga ★ ($10 fee), ShipShape Center	Deck 11
9:00 am	Daily Trivia Sheet available, Library	Deck 7
9:00 am – 11:00 am	Discover Shopping Guide available for questions, Gangway	Deck 1
9:00 am – 9:45 am	Aqua Tots Play Time, (Ages 18-36 months), On Air Club	Deck 3
9:30 am (approximate)	**First Tender Ashore**	

FOLD HERE ↑

AFTERNOON ACTIVITIES

1:00 pm	Crayola Art Workshop (Adult class, maximum of 25) Cloud Nine	Deck 14
3:00 pm	Brain Teasing Trivia, Pharaoh's Palace	Deck 5
3:00 pm	Family Floating Golf ⚹, Sports Pool	Deck 11
3:00 pm	World Cup, *Germany vs. Poland*, Schooner Bar	Deck 4
4:00 pm	Screening Room Movie: *Red Eye*, (Limited seating), Screening Room	Deck 2
4:00 pm	Shape Up, Pump Up ⚹, ShipShape Center	Deck 11
4:00 pm – 5:00 pm	How Merlot Can You Go? Wine Tasting (nominal fee), Vintages	Deck 5
4:30 pm	Cash Prize Bingo: *Cards on Sale*, Pharaoh's Palace	Deck 5
4:30 pm	Seminar: *Color Me Beautiful*, Freedom Day Spa	Deck 12
5:00 pm	Cash Prize Bingo: *Games Begin*, Pharaoh's Palace	Deck 5
5:00 pm	**Last Tender from Grand Cayman**	
5:00 pm	Friends of Bill W. Meeting, Skylight Chapel	Deck 15
5:00 pm	New Balance Pentathlon ⚹, Rock Climbing Wall	Deck 13
5:00 pm	PowerBox Ring Conditioning Class ⚹, ($10 fee), ShipShape Center	Deck 11
5:00 pm – 6:30 pm	Casual Portraits will be taken, Centrum	Decks 3,4
5:30 pm	**Freedom of the Seas sails for Montego Bay, Jamaica (210 nautical miles)**	
5:15 pm	Seminar: *"The Movers and Shakers in Art History,"* On Air Club	Deck 3

EVENING ACTIVITIES

7:00 pm	FREEDOM-ICE.COM Ice Show, (Tickets required), Studio B Center Ice	Deck 3
7:00 pm – 1:00 am	Cigar Aficionados, Connoisseur Club	Deck 5
7:00 pm – 10:00 pm	Classic Masters Exhibition, Art Gallery	Deck 3
7:30 pm – 8:45 pm	Casual Portraits will be taken, Centrum	Decks 3,4
7:45 pm	Majority Rules Gameshow, On Air Club	Deck 3
8:00 pm	NHL Hockey - Game 5, *Edmonton vs. Carolina*, Schooner Bar	Deck 4
8:00 pm – 8:30 pm	Krooze Komics Entertain, Royal Promenade	Deck 5
8:45 pm	Kids' Pirate Parade, Royal Promenade	Deck 5
9:00 pm	FREEDOM-ICE.COM Ice Show, (Tickets required), Studio B Center Ice	Deck 3
9:00 pm – 10:15 pm	Big Band Dancing, Pharaoh's Palace	Deck 5
10:00 pm – 10:30 pm	Summer Breeze Songs, Royal Promenade	Deck 5
10:00 pm – 11:00 pm	Family Karaoke, On Air Club	Deck 3
10:00 pm – 11:00 pm	Casual Portraits will be taken, Centrum	Decks 3,4
10:00 pm – 11:15 pm	Climbing Under The Stars, Rock Climbing Wall	Deck 13
10:00 pm – 11:15 pm	Surfing Under The Stars, FlowRider	Deck 13
10:15 pm – 11:00 pm	Dancing Margarita & Mojito Madness, Royal Promenade	Deck 5
10:30 pm	The Love & Marriage Game Show, (Adults only), Arcadia Theatre	Decks 3,4
11:15 pm – 12:45 am	Adult Karaoke, On Air Club	Deck 3
11:15 pm – 12:45 am	Surfin' Under The Stars Dance Party, Poolside	Deck 11
11:15 pm – 12:45 am	Fresh Pineapple Bar, Poolside	Deck 11
Midnight	Photo Art Exhibition, Photo Gallery	Deck 4

⚹ – Denotes ShipShape Dollars awarded for this activity.

Ship's Agent in Grand Cayman:
Bodden Shipping Agency
Tel. # 1-345-949-6254

MUSIC AND DANCING

Live Bands & DJs

Poolside Music	*With Caribbean Fusion* 3:00 pm – 4:00 pm & 4:15 pm – 5:00 pm Poolside, Deck 11
Sailaway Music	*With Celebrity DJ Smoke from the Scratch DJ Academy* 5:00 pm – 6:00 pm, Poolside, Deck 11 (Weather permitting)
Easy Listening Music	*With Viva Espressia* 5:00 pm – 6:00 pm, Royal Promenade, Deck 5 & 7:30 pm – 8:30 pm, Olive or Twist, Deck 14
Latin Dance Music	*With Sol y Arena* 7:30 pm – 8:30 pm & 10:00 pm – 12:30 am Boleros, Deck 4
Big Band Dancing	*With Musical Director Greg Carger and the Freedom of the Seas Orchestra* 9:00 pm – 10:15 pm Pharaoh's Palace, Deck 5
Standards, Dancing & Jazz	*With the Australian Jazz Quartet* 10:00 pm – 11:30 pm, Olive or Twist, Deck 14
Piano Bar Entertainment	*With Peter Ritz* 11:00 pm – 1:00 am, Schooner Bar, Deck 4
Nightclub Dancing	*With DJ Jamie* 11:00 pm – late, The Crypt, Decks 3 & 4 Adults only. Guests must be 18 years & older for admission and photo identification required if you appear to be under 18.
Surfin' Under The Stars Dance Party	*With the Stingrays* 11:15 pm – 12:45 am, Poolside, Deck 11 (Weather permitting)
New Spin on Jazz	*Join Celebrity DJ Smoke with the Australian Jazz Orchestra* Midnight – 1:15 am, Olive or Twist, Deck 14

Please note, the Viking Crown Lounge which includes Olive or Twist is for adults 18 and older only after 10:00 pm.

BAR SERVICE HOURS

Proof of Age *– Guests eighteen to twenty (18-20) years of age are welcome to enjoy beer and wine. Guests twenty-one (21) years of age and older are welcome to enjoy all alcoholic beverage.* *** Applicable regulatory age restrictions apply while the ship is in port and until the vessel enters international waters. Picture Identification required.*

DRINK OF THE DAY	MARGARITA AZUL	$5.95
Café Promenade Bar	7:00 am – 3:00 am	Deck 5
Squeeze Bar	9:00 am – 7:00 pm	Deck 11
Wipe Out Bar	9:00 am – 11:00 am 2:30 pm – 7:00 pm 10:00 pm – 11:30 pm	Deck 13
Pool Bar	9:00 am – 7:00 pm 11:00 pm – 1:00 am	Deck 11
Plaza Bar	11:30 am – 11:30 pm	Deck 11
Sorrento's Bar	11:30 am – 3:00 am	Deck 5
Bull & Bear Pub	1:00 pm – 2:00 am	Deck 5
Solarium Bar	2:00 pm – 6:00 pm	Deck 11
Boleros	2:00 pm – 2:00 am	Deck 4
Sky Bar	2:00 pm – 7:00 pm 11:00 pm – 1:00 am	Deck 12
Schooner Bar	2:00 pm – 2:00 am	Deck 4
On Air Club	3:00 pm – 1:00 am	Deck 3
Vintages Wine Bar	3:00 pm – 1:00 am	Deck 5
Olive or Twist	3:00 pm – 1:30 am	Deck 14
Pharaoh's Palace	4:00 pm – 6:00 pm 9:00 pm – 10:15 pm	Deck 5
Arcadia Theatre	4:00 pm – 5:30 pm 10:00 pm – midnight	Deck 3
Champagne Bar	4:00 pm – 1:00 am	Deck 5
Casino Bar	5:15 pm – late	Deck 4
Studio B	6:30 pm – 10:00 pm	Deck 3
Connoisseur Club Cigar Lounge	7:00 pm – 1:00 am	Deck 5
Fuel Bar (15-17 years)	8:00 pm – 1:00 am	Deck 12
The Crypt (Adults only)	11:00 pm – late	Decks 3,4

〔5日目〕ジャマイカのモンテゴベイ寄港日

CRUISE COMPASS

TEAR HERE

DAY 5

*Freedom of the Seas*SM
MONTEGO BAY, JAMAICA
Thursday, June 15, 2006

TODAY'S FORECAST
Partly Cloudy, 88°F

SUNRISE 6:35 am
SUNSET 7:47 pm

NEXT PORT-OF-CALL
Labadee, Haiti

It's Day 5 and it's time to explore Montego Bay. Today will be filled with exciting adventures - especially if you're taking advantage of one of our many exciting excursions. If you're not, it's time to book them now. These tours offer an insider's view of this beautiful island. So, don't miss out. And when you come back to the ship, you can keep the excitement going.

EXPLORATIONS!SM Shore Excursions and Port Information
Be sure to check with our Explorations!SM Desk for remaining availability on our shore excursions in Montego Bay. If you've already booked a shore excursion, make sure you check your tickets for departure times and locations. And, as always, don't forget your SeaPass card when going ashore. For detailed Explorations!SM descriptions, tune in to channel 15 on your stateroom TV and book right there through RCTV.

Discover Shopping Ashore – See the Shopping Team on the gangway from 8:00 am to 10:00 am, for buy-one-get-one free Tanzanite stud vouchers, find more, shop more, shop ashore in Montego Bay.

TEAR HERE

WIN-A-CRUISE BINGO
Final chance to buy cards!
7:45 pm – 8:10 pm, Arcadia Theatre, Deck 3
8:10 pm Game begins, Arcadia Theatre, Deck 3
If on second seating dining, the game will be over in time for dinner and main seating guests will have a great seat for the 9:00 pm show.

TONIGHT'S ENTERTAINMENT

FREEDOM-ICE.COM
Starring Our International Ice Cast
& Special Guest Stars ***George & Anna***
Showtimes: 5:00 pm & 7:00 pm, Studio B, Center Ice, Deck 3
Doors open 30 minutes prior to showtime.
Last chance to see this great show.
(Tickets are required for admission. Guests without tickets will be permitted entrance once the show begins if seats are available)

Your Cruise Director, Ken Rush, proudly presents

NOW YOU SEE IT
Starring the Magic of **Drew Thomas**
and featuring the Royal Caribbean Singers and Dancers

Showtime for Main Seating Guests 9:00 pm
Showtime for Second Seating Guests 10:45 pm
Arcadia Theatre, Decks 3 & 4

Please be reminded that the saving of seats and video taping of shows is strictly prohibited. Also, children in the first three rows of the theatre must be accompanied by a parent or guardian.

CALYPSO IN THE SKY
Enjoy our Calypso Band as they play island music.
10:00 pm – 11:30 pm, Flying Bridge, Royal Promenade, Deck 5

DINNER IS SERVED Formal Attire
Main Seating – 6:00 pm in Leonardo's, Isaac's & Galileo's Dining Rooms, Decks 3, 4 & 5
Second Seating – 8:30 pm in Leonardo's, Isaac's & Galileo's Dining Rooms, Decks 3, 4 & 5

TEAR HERE

INAUGURAL Season

ARRIVAL:	7:00 am (Ship's time)
ALL ABOARD:	3:30 pm (Ship's time)
DEPARTURE:	4:00 pm (Ship's time)
GANGWAY:	Deck 1

TODAY'S TIPS

- **SeaPass Boarding Card** – Don't forget to bring your SeaPass Boarding Card with you today when you leave for Montego Bay. You will not be allowed off or on the ship without it. A security checkpoint will be set up at the gangway for your safety.
- **Towels** – Beach Towels have been placed in your stateroom for your personal use when you go ashore. Should you desire an additional towel or replacement, please contact your Stateroom Attendant.
- **Moped & Motor Scooter Advisory** – Guests are strongly discouraged against renting mopeds or motor scooters in port(s)-of-call. Unfamiliar surroundings, unusual traffic and weather conditions, etc. may result in an accident. You may wish to consider an organized tour or taxi instead.
- **You're over halfway there!** – Start making a list of everything you still need to do before you leave the ship. Make those spa or specialty dining reservations now or pick up a duty-free souvenir in our Shops On Board.
- **Important Departure Information** – In order to ensure a smooth departure on Sunday, we kindly request all guests to please complete the Departure Information sheet in your stateroom and return to the Guest Relations Desk before 6:00 pm this evening.

Get ready for fun! Check out the Daily Planner for each day's activities.

CHECK YOUR DAILY PLANNER FOR EVENT/DINING TIMES AND LOCATIONS

FOOD & DRINK

Drink of the Day – Drink a *I Shot The Sheriff* for $5.95 and keep the glass. Yeah Mon! A blend of Rum, Vodka, Gin Tia Maria, then topped with Tripple Sec and Cranberry juice. Also, a souvenir sloozie glass for $8.95 with any frozen drink.

Champagne Bar – A little bubbly and a view. Grab your old and new friends and enjoy a drink in refined elegance. Get your Compass signed by your Cruise Director Ken and get 20% off in the Champagne Bar only.

Schooner Bar – Classic, dark teaks; oiled ropes; the smell of leather. The image that Royal Caribbean International had in mind when designing our original bar, the Schooner Bar. Create timeless memories as you have a drink and hum along to the piano melodies of Peter Ritz, Deck 4.

Cigar Aficionados – Enjoy a first-class cigar as you sip fine Cognac with friends in the Connoisseur Club. Munimum age 21.

Wine Trivia – Who is the famous Hollywood producer, who also make fine wine? For answers, visit Vintages.

ACTIVITIES

Crayola Art Workshop – Adult class - Join your Cruise Director's Staff and be a kid all over again! Maximum of 25 guests.

Family Disco Night – For an hour, we open up the adult nightclub to families with kids of all ages for disco music.

Team Trivia – Test your knowledge with your Cruise Director's Staff and fellow guests.

Screening Room Movie: *Charlie and the Chocolate Factory*. Comedy Adventure Starring Johnny Depp. Limited seating.

Lowe's Family Shipbuilding – Calling all families! Join the fun in building your own ship with supplies from our friends at Lowe's.

Family Name That Tune – Have fun with our staff and your own family.

Scrapbooking Workshop – You've escaped to a world of paradise. Join your Cruise Director's Staff and learn how to creatively highlight all those special moments of your formal evening through words and photos.

SPORTS & FITNESS

Polar BodyAge System – Live longer, grow younger, discover your body's real age and learn how to turn back the hands of time! Want to retire at 55 instead of 65, discover why 40 is the new 50 with Polar BodyAge technology.

Rock Climbing Wall – Feel a rush of adrenaline as you climb hand over hand up our 44-foot rock climbing wall. Instructional sessions start on the hour. Please bring socks. Children under the age of 18 must have a parent or legal guardian sign a waiver. No reservations required.

Golf Simulator – Play world-famous courses right here! Reserve a tee time on your stateroom TV. A nominal fee applies.

FlowRider – Join us for the ride of your life! The FlowRider simulates a real surfing and body boarding experience. 30,000 gallons of water per minute RUSH underneath the rider at 30 mph creating a five-foot ocean-like wave. Riders must be at least 58" to surf.

SPA & BEAUTY

The Freedom Day Spa is located on Deck 12 and is open today in port. Dial 6982 for an appointment.

Detox Seaweed Massage – Do you suffer from poor circulation, low energy levels, fluid retention, arthritis or fibromyalgia? If you answered yes to any of the above, sign up for our seaweed massage. This treatment includes a detoxifying seaweed wrap, scalp massage, foot and ankle massage, and culminates in either a 25 or 50-minute body massage. To enhance this treatment further, ask about our unique dry float bed. Today only $150.

Treat Yourself to Aroma-flex – Exploring all day or climbing the Rock Climbing Wall is a workout. And we've got the perfect way to reward you. Aroma-flex – it's the perfect combination of a 25-minute reflexology treatment for your feet and a back massage that will relax, rejuvenate and get you ready for another day of adventure. Today only $89.

Cellulite Reduction Program – Find out what cellulite is and how to reduce or eliminate it. You will learn that detoxifying the body is the first step in minimizing the appearance of cellulite and for losing unwanted inches off your body. After just one treatment, you will immediately notice firmer, smoother skin. Call or visit the Freedom Day Spa to learn more. Today only $99.

ONBOARD SHOPPING

14kt Gold at Massive Savings – Make room in your jewelry box because you can take up to 70% off regular retail prices on a wide selection of necklaces, chains, bracelets and anklets

Jewelry Extravaganza – Up to 50% off on selected diamond and designer costume jewelry.

Cosmetics – Learn new tricks from the world of cosmetics. Sign up for a professional makeover with Emma, our make up artist, in the Perfume Shop.

Formal Portrait Night – Keep this special vacation fresh in your memory with a professional formal portrait. These opportunities don't come along every day.

Old & Modern Masters Art Auction – Have you ever dreamed of owning an original Picasso, Dali, Chagall or Miro? Don't miss today's special art auction featuring the great masters and more. A true collector's auction at surprisingly affordable opening bids. $4000 art raffle, complimentary Champagne, and a free $95 artwork just for attending.

Australian Airbrushed Tattoos – Buy 2 tattoos and receive 1 of equal or lesser value for free!

GAMING

Earn The Nickname "Lucky" – Flip on RCTV in your stateroom to learn a new game. Then head down to the casino on Deck 4 to test your skills. Blackjack, Craps, Roulette, Texas Hold'em, Let It Ride, Caribbean Stud, and 3-Card Poker. Try the latest in Vegas-style slots including the very popular "Wheel of Fortune." Casino Royale has them all. So come and enjoy the fun and exciting atmosphere. Who knows – you could go home a big winner. It's that easy.

Cash Prize Bingo! – The Snowball Jackpot is snowballing! Get your cards before someone else screams BINGO! Jackpot stands at over $5,000.

Win-a-Cruise Bingo Pre-sales – Imagine all the Bingo you could play if you won a free cruise vacation. Beat the rush and get your lucky cards early.

Win-a-Cruise Bingo – Win a Caribbean cruise vacation for two! One game only at 8:10 pm tonight. 20 minutes later you could be the winner.

EVERYTHING ELSE

Someone is Missing You – Even though you've only been gone a few days, let them know you miss them, too, with one of our many ways to access your e-mail. Get online with royal caribbean online℠ workstations, or our new Wi-Fi wireless access.

Quiet Zone – Looking for a nice quiet place onboard? Try the Library on Deck 7. All guests are requested to respect silence in this area.

Crayola® – Coloring outside the lines – we encourage it! Adventure Art by Crayola® is a fun and creative activity normally offered to children 3-17 however today we offer an adult Crayola® class for adults so come bring out the kid in you!

OPEN HOURS

A Clean Shave/Barber 8:00 am – 8:00 pm Deck 5 Dial 2981
(Appointments advisable)
Adventure Ocean Back Deck 24 hours(Ages 12-17 only) Deck 12
Art Gallery 24 hours Deck 3
Australian Airbrushed Tattoos
3:00 pm – 5:30 pm Deck 11 Poolside
7:30 pm – 10:30 pm Deck 5, Royal Promenade
Casino Royale 4:15 pm – late Deck 4 Dial 3170
Discover Shopping Guide 8:00 am – 10:00 am Deck 1 Dial 3830
Doctor's Hours 9:00 am – 11:00 am
5:00 pm – 7:00 pm Deck 1 aft Dial 51
Emergency *(In case of emergency)* Dial 911
Explorations! Desk 7:00 am – 9:30 am Deck 5 Dial 3936
7:00 pm – 9:00 pm
Floral Cart 9:00 am – noon & 5:00 pm – 9:00 pm
Deck 5, Royal Promenade
Freedom Day Spa 8:00 am – 8:00 pm Deck 12 Dial 6982
Hair & Beauty Salon 8:00 am – 8:00 pm
Accupuncture 8:00 am – 8:00 pm
Teeth Whitening 8:00 am – 8:00 pm
Guest Relations 24 hours Deck 5 Dial 0
H2O Zone 8:00 am – 8:00 pm Deck 11
Per United States Public Health, pull ups/swimmer's or diapers are not permitted in the swimming pools including the H2O Zone. All children must be potty trained.
Language Assistance 10:00 am – 11:45 am
6:00 pm – 7:30 pm Deck 6
Library 24 hours Deck 7
Loyalty Ambassador 9:00 am – 11:00 am
3:00 pm – 7:00 pm Deck 6

Medical Facility 9:00 am – 11:00 am
5:00 pm – 7:00 pm Deck 1 aft Dial 51
Operator 24 hours Dial 0
Assistance for Overseas calls
8:00 am – 11:00 pm Dial 0
(All outside phone calls are $7.95 per minute)
Photo Gallery 4:00 pm – 11:00 pm Deck 4 Dial 2973
Photo Shop 4:30 pm – 11:00 pm Deck 4
royal caribbean onlineSM 24 hrs Deck 8
SeaTrek Dive Shop 7:30 am – 9:30 am Deck 11
4:00 pm – 6:00 pm
ShipShape Center 6:30 am – 10:00 pm Deck 11
Shops On Board 4:00 pm – 11:00 pm Deck 5 Dial 3867
Sports Outlet 4:00 pm – 7:00 pm & 10:00 pm – 11:30 pm
Deck 13
Swimming Pools 24 hours
Selected Pools and Hot Tubs are open 24 hours for your enjoyment. Guests are reminded there is no Lifeguard on duty. The Solarium Pool, Hot Tubs and the Solarium area are reserved for guests 16 and older only. Children must be supervised in the two Main Pools and the H2O Zone at all times. As a courtesy to other guests, reserving of deck chairs is not permitted. Chairs left vacant for 30 minutes or longer may be reassigned by the Pool Attendant to another waiting guest. Royal Caribbean International is not responsible for theft or loss of property left on deck chairs or pool areas. Reserved for Teens *(ages 12-14)* from 11:00 pm – 12:30 am.
Tuxedo Rentals 9:00 am – 11:00 am Dial 7094

SPORTS DECK ACTIVITIES & OPEN HOURS

(Adult sporting events are for guests 18 and older.)
Basketball Court - Open Play 21 hours Deck 13
(Reserved 11:00 am, 4:00 pm & 5:00 pm)
FlowRider Deck 13
Advanced Stand-Up Surfing 8:15 am – 9:00 am
Boogie Boarding 9:00 am – 11:00 am & 4:00 pm – 5:00 pm
Teen (12-14) Stand-up Surfing 5:00 pm – 6:00 pm
(Parents, please fill out a waiver for your teens (12-14 years) between 4:30 pm – 5:00 pm in the Living Room, Deck 12)
Stand-Up Surfing 6:00 pm – 7:00 pm
Please wear a t-shirt. Please remove all items of jewelry. A FlowRider waiver must be completed every day of the cruise.
Freedom Fairways Golf 24 hours Deck 13
Golf Simulator 9:00 am – 11:00 am Deck 13
5:00 pm – 7:00 pm
(By reservation, $25 fee)
You may reserve a Tee-Time for the Golf Simulator with one of your Sports Team at the Rock Wall Desk on Deck 13.
Ping Pong Tables 24 hours Deck 14
Rock Climbing Wall Deck 13
Open Climbing 9:00 am – 10:00 am & 4:00 pm – 7:00 pm
Please bring socks, t-shirt and shorts/pants. Clothing must be dry. No wet clothing please. A Rock Climbing Wall waiver must be completed once a cruise. Children must be 6 years of age or older to climb. Adults must be able to fit into XXL harness to climb. Please note, the Rock Climbing Wall will be closed if we are experiencing inclement weather.
Shuffleboard 24 hours Deck 4
Sports Pool Deck 11
Lap Swimming 6:00 am – 10:00 am
Adult Water Aerobics 10:00 am

For further information and descriptions of onboard services and policies, please read important information on the back of your Day 1 Cruise News.

DAILY PLANNER

FOLD HERE ↑

DAY 5

Freedom of the SeasSM
MONTEGO BAY, JAMAICA
Thursday, June 15, 2006
Arrival: 7:00 am (Ship's time)
All Aboard: 3:30 pm (Ship's time)
Departure: 4:00 pm (Ship's time)
Gangway Deck 1

DINING SCHEDULE & AFTERNOON SNACKS

BREAKFAST		
6:00 am – 8:00 am	Grab & Go Breakfast, Plaza Bar	Deck 11
7:00 am – 11:00 am	Breakfast, Jade/Windjammer Café	Deck 11
7:30 am – 9:30 am	Leonardo's Dining Room	Deck 3
LUNCH		
11:30 am – 3:00 pm	Jade/Windjammer Café	Deck 11
Noon – 2:00 pm	Leonardo's Dining Room	Deck 3
DINNER	**VENETIAN FEAST**	**FORMAL ATTIRE**
6:00 pm	Main Seating, All Dining Rooms	Decks 3, 4, 5
8:30 pm	Second Seating, All Dining Rooms	Decks 3, 4, 5
Please also note that bare feet, short pants and tank tops are not permitted at dinner.		
6:30 pm – 9:00 pm	Casual Dinner, Windjammer Café	Deck 11
6:30 pm – 9:00 pm	Casual Dinner, Jade	Deck 11
Casual Dinner in Jade and the Windjammer Café is buffet style.		
Please note that bare feet and tank tops are not permitted in these venues at dinner.		
SPECIALTY DINING		
6:00 pm – 9:45 pm	**Chops Grille,** The best steak on the high seas $20 dining fee per person applies, gratuities included (Reservations recommended, Dial 3055)	Deck 11
6:00 pm – 9:45 pm	**Portofino,** Intimate Italian Dining $20 dining fee per person applies, gratuities included (Reservations recommended, Dial 3035)	Deck 11
Dress Suggestion: Smart Casual. No short pants please. Guests 13 years and older are welcome. Please allow approximately two hours for dinner. Cancellations with less than 24 hours notice will be charged a fee of $10 per person.		
SNACKS		
24 hours	Café Promenade	Deck 5
11:30 am – 1:00 am	Johnny Rockets, ($3.95 cover charge per person, gratuities included) Reserved for Kids' Dinner from 5:30 pm – 7:00 pm.	Deck 12
11:30 am – 3:00 am	Sorrento's Pizza	Deck 5
2:00 pm – midnight	Ben & Jerry's Ice Cream (Nominal fee)	Deck 5
3:00 pm – 5:00 pm	Afternoon Tea, Windjammer Café	Deck 11
Midnight – 1:00 am	Midnight Delights, Boleros & Casino Royale	Deck 4

FOLD HERE ↑

MORNING ACTIVITIES

24 hours	Community Bulletin Board, Library	Deck 7
24 hours	royal caribbean online^{sm}	Deck 8
24 hours	Cards & Games available, Seven Hearts Card Room	Deck 14
24 hours	Discover Shopping Information	Channel 22
7:00 am	**Estimated Time of Arrival in Montego Bay, Jamaica**	
7:30 am	Balance in Motion ✱, ShipShape Center	Deck 11
8:00 am	Wheels in Motion ✱, ($10 fee), ShipShape Center	Deck 11
8:00 am	Shape Up, Step Up ✱, ShipShape Center	Deck 11
8:00 am – 8:45 am	Aqua Babies Play Time, (6-18 months), On Air Club	Deck 3
8:00 am – 10:00 am	Discover Shopping Guide Available, Gangway	Deck 1
8:30 am	New Balance Walk-a-Mile ✱, Sky Bar	Deck 12
9:00 am	Pathway to Yoga ✱, ($10 fee), ShipShape Center	Deck 11
9:00 am	Daily Trivia Sheet available, Library	Deck 7

FOLD HERE ↑

MORNING ACTIVITIES

9:00 am – 9:45 am	Aqua Tots Play Time, (18-36 months), On Air Club	Deck 3
9:00 am – 5:00 pm	Quiet Zone, Library	Deck 7
10:00 am	Adult Water Aerobics ✱, Sports Pool	Deck 11
10:00 am	Pathway to Pilates ✱, ($10 fee), ShipShape Center	Deck 11
11:00 am	Teens 3-on-3 Basketball ✱, Basketball Court	Deck 13
11:30 am	Team Trivia, Schooner Bar	Deck 4

AFTERNOON ACTIVITIES

Noon	World Cup, *England vs. Trinidad*, On Air Club	Deck 3
1:00 pm	Crayola Workshop, (Adult class, maximum of 25 guests), Cloud Nine	Deck 14
2:55 pm	World Cup, *Sweden vs. Paraguay*, On Air Club	Deck 3
3:00 pm	Family Name That Tune, Schooner Bar	Deck 4
3:30 pm	**All Aboard**	
3:30 pm	Old & Modern Masters Art Auction Preview, Pharaoh's Palace	Deck 5
3:30 pm	Cash Prize Bingo: Cards on Sale, Arcadia Theatre	Decks 3,4
3:45 pm – 5:30 pm	Calypso Sailaway Party, Poolside	Deck 11
4:00 pm	**Freedom of the Seas sails for Labadee, Haiti (345 nautical miles)**	
4:00 pm	Scrapbooking Workshop, Olive or Twist	Deck 14
4:00 pm	Old & Modern Masters Art Auction, Pharaoh's Palace	Deck 5
4:00 pm	Screening Room Movie: *Charlie & the Chocolate Factory*	Deck 2
4:00 pm	Seminar: Live Longer, Look Younger, ShipShape Center	Deck 11
4:00 pm	Cash Prize Bingo Games Begin, Arcadia Theatre	Decks 3,4
4:00 pm – 5:00 pm	Great Wine Tasting, (Nominal fee), Vintages Wine Bar	Deck 5
4:00 pm – 6:00 pm	Scratch DJ Class for Teens, (12-17), Fuel Disco	Deck 12
4:15 pm	Lowe's Family Shipbuilding, H2O Zone	Deck 11
4:30 pm	Seminar: Reflexology, Freedom Day Spa	Deck 12
5:00 pm	Aerobics in Motion ✱, ShipShape Center	Deck 11
5:00 pm	Friends of Bill W. Meeting, Skylight Chapel	Deck 15
5:00 pm	Adult Longest Drive Tournament ✱, Golf Simulator	Deck 13
5:00 pm	FREEDOM-ICE.COM, (Tickets required), Studio B Center Ice	Deck 3
5:00 pm – 6:30 pm	Formal Portraits will be taken, Centrum	Decks 3,4,5
5:45 pm – 7:00 pm	Win-a-Cruise Bingo Pre-sales, outside Café Promenade	Deck 5

EVENING ACTIVITIES

7:00 pm	FREEDOM-ICE.COM, (Tickets required), Studio B Center Ice	Deck 3
7:00 pm – 10:00 pm	An Evening with Picasso, Art Gallery	Deck 3
7:00 pm – 1:00 am	Cigar Aficionados, Connoisseur Club	Deck 5
7:30 pm – 8:45 pm	Formal Portraits will be taken, Centrum	Decks 3,4,5
7:45 pm – 8:10 pm	Win-a-Cruise Bingo Pre-sales, Arcadia Theatre	Decks 3,4
7:45 pm – 8:30 pm	Martini Tasting, *Learn how to make your own Martini*, (Nominal fee), Vintages Wine Bar	Deck 5
8:10 pm	Win-a-Cruise Bingo Game Begins, Arcadia Theatre	Decks 3,4
9:00 pm	Now You See It, (Main Seating Guests), Arcadia Theatre	Decks 3,4
9:00 pm	NBA Playoffs - Game 4, *Miami Heat vs. Dallas Maverics*, Schooner Bar	Deck 4
10:00 pm – 11:00 pm	Family Disco Hour, The Crypt	Decks 3,4
10:00 pm – 11:00 pm	Formal Portraits will be taken, Centrum	Decks 3,4,5
10:00 pm – 11:30 pm	Calypso in the Sky, Flying Bridge, Royal Promenade	Deck 5
10:45 pm	Now You See It, (Second Seating Guests), Arcadia Theatre	Decks 3,4
10:00 pm – 11:30 pm	Karaoke, (Adults only), On Air Club	Deck 3
11:45 pm – 12:45 am	Teen Karaoke, (15-17 years), On Air Club	Deck 3

✱ – Denotes ShipShape Dollars awarded for this activity.

Ship's Agent in Montego Bay, Jamaica:
Grace Kennedy Co. LTD, Telephone # 1 - 876 - 979 - 8077

MUSIC AND DANCING

Live Bands & DJs

Calypso Sailaway Party	*With our Island Band Caribbean Fusion* 3:45 pm – 5:30 pm, Poolside, Deck 11
Guitar Melodies	*With Leandro* 5:00 pm – 6:00 pm, Boleros, Deck 4; 7:30 pm – 8:30 pm, Schooner Bar, Deck 4
Cocktail Melodies	*With Pianist Tara Davis* 5:00 pm – 6:00 pm, Olive or Twist, Deck 14; 7:45 pm – 8:45 pm, Vintages Wine Bar, Deck 5
Live Dining Room Music	*With Viva Expressia* 6:00 pm – 7:00 pm & 8:30 pm – 9:30 pm All Dining Rooms
Cocktails and Dancing	*With the Australian Jazz Quartet* 7:45 pm – 8:45 pm, Olive or Twist, Deck 14
Latin Dance Music	*With Sol y Arena* 7:45 pm – 8:45 pm & 10:00 pm – 1:00 am Boleros, Deck 4
Calypso In The Sky	*With Caribbean Fusion* 10:00 pm – 11:30 pm Flying Bridge, Royal Promenade, Deck 5
Piano Bar Entertainment	*With Peter Ritz* 10:00 pm – 1:00 am, Schooner Bar, Deck 4
Standards, Dancing & Jazz	*With the Australian Jazz Quartet* 10:30 pm – 1:00 am, Olive or Twist, Deck 14
Nightclub Dancing	*With DJ Jamie* 11:00 pm – late, The Crypt, Decks 3 & 4. Guests must be 18 years & older for admission and photo identification will be required if you appear to be under 18.
Pub Entertainment	*Vocal Guitarist Jimmy Blakemore Plays Requests* 11:30 pm – 1:30 am, Bull & Bear, Deck 5
Dance Party	*With our Showband The Stingrays* 11:30 pm – 1:00 am, Pharaoh's Palace, Deck 5

Please note, the Viking Crown Lounge which includes Olive or Twist is for adults 18 and older only after 10:00 pm.

BAR SERVICE HOURS

***Proof of Age** – Guests eighteen to twenty (18-20) years of age are welcome to enjoy beer and wine. Guests twenty-one (21) years of age and older are welcome to enjoy all alcoholic beverage. ** Applicable regulatory age restrictions apply while the ship is in port and until the vessel enters international waters. Picture Identification required.*

DRINK OF THE DAY	I SHOT THE SHERIFF	$5.95
Café Promenade Bar	7:00 am – 3:00 am	Deck 5
Wipe Out Bar	9:00 am – 11:00 am 2:00 am – 7:00 pm	Deck 13
Squeeze Bar	9:00 am – 7:00 pm	Deck 11
Pool Bar	9:00 am – 7:00 pm	Deck 11
Plaza Bar	11:30 am – 11:30 pm	Deck 11
Sorrento's Bar	11:30 am – 3:00 am	Deck 5
Solarium Bar	1:00 pm – 6:00 pm	Deck 11
Vintages Wine Bar	1:00 pm – 1:00 am	Deck 5
Boleros	2:30 pm – 2:00 am	Deck 4
Schooner Bar	2:30 pm – 2:00 am	Deck 4
On Air Club	3:00 pm – 1:30 am	Deck 3
Olive or Twist	3:00 pm – 1:30 am	Deck 14
Sky Bar	2:00 pm – 7:00 pm	Deck 12
Bull & Bear Pub	2:00 pm – 2:00 am	Deck 5
Champagne Bar	3:00 pm – 1:00 am	Deck 5
Arcadia Theatre	3:30 pm – 5:00 pm 8:45 pm – midnight	Deck 3
Studio B	4:30 pm – 8:00 pm	Deck 3
Casino Bar	5:00 pm – late	Deck 4
Connoisseur Club Cigar Lounge	7:00 pm – 1:00 am	Deck 5
Fuel Bar (15-17 years)	8:00 pm – 1:00 am	Deck 12
Pharaoh's Palace	9:00 pm – 1:00 am	Deck 5
The Crypt (Family hour) **The Crypt** (Adults only)	10:00 pm – 11:00 pm 11:00 pm – late	Decks 3, 4

〔6日目〕ハイチのプライベートリゾート“ラバディ”寄港日

CRUISE COMPASS

TEAR HERE

DAY 6

Freedom of the Seas℠
LABADEE, HAITI
Friday, June 16, 2006

TODAY'S FORECAST
Partly Cloudy, 91°F

SUNRISE 6:13 am
SUNSET 7:33 pm

TOMORROW
At Sea

Today we arrive in Labadee, Haiti, our private destination reserved exclusively for Royal Caribbean International guests. Inside you'll find a list of exciting activities for your stay on shore and plenty for when you're back onboard, too. Have a fun day on the beach!

TENDERING
Please wait for announcements advising when the first guest tender departs. For guests on morning tours, please consult your tour tickets for meeting times and locations. Please use the portside elevators and staircases AFT ONLY to Deck 1 (below the Dining Rooms) to board the tenders **PLEASE DO NOT USE THE FORWARD ELEVATORS.**
Please remember that the last tender back to the ship leaves the dock at 4:30 pm. You must be on the pier by 4:15 pm in order to return to the Freedom of the Seas prior to sailing.

EXPLORATIONS!℠ Shore Excursions and Port Information – Be sure to check with our Explorations! Desk for remaining availability on our shore excursions for Labadee and Miami. If you've already booked a shore excursion, make sure you check your tickets for departure times and locations.

DISCOVER SHOPPING ASHORE – Win up to a $10,000 shopping spree, $50,000 in jewelry and watches, or a chance to drive a Formula 1 race car. Stop by the Discover Shopping Desk tonight and bring your receipts from purchases in guaranteed stores in Cozumel, Grand Cayman, Montego Bay and enter to win.

INAUGURAL Season

ARRIVAL	9:00 am
FIRST TENDER	9:30 am
LAST TENDER	4:30 pm
DEPARTURE	5:00 pm
GANGWAY	Deck 1

TEAR HERE

THE GREATEST SHOW AT SEA PARADE!
The Parade that brings out the kid in everyone!
5:45 pm, Royal Promenade, Deck 5

TONIGHT'S ENTERTAINMENT

ONCE UPON A TIME PRODUCTION SHOW
"A fairytale with a fun pop music twist"
Starring
The Royal Caribbean Singers and Dancers

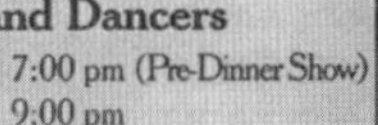

Showtime for Second Seating Guests	7:00 pm (Pre-Dinner Show)
Showtime for Main Seating Guests	9:00 pm

Arcadia Theatre, Decks 3 & 4

Once Upon A Time

Please be reminded that the saving of seats and video taping of shows is strictly prohibited. Also, children in the first three rows of the theatre must be accompanied by a parent or guardian.

STUDIO '84
Flashback to the 80's from Video, Music to the Clothing style. Get ready to laugh, smile and dance on the largest dance floor at sea.
Starring your Cruise Director's Staff and your fellow guests as the stars from the 80's.
10:15 pm Doors open
10:45 pm Show and Party begins
Studio B, Deck 3

Freedom of the Seas is proud to feature our outdoor Adults only club...

11:30 pm – 2:00 am, Solarium, Deck 11
Special twenty t-shirts available in our Shops On Board. Buy one and wear it tonight to the party!

CHEF'S DINNER Casual Attire
Main Seating – 6:00 pm in Leonardo's, Isaac's & Galileo's Dining Rooms, Decks 3, 4 & 5
Second Seating – 8:30 pm in Leonardo's, Isaac's & Galileo's Dining Rooms, Decks 3, 4 & 5

TEAR HERE

TODAY'S TIPS

- **SeaPass Boarding Card** – Don't forget to bring your SeaPass Boarding Card with you today when leaving for Labadee. You will not be allowed off or on the ship without it. A security checkpoint will be set up at the gangway for your safety.
- **Towels** – Beach Towels have been placed in your stateroom for your personal use when you go ashore. Should you desire an additional towel or replacement, please contact your Stateroom Attendant.
- Please note Labadee is a Marine reserve and the removal of aquatic life is not permitted.
- We know you never want to leave, but eventually you'll have to. To make it as easy and painless as possible, we suggest completing all cash transactions today or you may wait in long lines tomorrow. The Guest Relations Desk on Deck 5 or the Casino Royale on Deck 4 are available to cash traveler's checks or exchange large bills for smaller denominations.
- **Airport Transfers** – Final opportunity to purchase transfers to Miami and Fort Lauderdale airport. Guests wishing to purchase transfers must do so by 6:00 pm this evening at the Guest Relations Desk on Deck 5.

Get ready for fun! Check out the Daily Planner for each day's activities.

CROWN & ANCHOR SOCIETY AND LOYALTY AMBASSADOR

Royal Caribbean International offers you many fascinating and fun itineraries from breezy Maui to elegant Stockholm, from the stunning Parthenon to festive evening parties on colorful Aruba. See your Loyalty Ambassador located on Deck 6 and reserve your stateroom for your next cruise vacation today.

CHECK YOUR DAILY PLANNER FOR EVENT/DINING TIMES AND LOCATIONS

FOOD & DRINK

Drink of the Day – Drink a *Labadooze* for $5.95 and keep the souvenir glass. Enjoy the taste of this beautiful island, a blend of Dark Rum, pineapple mix, papaya, orange juice and a touch of Grenadine. Mmmmm.... Also a souvenir sloozie glass for $8.95 with any frozen drink.

Boleros – This is the place to let go. Enjoy the hot Latin rhythms and the ice-cold mojitos. It's always a fiesta at Boleros.

Chops Grille – If you're a meat lover, have we got the place for you! Choice meats are our specialty. So order up one of your favorite cuts and savor every tender bite. Don't miss out – make your reservation today.

Cook Books – Love the food? Well now you can eat like you're always on a Royal Caribbean International cruise vacation with our exclusive cook book. Available at the Guest Relations Desk or from your Dining Room Waiter.

ACTIVITIES

Adventure Family Game Show – Join us for The Family Quest. All ages are welcome.

Adventure Ocean Circus – Join the kids from Adventure Ocean for their very own show.

Friends Trivia Challenge – How much do you know about the television show *Friends?*

Family Shipboard Scavenger Hunt – How well do you know the Freedom? Bring the family and find out.

Kids' Karaoke and Joke Night – Kids are invited to have fun tonight in the On Air Club.

Teen Disco – Tonight the teens are allowed in the adult nightclub, The Crypt. Alcohol will not be served in The Crypt this evening.

Movie – Staying onboard today? Catch a movie in our very own Screening Room. Today's movie: *March of the Penguins.* Documentary. Limited seating.

Majority Rule Game Show II – Back by popular demand, join your Cruise Director's Staff for this fun event.

SPORTS & FITNESS

Polar BodyAge System – Take the BodyAge Challenge! Are you in your 40's but feel more like 60? Or maybe you're 70 and feel like a kid! Find out your body's true age with our unique system.

Fitness In Mind – Not in a hurry to go ashore? Or staying onboard? Join your Cruise Director's Staff for Fun Fitness.

Walk-a-Mile – Start out your morning with fresh air and breath-taking ocean views during a one-mile walk on deck.

SPA & BEAUTY

The Freedom Day Spa is located on Deck 12. Dial 6982 for an appointment.

Spa Taster – Today Only! – This is a limited-time special that will give you a taste of our lavish spa services. Come and enjoy a 25 minute aromatherapy massage followed by a 25 minute facial. Only $89.

Your Best Face Forward – Putting on your make-up shouldn't be a chore. Our salon professionals will show you the tricks of the trade and other shortcuts to make applying your make-up a pleasure.

Miracle Micro – Therapy Facial – This new, effective, non-invasive face therapy will soften fine lines, skin discoloration, enlarged pores, sun and age spots, and scars from chicken pox or acne. The result after just one treatment is youthful, more vibrant skin. Who doesn't need a miracle? Today only $120.

Acupuncture Seminar – Find out how acupuncture can help with pain management, weight loss, stress, detox, and can help you quit smoking. James, our Acupuncturist, will explain how he can help release your blocked energy, restore balance and improve your well-being today.

ONBOARD SHOPPING

Amber – It's all about amber. Hundreds of pieces from the 2003 Amber Collection will be available today on the Royal Promenade. Don't miss out on this great sale.

Inch of Gold – Today is your last chance to buy made-to-measure layered gold necklaces, anklets and more.

Photo Shop – Stock up on all your photo needs for this cruise vacation. Batteries and film for formal nights and underwater cameras for all of your underwater fun.

Photo Gallery – Do you have a photo you love? Cherish it for life and have it inscribed into crystal. Portraits, Anniversaries, Weddings, and Cruise Vacation memories.

Fine Art Exhibition – Tomorrow is the big day! Stop by and preview some of the art works that will be available at tomorrow's Final Art Auction Blowout. Tarkay, Erte, Dali, Chagall, Rembrandt, Picasso and more. Get your complimentary raffle tickets for our $10,000 art raffle just for stopping by.

Airbrushed Tattoss – Buy any extra large tattoo for the price of a medium tattoo.

GAMING

Slot Players – Make it a point to try our linked progressive machines. They have already paid out thousands of dollars; your turn could be next.

Wheel of Fortune – It's exciting. It's fun. It's a chance to win big. Come and take a spin at the Wheel of Fortune.

EVERYTHING ELSE

Community Bulletin Board – Looking for like minded interest? Stop by the Library on Deck 7 and fill out a card.

Quiet Zone – Looking for a nice quiet place onboard? Try the Library on Deck 7. All guests are requested to respect silence in this area.

royal caribbean online℠ – No matter where your adventure takes you, you can always stay in touch. Check e-mail and stocks, or e-mail a friend. This service is available 24 hours a day, at a nominal charge.

Smoking Policy – The Freedom of the Seas has designated smoking areas. Guests are asked to only smoke in specific areas.

Enjoy a juicy burger and thick shake from Johnny Rockets® or a delectable Ben & Jerry's® sundae made with your favorite, Chocolate Chip Cookie Dough Ice Cream. We've partnered with some great companies that are as dedicated as we are to making your cruise vacation the most memorable one ever. Your cruise vacation just wouldn't be complete without them!

DAILY PLANNER

FOLD HERE ↑

DAY 6

Freedom of the SeasSM
LABADEE, HAITI
Friday, June 16, 2006
Arrival: 9:00 am
First Tender: 9:30 am
Last Tender: 4:30 pm
Departure: 5:00 pm
Gangway: Deck 1

OPEN HOURS

A Clean Shave/Barber 8:00 am – 8:00 pm Deck 5
(Appointments advisable)
Adventure Ocean Back Deck 24 hours(Ages 12-17 only) Deck 12
Art Gallery 7:00 pm – 10:00 pm Deck 3
Australian Airbrushed Tattoos 3:30 pm – 5:30 pm Deck 11 Poolside
7:30 pm – 9:00 pm Deck 5 Royal Promenade
11:00 pm – 12:30 am Deck 11 Poolside
Casino Royale Slots 9:00 am – late
Tables 2:00 pm – late Deck 4 Dial 3171
Discover Shopping Desk 5:00 pm – 6:00 pm Deck 5 Dial 3830
Doctor's Hours 9:00 am – 11:00 am
5:00 pm – 7:00 pm Deck 1 aft Dial 51
Emergency ***(In case of)*** Dial 911
Explorations! Desk 8:00 am – 9:00 am
5:00 pm – 6:00 pm Deck 5 Dial 3936
Floral Cart 5:00 pm – 9:00 pm Deck 5 Royal Promenade
Freedom Day Spa 8:00 am – 8:00 pm Deck 12 Dial 6982
Accupuncture 8:00 pm – 8:00 pm
Hair & Beauty Salon 8:00 am – 8:00 pm
Teeth Whitening 8:00 am – 8:00 pm
Guest Relations 24 hours Deck 5 Dial 0
H2O Zone 8:00 am – 8:00 pm Deck 11
Per United States Public Health, pull ups/swimmer's or diapers are not permitted in the swimming pools including the H2O Zone. All children must be potty trained.
Language Assistance 10:00 am – 11:45 am
6:00 pm – 7:30 pm Deck 6
Library 24 hours Deck 7
Loyalty Ambassador 9:00 am – 11:00 am
3:00 pm – 7:00 pm Deck 6
Medical Facility 9:00 am – 11:00 am
5:00 pm – 7:00 pm Deck 1 aft Dial 51
Operator 24 hours Dial 0
Assistance for Overseas calls
8:00 am – 11:00 pm Dial 0
(All outside phone calls are $7.95 per minute)
Photo Gallery & Shop 4:00 pm – 11:00 pm Deck 3 Dial 3870
royal caribbean onlineSM 24 hrs Deck 8
SeaTrek Dive Shop 5:00 pm – 6:00 pm Deck 11 Dial 3889
ShipShape Center 6:30 am – 10:00 pm Deck 11
Shops On Board Deck 5 Dial 3867
General Store & Logo Store 8:30 am – 11:30 am
All Stores 5:00 pm – 11:00 pm
Sports Outlet 8:15 am – 11:00 am & 3:00 pm – 5:00 pm, Deck 13
Swimming Pools 24 hours
Selected Pools and Hot Tubs are open 24 hours for your enjoyment. Guests are reminded there is no Lifeguard on duty. The Solarium Pool, Hot Tubs and the Solarium area are reserved for guests 16 and older only. Children must be supervised in the two Main Pools and the H2O Zone at all times. As a courtesy to other guests, reserving of deck chairs is not permitted. Chairs left vacant for 30 minutes or longer may be reassigned by the Pool Attendant to another waiting guest. Royal Caribbean International is not responsible for theft or loss of property left on deck chairs or pool areas.
Tuxedo Rentals Dial 7094

SPORTS DECK ACTIVITIES & OPEN HOURS

(Adult sporting events are for guests 18 and older.)
Basketball Court - Open Play 21 hours Deck 13
(Reserved 5:00 pm, 6:00 pm, 7:30 pm)
FlowRider Deck 13
Advanced Stand-Up Surfing 8:15 am – 9:00 am
Boogie Boarding 9:00 am – 11:00 am
Stand-up Surfing 4:00 pm – 5:00 pm & 6:00 pm – 7:00 pm
Teen (15-17 years) Stand-Up 5:00 pm – 6:00 pm
Parents, please fill out a waiver for your teens (15-17 years) between 4:30 pm – 5:00 pm in the Living Room, Deck 12.
Please wear a t-shirt. Please remove all items of jewelry. A FlowRider waiver must be completed every day of the cruise.
Freedom Fairways Golf 23 hours Deck 13
(Reserved 9:00 am)
Golf Simulator 9:00 am – 11:00 am, 4:00 pm – 5:00 pm &
6:00 pm – 7:00 pm
(By reservation, $25 fee), Deck 13
Ping Pong Tables 24 hours Deck 14
Rock Climbing Wall Deck 13
Open Climbing Session 9:00 am – 11:00 am, 4:00 pm – 5:00 pm &
6:00 pm – 7:00 pm
Family Climbing Sessions 5:00 pm – 6:00 pm
Please bring socks, t-shirt and shorts/pants. Clothing must be dry. No wet clothing please. A Rock Climbing Wall waiver must be completed once a cruise. Children must be 6 years of age or older to climb. Adults must be able to fit into XXL harness to climb. Please note, the Rock Climbing Wall will be closed if we are experiencing inclement weather.
Shuffleboard 24 hours Deck 4
Sports Pool Deck 11
Lap Swimming 6:00 am – 10:00 am
Open Play Water Polo 5:00 pm – 7:00 pm
Volleyball Court – Open Play 5:00 pm – 7:00 pm Deck 13

For further information and descriptions of onboard services and policies, please read important information on the back of your Day 1 Cruise News.

DINING SCHEDULE & AFTERNOON SNACKS

BREAKFAST

6:00 am – 8:00 am	Grab and Go Breakfast, Plaza Bar	Deck 11
7:00 am – 11:00 am	Breakfast, Jade & Windjammer Café	Deck 11
7:30 am – 9:30 am	Leonardo's Dining Room	Deck 3

LUNCH

Noon – 2:00 pm Lunch on the Island - 3 Locations.
Columbus Cove Café, Dragon's Rock and Café Labadee
Please Note: All food and drinks served on the Island are provided from the Freedom of the Seas' onboard galley.
Noon – 2:00 pm Leonardo's Dining Room Deck 3

DINNER	**CHEF'S DINNER**	**CASUAL**
6:00 pm	Main Seating, All Dining Rooms	Decks 3, 4, 5
8:30 pm	Second Seating, All Dining Rooms	Decks 3, 4, 5

Please also note that bare feet and tank tops are not permitted at dinner.

6:30 pm – 9:00 pm	Casual Dinner, Windjammer Café	Deck 11
6:30 pm – 9:00 pm	Casual Dinner, Jade	Deck 11

Casual Dinner in Jade and the Windjammer Café is buffet style.
Please note that bare feet and tank tops are not permitted in these venues at dinner.

FOLD HERE ↑

SPECIALTY DINING

6:00 pm – 9:45 pm **Chops Grille,** The best steak on the high seas Deck 11
$20 dining fee per person applies, gratuities included
(Reservations recommended, Dial 3055)
6:00 pm – 9:45 pm **Portofino,** Intimate Italian Dining Deck 11
$20 dining fee per person applies, gratuities included
(Reservations recommended, Dial 3035)
Dress Suggestion: Smart Casual. No short pants please. Guests 13 years and older are welcome. Please allow approximately two hours for dinner. Cancellations with less than 24 hours notice will be charged a fee of $10 per person.

SNACKS

24 hours	Café Promenade	Deck 5
11:00 am – 4:00 pm	Ben & Jerry's Ice Cream (Nominal fee)	
	Satellite Station	Labadee
11:30 am – 1:00 am	Johnny Rockets ($3.95 cover charge per person, gratuities included)	Deck 12
11:30 am – 3:00 am	Sorrento's Pizza	Deck 5
2:00 pm – midnight	Ben & Jerry's Ice Cream (Nominal fee)	Deck 5
2:45 pm – 5:00 pm	Afternoon Tea, Windjammer Café	Deck 11
11:30 pm – midnight	Culinary Gala Sensation, Windjammer Cafe	Deck 11

(This is a perfect opportunity for guests to bring their cameras and take photos of the magnificent food displayed, eagerly prepared by our talented onboard Culinary Team.)
12:15 am – 1:00 am Culinary Gala Sensation, Open for Dining, Windjammer Cafe Deck 11

MORNING ACTIVITIES

24 hours	Community Bulletin Board, Library	Deck7
24 hours	royal caribbean online℠	Deck 8
24 hours	Discover Shopping Talk, Continuous	Channel 22
24 hours	Cards & Games available, Seven Hearts Card Room	Deck 14
7:30 am	Balance in Motion ★, ShipShape Center	Deck 11
8:00 am	Aerobics in Motion ★, ShipShape Center	Deck 11
8:00 am	Wheels in Motion ★,($10 fee), ShipShape Center	Deck 11
8:00 am – 8:45 am	Aqua Babies Play Time, (Ages 6-18 months) On Air Club	Deck 3
8:30 am	New Balance Walk-a-Mile ★, Sky Bar	Deck 12
9:00 am	**Estimated Time of Arrival in Labadee, Haiti**	
9:00 am	Pathway to Yoga on the Beach ★, ($10 fee), Meet in the Aerobics Studio @ 8:30 am	Deck 11
9:00 am	Family Gilmore Golf, Freedom Fairways	Deck 13
9:00 am	Daily Trivia Sheet available, Library	Deck 7
9:00 am – 9:45 am	Aqua Tots Play Time, (Ages 18-36 months), On Air Club	Deck 3
9:30 am (approximate)	**First Tender to Labadee Departs**	
11:00 am	Adult Beach Volleyball, Volleyball Court	Labadee
11:30 am	Team Trivia, Schooner Bar	Deck 4

FOLD HERE ↑

AFTERNOON ACTIVITIES

Noon	Adventure Ocean Family Water Balloon Toss, Kids Adventure Beach	Labadee
1:30 pm	Folkloric Show, Café Labadee	Labadee
2:00 pm	Line Dance Class ⚓, Pharaoh's Palace	Deck 5
2:55 pm	World Cup, *Mexico vs. Angola*, On Air Club	Deck 3
3:00 pm	Family Beach Olympics, Volleyball Court	Labadee
3:00 pm	*Friends* Trivia Challenge, Schooner Bar	Deck 4
3:30 pm – 5:30 pm	80's Sailaway Party, Poolside	Deck 11
4:00 pm	Family Shipboard Scavenger Hunt, Schooner Bar	Deck 4
4:00 pm	Screening Room Movie: *March of the Penguin*, (Limited seating), Screening Room	Deck 2
4:00 pm	Pathway to Pilates ⚓, ($10 fee), ShipShape Center	Deck 11
4:00 pm	Seminar: *Put your Best Face Forward*, Freedom Day Spa	Deck 12
4:15 pm	Cash Prize Bingo: *Cards on Sale*, Pharaoh's Palace	Deck 5
4:30 pm	**Last Tender from Labadee, Haiti**	
4:30 pm	Seminar: *Accupuncture*, *The Point of Well Being*, Freedom Day Spa	Deck 12
4:45 pm	Cash Prize Bingo: *Games Begin*, Pharaoh's Palace	Deck 5
5:00 pm	**Freedom of the Seas sails for Miami, Florida (616 nautical miles)**	
5:00 pm	Friends of Bill W. Meeting, Skylight Chapel	Deck 15
5:00 pm	Wheels in Motion ⚓ ($10 fee), ShipShape Center	Deck 11
5:00 pm – 6:00 pm	Scratch DJ Class for Teens, (12-17 years), Fuel Drive	Deck 12
5:00 pm – 6:00 pm	"Sweepstakes Hour," Discover Shopping Desk	Deck 5
5:00 pm – 6:30 pm	Casual Portraits will be taken, Centrum	Decks 3,4
5:15 pm	Jewish Sabbath Service, (minimum of 10 needed), Aruba Conference Room	Deck 2
5:45 pm	The Greatest Show at Sea Parade, Royal Promenade	Deck 5

EVENING ACTIVITIES

7:00 pm	Once Upon A Time Production Show, (Second Seating Guests Pre-Dinner Show), Arcadia Theatre	Decks 3,4
7:00 pm – 10:00 pm	Fine Art Exhibition, Art Gallery	Deck 3
7:30 pm – 8:45 pm	Casual Portraits will be taken, Centrum	Decks 3,4
8:00 pm	Majority Rule Game Show II, On Air Club	Deck 3
8:45 pm	Adventure Ocean Circus, Royal Promenade	Deck 5
9:00 pm	Once Upon A Time Production Show, (Main Seating Guests), Arcadia Theatre	Decks 3,4
9:30 pm – 10:30 pm	Dessert Wine, Port & Chocolate Tasting, (Nominal fee), Vintages Wine Bar	Deck 5
10:00 pm	Family Quest, Pharaoh's Palace	Deck 5
10:00 pm – 11:00 pm	Casual Portraits will be taken, Centrum	Decks 3,4
10:15 pm	Doors Open for Studio '84 Dance Party, Studio B	Deck 3
10:30pm – 11:30 pm	Kids' Karaoke & Joke Night, On Air Club	Deck 3
10:45 pm – 11:30 pm	Studio '84 Dance Party, Studio B	Deck 3
11:00 pm – 12:15 am	Teen Disco, (12-14 years), The Crypt	Decks 3,4
11:30 pm – 2:00 am	twenty, (Adults only), Solarium	Deck 11
11:30 pm – midnight	Culinary Gala Sensation open for viewing, Windjammer Café	Deck 11
12:15 am – 1:00 am	Culinary Gala Sensation open for dining, Windjammer café	Deck 11
12:30 am – late	Teen Disco, (15-17 years), The Crypt	Decks 3,4
2:00 am – late	After Hours Party, (Adults only), Olive or Twist	Deck 14

⚓ – Denotes ShipShape Dollars awarded for this activity.

Ship's Labadee On-Site Manager:
Telephone # 305 - 982 - 2417

MUSIC AND DANCING

Live Bands & DJs

80's Sailaway Party	*With the Stingrays* 3:30 pm – 5:30 pm, Poolside, Deck 11
Classical Concert	*With Viva Expressia* 5:00 pm – 6:00 pm, Olive or Twist, Deck 14
Dining Room Music	*With Tara Davis* 6:00 pm – 7:00 pm & 8:30 pm – 9:30 pm All Dining Rooms
Guitar Melodies	*With Leonardo* 7:30 pm – 8:30 pm & 10:00 pm – 11:00 pm Olive or Twist, Deck 14
Latin Dance Music	*With Sol y Arena* 7:30 pm – 8:30 pm & 10:00 pm – 12:30 am Boleros, Deck 4
Piano Bar Entertainment	*With Peter Ritz* 10:00 pm – 1:00 am, Schooner Bar, Deck 4
Pub Entertainment	*Vocal Guitarist Jimmy Blakemore Plays Request* 10:30 pm – 1:00 am, Bull & Bear, Deck 5
Teen Nightclub	*With DJ Jamie* 11:00 pm – late, The Crypt, Decks 3 & 4
Dancing, Standards & Jazz	*With the Australian Jazz Quartet* 11:15 pm – 1:30 am, Olive or Twist, Deck 14
twenty	*With Celebrity DJ Smoke from the Scratch DJ Academy* 11:30 pm – 2:00 am, Solarium, Deck 11 Guests must be 18 years or older for admission and photo identification will be required if you appear to be under 18.
After Hours Party	*With the Freedom of the Seas Musicians* 2:00 am – late, Olive or Twist, Deck 14 Guests must be 18 years or older for admission and photo identification will be required if you appear to be under 18.

Please note, the Viking Crown Lounge which includes Olive or Twist is for adults 18 and older only after 10:00 pm.

BAR SERVICE HOURS

***Proof of Age** – Guests eighteen to twenty (18-20) years of age are welcome to enjoy beer and wine. Guests twenty-one (21) years of age and older are welcome to enjoy all alcoholic beverage. ** Applicable regulatory age restrictions apply while the ship is in port and until the vessel enters international waters. Picture Identification required.*

DRINK OF THE DAY	LABADOOZE	$5.95
Café Promenade Bar	7:00 am – 3:00 am	Deck 5
Squeeze Bar	9:00 am – 7:00 pm	Deck 11
Wipe Out Bar	9:00 am – 11:00 am 2:00 pm – 7:00 pm	Deck 13
Pool Bar	9:00 am – 7:00 pm 11:00 pm – 2:00 am	Deck 11
Plaza Bar	11:30 am – 1:30 am	Deck 11
Sorrento's Bar	11:30 am – 3:00 am	Deck 5
Bull & Bear Pub	1:00 pm – 2:00 am	Deck 5
Sky Bar	2:00 pm – 7:00 pm	Deck 12
Solarium Bar	2:00 pm – 6:00 pm 11:00 pm – 2:00 am	Deck 11
Boleros	2:00 pm – 2:00 am	Deck 4
Vintages Wine Bar	3:00 pm – 1:00 am	Deck 5
Champagne Bar	3:00 pm – 1:00 am	Deck 5
Schooner Bar	3:00 pm – 2:00 am	Deck 4
Olive or Twist	3:00 pm – late	Deck 14
Pharaoh's Palace	4:00 pm – 5:30 pm 9:00 pm – 11:30 pm	Deck 5
On Air Club	4:00 pm – 1:00 am	Deck 3
Casino Bar	5:15 pm – late	Deck 4
Connoisseur Club Cigar Lounge	7:00 pm – 1:00 am	Deck 5
Fuel Bar (15-17 years)	8:00 pm – 1:00 am	Deck 12
Arcadia Theatre	8:30 pm – midnight	Deck 3
Studio B	10:15 pm – 11:30 pm	Deck 3
The Crypt (Teens only)	11:00 pm – late	Decks 3,4

〔7日目〕終日航海日

CRUISE COMPASS

TEAR HERE

DAY 7

Freedom of the Seas℠
AT SEA
Saturday, June 17, 2006

TODAY'S FORECAST
Scattered Thunderstorms, 89°F

SUNRISE 6:21 am
SUNSET 8:05 pm

NEXT PORT-OF-CALL
Miami, Florida

This is Day 7. And even though all great things must come to an end, we want you to know one thing: IT'S NOT OVER YET! Today is the day to do the things that you promised yourself you'd do all week and see the things you haven't seen. Basically, we want you to have another unforgettable day. Listed on these pages is just a sample of today's highlights. For a complete list of activities, turn to your Daily Planner.

EXPLORATIONS! Shore Excursions & Port Information

Just because today's the last day of your cruise vacation, it doesn't mean the adventure has to stop. Miami offers many exciting shore excursions like a Everglades Safari Tour and Miami Highlights Tour. Some tours will drop you off at the Miami International Airport at the conclusion of the tours. Consult our Interactive TV system for full tour descriptions. The Explorations! Desk will be open today for additional information and sales.

CAPTAIN'S CORNER

Captain Bill and Cruise Director Ken answer any questions you may have about shipboard life and operations.
9:45 am – 10:30 am, Pharaoh's Palace, Deck 5

TEAR HERE

SNOWBALL JACKPOT BINGO!

The Jackpot must go today!
10:30 am cards on sale, 11:00 am games begin
This morning's session is your last chance to use any coupons.

FREEDOM LIVE TELEVISION SHOW

Join Ken and special guests and be a part of a LIVE television show.
11:15 am, FlowRider, Deck 13

TONIGHT'S ENTERTAINMENT

FAREWELL VARIETY SHOW

Hosted by your Cruise Director, ***Ken Rush***
Starring the Comedy of ***Eric Lyden***
also featuring the Royal Caribbean Singers and Dancers

Arcadia Theatre, Decks 3 & 4
Showtime for Second Seating Guests 7:00 pm (Pre-dinner show)
Showtime for Main Seating Guests 9:00 pm

Don't miss the premier of the Cruise in Review.

Please be reminded that the saving of seats and video taping of shows is strictly prohibited. Also, children in the first three rows of the theatre must be accompanied by a parent or guardian.

FAREWELL 50's & 60's ROCK N' ROLL DANCE PARTY

Join our Rockin' Showband, the Stingrays with 50's and 60's music.
10:30 pm – 12:30 am, Poolside, Deck 11

INTIMATE JAZZ CABARET

Starring the Royal Caribbean Singers and featuring the Australian Jazz Quartet and Musical Director Greg Carger.
11:00 pm, Olive or Twist, Deck 14

FEAST OF NATIONS — **Casual Attire**
Main Seating – 6:00 pm in Leonardo's, Isaac's & Galileo's Dining Rooms, Decks 3, 4 & 5
Second Seating – 8:30 pm in Leonardo's, Isaac's & Galileo's Dining Rooms, Decks 3, 4 & 5

TEAR HERE

TODAY'S TIPS

- **Last Day!** – Today is your last day onboard so be sure to try those things you don't want to miss – winning at Bingo, climbing the rock wall, try the FlowRider, buying souvenirs and catching that final sunset over the Caribbean.
- **Thanks!** – You may already be planning your next cruise vacation but tonight is the night to thank those people who made this one so special. Show them how much you appreciated their Gold Anchor Service by extending a gratuity. And don't forget to tell them you'll see them on your next cruise vacation.
- **We care what you think** – Please complete your Guest Satisfaction Survey and let us know how you enjoyed your vacation. When finished, be sure to place them in the light green boxes at the Guest Relations Desk. Please do not leave them in your stateroom.
- Your SeaPass statement will be delivered to your stateroom upon arrival in Miami. Charges will be billed directly to your credit card. Remember, you can review your charges at any time on RCTV.
- **Sunday Morning Purchases** – Guests with SeaPass accounts secured with a credit card will be able to make purchases in our Video Arcade, Photo Gallery, Medical Facility and selected Bar venues on Sunday morning. Charges will be billed directly to your credit card. Remember, you can review your current charges at any time on RCTV.

Get ready for fun! Check out the Daily Planner for each day's activities.

CROWN & ANCHOR SOCIETY AND LOYALTY AMBASSADOR

What's better than a Royal Caribbean cruise vacation? Another Royal Caribbean cruise vacation. Make your reservations while you're still onboard and receive an onboard credit of up to $200 per stateroom – a special offer only while you are on this cruise vacation. See your Loyalty Ambassador located on Deck 6 to reserve your stateroom today.

CHECK YOUR DAILY PLANNER FOR EVENT/DINING TIMES AND LOCATIONS

FOOD & DRINK

Drink of the Day – Drink a *Freedom Breezer* for $5.95 and keep the souvenir glass. A blend of Bacardi Razz, Blue Curacao, Peach Snapps and Pineapple Juice. Also a souvenir sloozie glass for $8.95 with any frozen drink.

Bull & Bear Pub – Relax with an ice-cold English pint in a traditional pub atmosphere anytime you want. Remember – you're on vacation.

Olive or Twist – Born in New Orleans, and alive and kickin' right here onboard. Sit back with friends and your favorite drink and just soak in the cool sounds of Jazz.

The Crypt – If you're more a creature of the night, things come alive once the sun goes down at this hot night-time spot. Dance till the wee hours of the night to all your favorite tunes.

ACTIVITIES

Mr. Sexy Legs Contest – Ooh, la, la... Show the women what you've got at the Men's Sexy Legs Contest. This will be one of the funniest contest you'll ever see.

Final Team Trivia – Last chance to join your Cruise Director's Staff and test your knowledge.

Back Stage Tour – You sat in the audience before, but haven't you always wondered what went on behind the scenes? Find out on this fun back sage tour.

Ice Skating – Show off your best figure eight or pirouette on one of the only ice-skating rinks at sea. Long pants and socks are required. Children under the age of 18 must have a parent or guardian to sign a waiver for them.

Towel Folding Demonstration – Dying to know how our Stateroom Attendants fold the towels into so many different surprises? You're not alone! Join us as we find out how they do it.

Cry Baby Cinema – Parents, tired of missing the latest movies because of your infant or toddler? Welcome to *Cry Baby Cinema* - a new movie experience designed with you and your baby in mind, where cry babies are welcome! Bring your family, gather the baby and enjoy the film.

Adventure Ocean Family Talent Show – A human pyramid, a barbershop quartet. What's your family talent? Let everyone see it at the Family Talent show. It's a great time for everyone. Rehearsal required.

Learn How to Skate – Join our International Ice Cast to learn the basic skills of ice skating. These sessions are for those guests who previously signed up only. Long pants and socks are required.

Screening Room Movie – *Good Night and Good Luck. Drama/History starring David Strathairn*

Scrapbook Workshop – You don't want to forget the pristine playground you discovered or the relaxing moments you experienced. Preserve these carefree days using a "tear-ific" technique demonstrated by your Cruise Director's Staff.

Joke & Karaoke Night – Adults only. Tell everyone your favorite joke or sing a song but please remember no crude or rude jokes allowed. If you have any questions, please screen your joke with the Cruise Director's Staff beforehand.

SPORTS & FITNESS

Polar BodyAge System – Take the BodyAge Challenge! Are you in your 40's but feel more like 60? Or maybe you're 70 and feel like a kid! Find out your body's true age with our unique system.

FlowRider – Last chance to enjoy the ride of your life! The FlowRider simulates a real surfing and body boarding experience. 30,000 gallons of water per minute RUSH underneath the rider at 30 mph creating a five-foot ocean-like wave. Riders must be at least 58" to surf.

ShipShape Dollar Exchange – These are probably the only things you don't want to take home from this vacation. Exchange your ShipShape Dollars for prizes at the Wipe Out Bar.

SPA & BEAUTY

The Freedom Day Spa is located on Deck 12. Dial 6982 for an appointment.

Puffy Eye Seminar – Put the twinkle and sparkle back in your eyes! Join us in the Beauty Salon to find out how to remove dark circles, puffiness and fine lines. FREE eye spa treatment for all who attend.

Partners Massage Class – Bring your friend or partner and learn some basic massage techniques to use at home. Leave with some practical experience and a bottle of aromatherapy oil to get you started. $52 per session per couple – sign-up required.

Cellulite Reduction Program – Pssst, we've got the scoop to help you battle cellulite. Find out what it is, and how to reduce and get rid of it from our Body Correction Specialist.

Three-Minute Makeovers – Make the best of three minutes. Please come and join us as we demonstrate for the last time the fastest makeover on the high seas.

ONBOARD SHOPPING

Shop Till You Drop – Today is your last chance to shop tax and duty free onboard. So get shopping and be sure to keep an eye out for our special sales on the Royal Promenade.

T-shirt Blowout – Cool Caribbean T-shirts, 2 for $25 and Royal Caribbean Pocket T-shirts that are 50% off.

Cruise in Review Video and DVD – Your friends might not believe you when you tell them all the incredible things you did on your cruise. So prove it to them with your own Cruise in Review DVD. On sale in the photo shop.

Final Art Auction – Over $10,000 worth of art and free prizes to be given away. This auction features the best of the Freedom of the Seas collections. New works up for bid, save 40-80% below retail gallery prices. Free Champagne.

Airbrushed Tattoos – This is your last chance to impress your friends and family with a realistic temporary tattoo, airbrushed tattoos are fun for all ages.

GAMING

Casino Royale – It's your last night to win big. Try your luck tonight with Caribbean Stud Poker, Let it Ride, American Roulette, Blackjack, Craps, Texas Hold'em, plus 306 slot machines. Casino Royale, Deck 4.

Wheel of Fortune – It's exciting. It's fun. It's a chance to win big. Come and take a spin at the Wheel of Fortune.

Snowball Jackpot Bingo – Someone has to win all this money - don't you want it to be you? Last chance to use any bingo coupons! Please note that if the jackpot is won this morning there will be no session this afternoon.

Winner Takes All – You want to win big money right? Well, this round of bingo is for you. There are thousands of dollars to be won. This session will only be played if the jackpot was not won this morning.

EVERYTHING ELSE

Community Bulletin Board – Looking for like minded interest? Stop by the Library on Deck 7 and fill out a card.

Smoking Policy – The Freedom of the Seas has designated smoking areas. Guests are asked to only smoke in specific areas.

Fathers' Day – Don't forget tomorrow is Fathers's Day so on behalf of all of us onboard Freedom of the Seas, Happy Early Fathers' Day to all fathers onboard.

FAREWELL MESSAGE

As our cruise vacation draws to a close, we wish to express on behalf of the Master, Hotel Director, Officers, Staff and Crew, our best wishes for a safe return home. We trust you will have pleasant memories when reflecting on your sailing experience s and sincerely hope to see you on one of our Royal Caribbean International ships in the near future. We have been friends and shipmates for 2,040 nautical miles. Godspeed.

OPEN HOURS

A Clean Shave/Barber (Appointments advisable)	8:00 am – 8:00 pm	Deck 5	
Adventure Ocean Back Deck	24 hours(Ages 12-17 only)		Deck 12
Art Gallery	24 hours	Deck 3	
Australian Airbrushed Tattoos	10:30 am – 5:30 pm	Deck 11	Poolside
	8:00 pm – 11:00 pm	Deck 5	Royal Promenade
Casino Royale Slots	9:00 am – late		
Tables	1:00 pm – late	Deck 4	Dial 3171
Discover Shopping Desk	1:00 pm – 2:00 pm	Deck 5	Dial 3830
	5:00 pm – 6:00 pm		
	7:30 pm – 8:30 pm		
Doctor's Hours	9:00 am – 11:00 am		
	5:00 pm – 7:00 pm	Deck 1 aft	Dial 51
Emergency	***(In case of)***		Dial 911
Explorations! Desk	11:00 am – noon		
	5:30 pm – 6:30 pm	Deck 5	Dial 3936
Floral Cart	5:00 pm – 9:00 pm	Deck 5	Royal Promenade
Freedom Day Spa	8:00 am – 8:00 pm	Deck 12	Dial 6982
Accupuncture	8:00 pm – 8:00 pm		
Hair & Beauty Salon	8:00 am – 8:00 pm		
Teeth Whitening	8:00 am – 8:00 pm		
Guest Relations	24 hours	Deck 5	Dial 0
H2O Zone	8:00 am – 8:00 pm	Deck 11	

Per United States Public Health, pull ups/swimmer's or diapers are not permitted in the swimming pools including the H2O Zone. All children must be potty trained.

Language Assistance	10:00 am – 11:45 am		
	6:00 pm – 7:30 pm	Deck 6	
Library	24 hours	Deck 7	
Loyalty Ambassador	9:00 am – 7:30 am	Deck 6	
Medical Facility	9:00 am – 11:00 am		
	5:00 pm – 7:00 pm	Deck 1 aft	Dial 51
Operator	24 hours		Dial 0
Assistance for Overseas calls			
	8:00 am – 11:00 pm		Dial 0
	(All outside phone calls are $7.95 per minute)		
Photo Gallery & Shop	9:00 am – midnight	Deck 4	Dial 3870
royal caribbean onlineSM	24 hrs	Deck 8	
SeaTrek Dive Shop	9:00 am – 5:00 pm	Deck 11	Dial 3889
ShipShape Center	6:30 am – 10:00 pm	Deck 11	
Shops On Board	9:00 am – 11:00 pm	Deck 5	Dial 3867
Sports Outlet	8:15 am – 6:00 pm	Deck 13	
Swimming Pools	24 hours		

Selected Pools and Hot Tubs are open 24 hours for your enjoyment. Guests are reminded there is no Lifeguard on duty. The Solarium Pool, Hot Tubs and the Solarium area are reserved for guests 16 and older only. Children must be supervised in the two Main Pools and the H2O Zone at all times. As a courtesy to other guests, reserving of deck chairs is not permitted. Chairs left vacant for 30 minutes or longer may be reassigned by the Pool Attendant to another waiting guest. Royal Caribbean International is not responsible for theft or loss of property left on deck chairs or pool areas.

SPORTS DECK ACTIVITIES & OPEN HOURS

(Adult sporting events are for guests 18 and older.)

Basketball Court - Open Play	22 hours (Reserved 10:00 am, 4:00 pm)	Deck 13
FlowRider	Deck 13	
Advanced Stand-Up Surfing	8:15 am – 9:00 am	
Stand-up Surfing	9:00 am – 11:00 am & 1:00 pm – 3:00 pm	
Boogie Boarding	11:00 am – 1:00 pm & 3:00 pm – 5:00 pm	

Please wear a t-shirt. Please remove all items of jewelry. A FlowRider waiver must be completed every day of the cruise.

Freedom Fairways Golf	24 hours	Deck 13
Golf Simulator	9:00 am – 6:00 pm (By reservation, $25 fee), Deck 13	
Ping Pong Tables	24 hours	Deck 14
Rock Climbing Wall		Deck 13
Advanced Session	9:00 am – 10:00 am	
Sessions	10:00 am, 11:00 am, 2:00 pm, 3:00 pm, 4:00 pm & 5:00 pm	

Please bring socks, t-shirt and shorts/pants. Clothing must be dry. No wet clothing please. A Rock Climbing Wall waiver must be completed once a cruise. Children must be 6 years of age or older to climb. Adults must be able to fit into XXL harness to climb. Please note, the Rock Climbing Wall will be closed if we are experiencing inclement weather.

Shuffleboard	24 hours	Deck 4
Sports Pool		Deck 11
Lap Swimming	6:00 am – 10:00 am	
Pool Games	1:30 pm	
Family Wacky Water Races	3:00 pm	

For further information and descriptions of onboard services and policies, please read important information on the back of your Day 1 Cruise News.

DAILY PLANNER

FOLD HERE ↑

DAY 7

Freedom of the SeasSM
AT SEA
Saturday, June 17, 2006

DINING SCHEDULE & AFTERNOON SNACKS

BREAKFAST		
6:00 am – 8:00 am	Grab and Go Breakfast, Plaza Bar	Deck 11
7:30 am – 11:00 am	Breakfast, Jade & Windjammer Café	Deck 11
8:00 am – 10:00 am	Leonardo's Dining Room	Deck 3
LUNCH		
11:30 am – 3:00 pm	Jade & Windjammer Café	Deck 11
Noon – 2:00 pm	Leonardo's Dining Room	Deck 2
DINNER	**FEAST OF NATIONS**	**CASUAL**
6:00 pm	Main Seating, All Dining Rooms	Decks 3,4,5
8:30 pm	Second Seating, All Dining Rooms	Decks 3,4,5

Please also note that bare feet and tank tops are not permitted at dinner.

6:30 pm – 9:00 pm	Casual Dinner, Jade & Windjammer Café	Deck 11

Casual Dinner in Jade and the Windjammer Café is buffet style.
Please note that bare feet and tank tops are not permitted in these venues at dinner.

SPECIALTY DINING		
6:00 pm – 9:45 pm	**Chops Grille,** The best steak on the high seas $20 dining fee per person applies, gratuities included (Reservations recommended, Dial 3055)	Deck 11
6:00 pm – 9:45 pm	**Portofino,** Intimate Italian Dining $20 dining fee per person applies, gratuities included (Reservations recommended, Dial 3035)	Deck 11

FOLD HERE ↑

Dress Suggestion: Smart Casual. No short pants please. Guests 13 years and older are welcome. Please allow approximately two hours for dinner. Cancellations with less than 24 hours notice will be charged a fee of $10 per person.

SNACKS		
24 hours	Café Promenade	Deck 5
11:00 am – midnight	Ben & Jerry's Ice Cream (Nominal fee)	Deck 5
11:30 am – midnight	Johnny Rockets ($3.95 cover charge per person, gratuities included)	
11:30 am – 2:00 am	Sorrento's Pizza	Deck 5
3:00 pm – 5:00 pm	Afternoon Tea, Windjammer Café	Deck 11
Midnight – 1:00 am	Midnight Delights, Boleros & Casino Royale	Deck 4

MORNING ACTIVITIES

24 hours	Departure Video, Stateroom Television	Channel 41
24 hours	Community Bulletin Board, Library	Deck7
24 hours	royal caribbean online^sm	Deck 8
24 hours	Cards & Games available, Seven Hearts Card Room	Deck 14
7:00 am	Trekk, Indoor Walking ☆, ShipShape Center	Deck 11
7:30 am	Balance in Motion ☆, ShipShape Center	Deck 11
8:00 am	Aerobics in Motion ☆, ShipShape Center	Deck 11
8:00 am – 8:45 am	Aqua Babies Play Time, (Ages 6-18 months) On Air Club	Deck 3
8:30 am	New Balance Walk-a-Mile ☆, Sky Bar	Deck 12
9:00 am	Advanced Ice Skating, (Must have your own skates), Studio B	Deck 3
9:00 am	Pathway to Yoga ☆,($10 fee), ShipShape Center	Deck 11
9:00 am – 9:45 am	Aqua Tots Play Time, (Ages 18-36 months), On Air Club	Deck 3
9:30 am	Backstage Tour, Arcadia Theatre	Decks 3,4
9:45 am	Adventure Ocean Talent Show Rehearsal, On Air Club	Deck 3
9:45 am	Captain's Corner, Pharaoh's Palace	Deck 5
9:45 am – 10:30 am	Learn to Skate, (for those who previously signed up, 6-17 years), Studio B	Deck 3
10:00 am	Seminar: *Eat More to Weigh Less*, ShipShape Center	Deck 11
10:00 am	Adult Volleyball Tournament ☆, Basketball Court	Deck 13
10:00 am	Cry Baby Cinema: *Good Night and Good Luck*, Screening Room	Deck 2
10:30 am – 11:00 am	Snowball Jackpot Bingo Cards on Sale, Arcadia Theatre Lobby	Deck 3
10:30 am – 11:15 am	Learn to Skate, (for those who previously signed up, 18+ years), Studio B	Deck 3

FOLD HERE ↑

MORNING ACTIVITIES

11:00 am	PowerBox Ring Conditioning ✗, ($10 fee), ShipShape Center	Deck 11
11:00 am	Adventure Ocean Family Talent Show, OnAir Club	Deck 3
11:00 am	Snowball Jackpot Bingo Games Begin, Arcadia Theatre	Decks 3,4
11:00 am	Seminar: *Detox for Weight Loss*, ShipShape Center	Deck 11
11:15 am	Freedom LIVE Television Show, FlowRider	Deck 13
11:15 am – noon	Ice Skating, Studio B	Deck 3
11:45 am	Seminar: *Puffy Eyes, Put the Twinkle Back*, Freedom Day Spa	Deck 12

AFTERNOON ACTIVITIES

Noon	Meet the Krooze Komics, Royal Promenade	Deck 5
Noon	Pilates Reformer Group Class ✗, ($15 fee), ShipShape Center	Deck 11
Noon	Teen Speed Climb Challenge, (12-17 years), Rock Climbing Wall	Deck 13
12:30 pm	Final Art Auction Preview, Pharaoh's Palace	Deck 5
1:00 pm	Towel Folding, On Air Club	Deck 3
1:00 pm	Scrapbooking, (Adult class, maximum of 50 guests), Olive or Twist	Deck 14
1:00 pm	Screening Room Movie: *Good Night and Good Luck*, (Limited seating)	Deck 2
1:00 pm	Partner Massage Class, ($52 per couple), Freedom Day Spa	Deck 12
1:15 pm	Final Champagne Art Auction, Pharaoh's Palace	Deck 5
1:15 pm – 2:00 pm	Ice Skating, Studio B	Deck 3
1:30 pm	Pool Games, Sports Pool	Deck 11
1:30 pm – 2:30 pm	Winner Takes All Bingo pre-sales, outside Windjammer Café	Deck 11
1:30 pm – 2:30 pm	Awesome Wine Tasting, (Nominal fee), Vintages Wine Bar	Deck 5
2:00 pm	Mr Sexy Legs Contest, Poolside	Deck 11
2:00 pm	Seminar: *Discover the Burn Fat Fast Zone*, ShipShape Center	Deck 11
2:00 pm – 2:45 pm	Ice Skating, Studio B	Deck 3
2:30 pm	Seminar: *Cellulite Reduction*, Freedom Day Spa	Deck 12
2:45 pm – 3:30 pm	Winner Takes All Bingo cards on sale, (Final chance), Arcadia Theatre	Decks 3,4
3:00 pm	World Cup, *Italy vs. USA*, On Air Club	Deck 3
3:00 pm	Seminar: *Three Minute Makeovers*, Freedom Day Spa	Deck 12
3:00 pm	Family Wacky Water Races ✗, Sports Pool	Deck 11
3:15 pm – 4:00 pm	Ice Skating, Studio B	Deck 3
3:30 pm	Seminar: *Live Longer, Look Younger*, ShipShape Center	Deck 11
3:30 pm	Winner Takes All Bingo, Games begin, Arcadia Theatre	Decks 3,4
4:00 pm	Screening Room Movie: *Good Night and Good Luck*, (Limited seating)	Deck 2
4:00 pm	Wheels in Motion ✗, ($10 fee), ShipShape Center	Deck 11
4:00 pm – 4:45 pm	Ice Skating, Studio B	Deck 3
4:00 pm – 5:00 pm	Connoisseur Wine Tasting, (Nominal fee), Vintages Wine Bar	Deck 5
4:30 pm – 6:00 pm	ShipShape Dollar Exchange, Wipe Out Bar	Deck 13
4:45 pm – 5:30 pm	Teen Ice Skating, (12-17 years), Studio B	Deck 3
5:00 pm	Friends of Bill W. Meeting, Skylight Chapel	Deck 15
5:00 pm	Boot Camp Circuit Class ✗, ShipShape Center	Deck 11
5:00 pm	Adult Soccer Tournament ✗, Basketball Court	Deck 13
5:00 pm – 6:00 pm	Scratch DJ Graduation for Teens, (12-17 years), Fuel	Deck 12

EVENING ACTIVITIES

6:00 pm	Final Book Return, Library	Deck 7
7:00 pm	Farewell Variety Show, (Second Seating Guests Pre-Dinner Show), Arcadia Theatre	Decks 3,4
7:00 pm – 1:00 am	Cigar Aficionados, Connoisseur Club	Deck 5
8:00 pm	Final Team Trivia, On Air Club	Deck 3
8:00 pm	NHL Hockey - Game 6, *Edmonton vs. Carolina*, Pharaoh's Palace	Deck 5
9:00 pm	Farewell Variety Show, (MainSeating Guests), Arcadia Theatre	Decks 3,4
9:30 pm – 10:30 pm	Dessert Wine, Port & Chocolate Tasting, Vintages Wine Bar	Deck 5
10:30 pm – midnight	Farewell 50's & 60's Dance Party, Poolside	Deck 11
10:30 pm – midnight	Joke & Karaoke Night, (Adults only), On Air Club	Deck 3
11:00 pm	Intimate Jazz Cabaret, Olive or Twist	Deck 14

✗ – Denotes ShipShape Dollars awarded for this activity.

MUSIC AND DANCING

Live Bands & DJs

Island Music	*With Caribbean Fusion* 12:30 pm – 1:30 pm & 3:45 pm – 5:30 pm Poolside, Deck 11
Afternoon Jazz	*With the Australian Jazz Quartet* 2:00 pm – 3:30 pm, Royal Promenade, Deck 5
Easy Listening Music	*With Viva Expressia* 4:00 pm – 5:00 pm, Royal Promenade, Deck 5
Guitar Melodies	*With Leandro* 5:00 pm – 6:00 pm & 7:30 pm – 8:30 pm Olive or Twist, Deck 14
Dining Room Music	*With Tara Davis* 6:00 pm – 7:00 pm & 8:30 pm – 9:30 pm All Dining Rooms
Latin Dance Music	*With Sol y Arena* 7:30 pm – 8:30 pm & 10:30 pm – 1:00 am Boleros, Deck 4
Strings In The Sky	*With Viva Expressia* 7:45 pm – 9:00 pm Flying Bridge, Royal Promenade, Deck 5
Piano Bar Entertainment	*With Peter Ritz* 10:00 pm – 1:00 am, Schooner Bar, Deck 4
Pub Entertainment	*Vocal Guitarist Jimmy Blakemore Plays Request* 10:30 pm – 1:00 am, Bull & Bear, Deck 5
Nightclub Dancing	*With DJ Jamie* 10:30 pm – midnight & 2:00 am – late The Crypt, Decks 3 & 4 Guests must be 18 years or older for admission and photo identification will be required if you appear to be under 18.
Farewell 50's & 60's Party	*With our Showband The Stingrays* 10:30 pm – midnight, Poolside, Deck 11
Classic Jazz	*With the Australian Jazz Quartet* 10:15 pm – 10:45 pm, Olive or Twist, Deck 14
Intimate Jazz Cabaret	*With the Royal Caribbean Singers* 11:00 pm, Olive or Twist, Deck 14
Celebrity DJ	*With Celebrity DJ Smoke from the Scratch DJ Academy* Midnight – 2:00 am, The Crypt, Decks 3 & 4 Guests must be 18 years or older for admission and photo identification will be required if you appear to be under 18.
Jazz Jam Session	*With the Australian Jazz Quartet & Special Guests* Midnight – 1:00 am, Olive or Twist, Deck 14

Please note, the Viking Crown Lounge which includes Olive or Twist is for adults 18 and older only after 10:00 pm.

BAR SERVICE HOURS

***Proof of Age** – Guests eighteen to twenty (18-20) years of age are welcome to enjoy beer and wine. Guests twenty-one (21) years of age and older are welcome to enjoy all alcoholic beverage. ** Applicable regulatory age restrictions apply while the ship is in port and until the vessel enters international waters. Picture Identification required.*

DRINK OF THE DAY	FREEDOM BREEZER	$5.95
Café Promenade Bar	7:00 am – 3:00 am	Deck 5
Wipe Out Bar	9:00 am – 6:00 pm	Deck 13
Squeeze Bar	9:00 am – 7:00 pm	Deck 11
Solarium Bar	9:00 am – 7:00 pm	Deck 11
Pool Bar	9:00 am – 7:00 pm 10:30 pm – 12:30 am	Deck 11
Arcadia Theatre	10:00 am – noon 3:30 pm – 5:00 pm	Deck 3
Sky Bar	11:00 am – 7:00 pm 10:30 pm – 12:30 am	Deck 12
Plaza Bar	11:30 am – 11:30 pm	Deck 11
Sorrento's Bar	11:30 am – 3:00 am	Deck 5
Boleros	Noon – 2:00 am	Deck 4
Casino Bar	Noon – late	Deck 4
Vintages Wine Bar	1:00 pm – 1:00 am	Deck 5
Olive or Twist	1:00 pm – late	Deck 14
Bull & Bear Pub	2:00 pm – 2:00 am	Deck 5
Schooner Bar	2:00 pm – 2:00 am	Deck 4
On Air Club	3:00 pm – 1:00 am	Deck 3
Champagne Bar	3:00 pm – 1:00 am	Deck 5
Connoisseur Club Cigar Lounge	7:00 pm – 1:00 am	Deck 5
Pharaoh's Palace	8:00 pm – 10:00 pm	Deck 5
Fuel Bar (15-17 years)	8:00 pm – 1:00 am	Deck 12
The Crypt (Adults only)	11:00 pm – late	Decks 3,4

〔8日目〕下船日

AS YOU DEPART
CRUISE COMPASS

Sunday Morning　　Miami

Dining Schedule

BREAKFAST

6:00 am – 8:30 am	Leonardo's Dining Room (Continuous service)	Deck 3
6:00 am – 9:00 am	Buffet Breakfast, Windjammer Café	Deck 11

Please note: Room Service will not be available after 1:00 am Sunday morning.

Activities

6:30 am	**Express Departure** – Those guests who decide to carry their own luggage off the ship.
7:00 am – 9:30 am	Photographs available for purchase, Photo Gallery, Deck 4.
7:00 am – 10:00 am	We request all guests to wait comfortably in the Arcadia Theatre on Deck 3 forward or Studio B and the On Air Club on Deck 3 aft.
8:00 am – 10:00 am	Basketball Open Play, Basketball Court, Deck 13.
8:00 am – 10:00 am	Freedom Fairways Golf Open, Sports Deck, Deck 13.

Guest Services

7:00 am – 10:00 am	CNN will be shown, Studio B, Deck 3
7:00 am – 10:00 am	The Today Show featuring the Freedom of the Seas, Arcadia Theatre, Deck 3.
7:00 am – 10:00 am	ESPN International, On Air Club, Deck 3
Open 24 Hours	**Medical Emergency (Dial 911),** Deck 1 aft.
Open 24 Hours	Guest Relations Desk (Dial 0), Deck 5.

BAR SERVICE HOURS

CAFÉ PROMENADE	6:30 am – 9:30 am
ON AIR	7:00 am – 9:30 am

Guests twenty-one (21) years of age and older are welcome to enjoy all alcoholic beverages. Picture identification required.

IMPORTANT NOTICE

Please complete your Guest Satisfaction Survey and deposit it in the box at the Guest Relations Desk or on the gangway as you depart. Please do not leave the survey in your stateroom.

- **Gangway is located on Deck 4 forward and midship.**
- **6:30 am Express Departure**
  For guests who decide to carry their own luggage off the ship.
- **7:15 am (approximately)**
  Departure will commence. Please listen for announcements. Do not congregate in the gangway lobbies but wait comfortably in the Arcadia Theatre, Studio B, the On Air Club or on the outside open decks until your color tag has been announced. Please note that these will be the ONLY locations where announcements will be made for departure.
- **7:00 am (continuous)**
  Enjoy CNN in Studio B, the Today Show featuring the Freedom of the Seas in the Arcadia Theatre, Deck 3. and ESPN International in the On Air Club. *(Showing continuously)*
- **7:00 am (Beginning at)**
  Guests holding tickets for Miami tours, please meet outside in the terminal. Please be in the terminal 10 minutes before your departure.
- **8:00 am**
  We would appreciate it if you would vacate your staterooms as soon as possible. We suggest 8:00 am at the very latest as we must prepare for our next cruise vacation and would like to extend the same courtesy that was extended to you in ensuring the ship being ready. **Please leave your stateroom safe OPEN and UNLOCKED before vacating your stateroom.**
- **Prior to 8:00 am**
  Any SeaPass inquiries must be presented to the Guest Relations Desk before 8:00 am.
- ♿ **Guests with disabilities**
  Guests who require wheelchair assistance, please meet in Boleros, Deck 4, when your color tag is called. Please be advised that this service is limited and has an approximately waiting time of 45 minutes.

On behalf of Royal Caribbean International, we would like to thank you for cruising onboard the Freedom of the Seas. We look forward to welcoming you aboard another Royal Caribbean International ship. Have a safe journey home!

As You Depart

Five easy steps to check out

Express Departure

BEAT THE RUSH!

For those guests wishing to expedite their departure on Sunday morning, we are pleased to offer you express departure should you decide to participate in carrying your own luggage off the ship. To participate in the express departure process, DO NOT LEAVE YOUR LUGGAGE OUTSIDE YOUR STATEROOM ON SATURDAY EVENING. Instead, on Sunday morning after the ship has been cleared by Customs and Border Protection officials, an announcement will be made advising all guests participating in this program that they may depart the ship carrying their own luggage (approximately 6:30 am). Should you have any questions regarding the Express Departure process please contact your Stateroom Attendant.

STEP 1 *The Night Before*

- Remove all old tags from your luggage except for personal identification.
- Attach one of the colored tags, delivered by your Stateroom Attendant, to each piece of luggage and one tag on your carry-on luggage to serve as a reminder as to what color tag you received. Please note that if you are participating in an Express Departure you do not need any luggage tags.
- Place each piece of luggage outside your stateroom between 7:00 pm and 11:00 pm on Saturday evening. ***Please make sure that you do not pack your flight tickets, passport/proof of citizenship, medication and that you keep some clothes for Sunday.*** It is imperative that your luggage be placed outside your stateroom before retiring Saturday evening to ensure that your luggage is received in a timely manner in the terminal.
- Under no circumstances should you accept a parcel or piece of luggage that does not belong to you.
- Please take a moment to view the departure video on channel **41** for an overview of the departure process.

STEP 2 *SeaPass*

SeaPass accounts are automatically billed to your credit card. A statement of your account will be delivered to your stateroom by 6:30 am Sunday. However, accounts established with a credit card will remain active on Sunday morning for any last-minute purchases. If you have any questions regarding your account, please contact the Guest Relations Desk on Sunday morning. Remember, you can review your folio at any time prior to this by using the RCTV system. Stateroom refrigerators will be checked Sunday morning prior to your departure and any consumed items will be billed in addition to your statement received on Sunday morning.

STEP 3 *Mandatory Customs and Border Protection Inspection*

The United States Customs and Border Protection forms can be easily and quickly filled out. However, should you need assistance, please do not hesitate to ask.

United States Residents who have exceeded their exemptions in merchandise, liquor or are carrying monies in excess of $10,000 must make themselves known to a United States Customs and Border Protection Officer in the terminal building as they depart the ship. BY LAW IT IS IMPERATIVE THAT YOU DECLARE THESE ITEMS TO THE OFFICIALS.

Your Customs exemption of $800 allows you Duty-Free status on:

- $800 in merchandise from any of our ports or purchased onboard.
- 1 carton of 200 cigarettes – must be 18 years or older. Excess United States-manufactured cigarettes made for export only will be seized. Foreign-manufactured tobacco products will be subject to duty and internal revenue taxes.
- 100 cigars – must be 18 years or older. No Cuban cigars are permitted into the United States.
- 1 liter of alcohol – must be 21 years or older. One additional liter is permitted if it was manufactured and purchased in Haiti or Jamaica. Applicable internal revenue taxes and duties will be assessed on overages.

DEPARTURE ARRANGEMENTS　*Freedom of the Seas*

All guests must present themselves personally to a United States Customs and Border Protection Officer for an immigration inspection. This includes United States Citizens and Residents. This inspection will take place in the terminal after leaving the ship. United States Citizens/Residents and Canadians must show their passport or proof of citizenship (e.g. citizenship card or birth certificate AND photo identification, or A.R.C. card). Non-United States Citizens must show their passport and a completed immigration form (Form I-94), if holding a visa. Please have all your documents in hand and your passport open to the photo page.

United States Customs and Border Protection prohibits the transportation of any agricultural products, such a fruits, vegetables, plants or meats into the United States. Any prohibited items taken off the ship will be seized and a fine may be imposed.

STEP 4 *Final Departure*

Please wait in the Arcadia Theatre, Studio B, On Air Club or on the outside open decks until your color tag is called to the gangway. Please note that these are the ONLY locations where announcements will be made for departure. We request that you keep the gangway areas clear at all times. **Your luggage will not be available until your color is called.**

Guests requiring wheelchair assistance should meet in Boleros on Deck 4 when your color tag is called. You will be escorted off as your color is called. **(Please be advised this is a limited service and approximate waiting time is 45 minutes.)**

- **Guest Satisfaction Surveys** – As your hosts for this cruise vacation, we would sincerely appreciate you sharing your personal opinion about your vacation onboard the Freedom of the Seas. We kindly request that you complete the Guest Satisfaction Survey card and drop it in the sealed Guest Satisfaction Survey boxes by the Guest Relations Desk, Deck 5, or on the gangways as you depart the ship.

  Please do not leave the card in your stateroom as only cards deposited by you in the sealed boxes are reviewed for their comments and suggestions.

  Thank you in advance for your comments and we hope to see you sailing again with Royal Caribbean International.

STEP 5 *Onshore Connections*

- Royal Caribbean representatives will be in the port terminal to assist you.

Departure Order

The first color will be called off the ship at approximately 7:15 am. The last color will be called at approximately 10:00 am.

Please note this order is subject to the flow of guests and luggage into and out of the terminal and may change slightly.

7:15 am	Beige, Lavender, Gray	(All numbers)
7:45 am – 8:15 am	Yellow, Red	(All numbers)
8:15 am – 8:45 am	Brown, Orange	(All numbers)
9:00 am – 9:30 am	Blue, Light Blue 1, Light Blue 2	(All numbers)
9:30 am – 10:00 am	Green 1, Green 2, Purple 1, Purple 2, Pink 1, Pink 2	

Get out there.℠

Final Checklist

Guest Satisfaction Surveys
We value your opinion and encourage you to complete the survey and drop it in the specially designated boxes located by the Guest Relations Desk on Deck 5 or on the gangway as you depart the ship.

Tipping
We hope that you have had the vacation of a lifetime onboard the Freedom of the Seas and that you have especially enjoyed our unique brand of service onboard. We call it Gold Anchor Service. For our service staff we provide the following gratuity suggestions:

Guests with Stateroom accommodations	PER CRUISE / PER PERSON
Head Waiter	$5.25
Dining Room Waiter	$24.50
Assistant Waiter	$14.00
Stateroom Attendant Team	$24.50
Alternative Dining at your Discretion	

Guests with Suite accommodations	PER CRUISE / PER PERSON
Head Waiter	$5.25
Dining Room Waiter	$24.50
Assistant Waiter	$14.00
Sateroom Attendant Team	$40.25
Alternative Dining at your Discretion	

Gratuities are extended on the final evening. Room Service Attendants are tipped as service is rendered. A 15% gratuity will be added to all beverage charges.

Please note that guests who opted to prepay their gratuities on their SeaPass account will receive their gratuity vouchers on Saturday afternoon from their Stateroom/Suite Attendant.

Staterooms
Please check your stateroom thoroughly before departing making sure that you take all of your personal belongings with you. All lost and found items must be claimed by guests during each cruise vacation and prior to departure. All unclaimed items will be donated to local charities at the end of each cruise vacation.

Please vacate your stateroom by 8:00 am and wait comfortably in the Arcadia Theatre, Studio B, the On Air Club or on the outside open decks after breakfast, avoiding the gangway areas and surrounding vicinity on Deck 4.

In-room Safe
Please leave your safe open and unlocked before vacating your stateroom.

ディナーのメニュー

Captain's Gala Dinner

APPETIZERS

V/8 Juice　Pear Nectar　Pineapple Juice
Melon with Prosciutto Ham　Baltic Caviar
✓ Shrimp Cocktail, Brandy Sauce　Quiche Florentine
Egg Skabeloff garnished with Smoked Salmon

SOUPS

Cream of Leek Parisienne　✓ Clear Oxtail Amontillado
Chilled Carrot with Dill

SALADS

Caesar Salad
Romaine Lettuce, Anchovies, Romano Cheese and Croutons garnished with Hearts of Palm and Parsley

✓ Hearts of Lettuce with Artichoke
Your choice of Pimento Vinaigrette, Russian or Caesar Dressing

ENTREES

✓ Fillet of Halibut Veronique
Presented with a light Chardonnay Sauce and seedless Grapes

Cuisses de Grenouilles
Frog Legs sauteed in a Garlic Herb Butter

Roasted Rack of Lamb Provencale
Spiced with Garlic, Thyme and Rosemary, served with Lorette Potatoes

Roast Duckling
Complemented by a sweet Blueberry Sauce

Filet Mignon
Broiled Filet of Prime Beef accompanied by Sauce Bearnaise

✓ Brussel Sprouts　✓ Vegetables Jardiniere　Yellow Squash Rissolees
Lorette Potatoes　Baked Idaho Potato

CHEESES

Cream of Gruyere　Norwegian Jarlsberg　Port Salut　Cheddar　Cream
Edam　Swiss　Gorgonzola　Camembert　Bel Paese
Fresh Fruits

DESSERTS

Black Forest Cake　✓ Pineapple Meringue Pie
Charlotte Royal au Cointreau　Cherries Jubilee
Banana or Pistachio Ice Cream　Champagne Sherbet

キャプテン・ガラディナーのメニュー・その1

Captain's Gala Dinner

The Chef Suggests

The Chef has selected a gala menu to celebrate the Captain's Gala Dinner.
We are pleased to recommend these courses that complement each other
and are designed to provide a delightful dining experience.

Melon with Prosciutto Ham

Clear Oxtail Amontillado
Light Consomme with a splash of Sherry

Caesar Salad
Romaine Lettuce, Anchovies, Romano Cheese and Croutons
combine to make the classic Caesar, tossed with Caesar Dressing

Fillet of Halibut Veronique
Presented with a light Chardonnay Sauce and seedless Grapes

Vegetables Jardiniere　Lorette Potatoes

Cream of Gruyere Cheese　Cherries Jubilee

Wine Steward Recommendations

Chablis
France
A full-bodied White Wine
from Burgundy, superb with Seafood

Chateauneuf du Pape
Rhone
A soft round rich Red Wine from the Rhone,
ideal with Beef or Lamb

Beverages

Freshly brewed Regular or Decaffeinated Coffee　Sanka　Postum
Tea　Herbed Tea　Milk　Hot Chocolate　Soft Drinks

キャプテン・ガラディナーのメニュー・その２

演習

あなたが「フリーダム・オブ・ザ・シーズ」の1週間クルーズに乗船するとして，以下の各項目について計画を立てなさい。

⑴ 各寄港地で参加するオプショナルツアーの計画を立てなさい。
⑵ どのような服を持参すべきか考えなさい。
⑶ 2日目は航海日である。この日のスケジュールを考えなさい。
⑷ 4日目はグランドケイマン島に寄港する。この日のスケジュールを考えなさい。
⑸ 下船のためにするべき行動についてまとめなさい。
⑹ キャプテン・ガラディナーで食べる料理を選びなさい。

おわりに

今，世界中でブームになっている「客船による周遊観光」すなわち「クルーズ」について，本書で学んでいただきました。物流を中心とする海事産業では，限られた海上輸送需要の中で，過激な需要獲得競争が世界規模で繰り広げられていますが，唯一，クルーズの分野は，需要自体が急拡大している成長分野です。

しかし，古い歴史をもつクルーズがそのままの形で成長したわけではありません。客船が飛行機の登場で役割を終えた 50 年ほど前に，それまでの伝統的クルーズから脱皮した新しい現代クルーズというビジネスモデルが登場して，それが牽引して，クルーズを巨大な産業にまで成長させ，世界的な広がりを見せているのです。ここで学ぶべき大事なことは，クルーズの世界に大きな「イノベーション」が起こったからこそ，今のクルーズの成長があるということです。

クルーズの世界には夢があります。本書でクルーズの疑似体験をした後には，ぜひ，本当のクルーズを体験してみてください。東アジアでもクルーズが急成長していて，日本にもたくさんの大型クルーズ客船が来航するようになりました。また，日本発着のクルーズも急増しています。

日本では，クルーズに対して，値段が高く，長い休みが必要で，たいくつで堅苦しいといったイメージが浸透していますが，今のクルーズは，価格も安く，短い期間で楽しめる究極のレジャーです。ぜひ，クルーズを楽しんでみてください。クルーズに魅了され，病みつきになること請け合いです。

では，読者のみなさまの新しい船出を祈念して，ボン・ボヤージュ（ご安寧な航海を）！

索 引

【著者紹介】

池田 良穂（いけだ よしほ）
大阪府立大学名誉教授，大阪経済法科大学客員教授
船舶工学，海洋工学，クルーズビジネス等が専門。専門分野での学術研究だけでなく，船に関する啓蒙書を多数執筆し，雑誌等への寄稿，テレビ出演も多く，わかり易い解説で定評がある。

イラスト

中山 美幸（なかやま みゆき）
大阪芸術大学でデザインを学び，日本クルーズ＆フェリー学会の学会誌「Cruise & Ferry」の編集，各種パンフレット，ポスター等のデザインに従事している。

ISBN978-4-303-16412-6

クルーズビジネス

2018年4月25日 初版発行

著 者 池田良穂
発行者 岡田節夫
発行所 海文堂出版株式会社

検印省略

本 社 東京都文京区水道2-5-4（〒112-0005）
電話 03（3815）3291㈹ FAX 03（3815）3953
http://www.kaibundo.jp/
支 社 神戸市中央区元町通3-5-10（〒650-0022）

日本書籍出版協会会員・工学書協会会員・自然科学書協会会員

PRINTED IN JAPAN 印刷 東光整版印刷／製本 誠製本